McGRAW-HILL
BOOK COMPANY
New York
St. Louis
San Francisco
Düsseldorf
Johannesburg
Kuala Lumpur
London
Mexico
Montreal
New Delhi
Panama
Paris
São Paulo
Singapore
Sydney
Tokyo
Toronto

DAVID I. STEINBERG
Associate Professor of Mathematics
Southern Illinois University
Edwardsville, Illinois

Computational Matrix Algebra

This book was set in Times New Roman.
The editors were Jack L. Farnsworth and Barry Benjamin;
the cover was designed by Pedro A. Noa;
and the production supervisor was Sam Ratkewitch.
The drawings were done by John Cordes, J & R Technical Services, Inc.
Kingsport Press, Inc., was printer and binder.

Library of Congress Cataloging in Publication Data

Steinberg, David.
Computational matrix algebra.

1. Matrices. I. Title.
QA188.S68 512.9'43 73-20451
ISBN 0-07-061110-6

COMPUTATIONAL MATRIX ALGEBRA

1234567890KPKP7987654

CONTENTS

PREFACE

An undergraduate course in elementary matrix algebra has become an integral part of the curriculum in a great number of engineering and applied science programs. Unfortunately, most elementary texts on matrix algebra are purely theoretical; that is, computational considerations are largely ignored. With the growing use of high-speed electronic digital computers to solve large-scale algebraic systems (simultaneous equations or eigenvalue problems) it is becoming more and more important for the student to be exposed to such concepts as numerical methods, computer arithmetic, convergence of an algorithm, and efficiency of a computational method. It is the purpose of this text to include these ideas in the development of the theory of matrix algebra.

Most universities now have some type of computer facility, and it is hoped that where such a facility is available, the student will be given the opportunity to gain access to it. In such a case, the student is urged to write his own programs to try out the methods presented in the text. Some of the exercises are designed with this in mind.

Computational Matrix Algebra is primarily intended for sophomores and juniors majoring in engineering or other applied sciences. No mathematics background beyond high school algebra and trigonometry is assumed.

However, since a certain level of sophistication in mathematical concepts would be beneficial to the understanding of the material in this text, it is recommended for use by students who have studied at least one year of college-level mathematics.

It has been the author's experience that most of the material in *Computational Matrix Algebra* can be covered in a one-semester course; the entire text can certainly be covered quite thoroughly in a two-quarter course. This text could also be used for a full two-semester (three-quarter) sequence, with the inclusion of some supplementary material, such as advanced linear algebra, more numerical methods (e.g., the algorithms mentioned in Sec. 10-6), or a more detailed discussion of some of the computer programming aspects of the various methods presented in the text.

The author wishes to express his gratitude to Professors Philip J. Davis, Daniel A. Prelewicz, and Robert A. Walsh for their invaluable suggestions, and to Mr. John C. Lamb for the remarkably thorough assistance he provided in the proofreading of the manuscript. The author is also grateful for the extreme patience displayed by each of the following persons, who tolerated his abominable handwriting and other personality quirks, and who overcame these severe handicaps to produce a magnificently typed manuscript: Judy Forman, Mildred Gale, Gloria Horkits, Paulette Luter, Peggy Reask, Sharon Schaefer, and Chris Thiemann.

DAVID I. STEINBERG

1
INTRODUCTION

1-1 AN OVERVIEW

A wide variety of mathematical models arising from problems in engineering, the physical sciences, and the social sciences involves the treatment and solution of systems of linear algebraic equations. Frequently, such systems contain a large† number of variables and equations, and the problem of solving a large system is not a trivial one, even with the aid of high-speed computers. Moreover, the use of digital computers in solving problems introduces several additional factors which must be considered in the development of solution techniques, such as computational efficiency and computational accuracy.

If the reader has access to a digital computer, he is urged to try out the methods discussed in this text on a computer, where appropriate. Many modern computer centers have programs already written, and available to the user, for most of the methods which we shall discuss. Of course, the reader is also encouraged to write his own computer programs, in the interests of

† By "large" we mean as many as 50 to 100—or even hundreds of—equations and variables. In fact, in Chap. 8 we shall discuss a type of problem which in many applications involves several hundred equations and thousands of variables.

enhancing his understanding and appreciation of these methods, and perhaps also increasing his skills as a programmer. However, a prior knowledge of computer programming is not necessary, nor is it assumed.

In this text we shall endeavor to maintain a proper balance between theoretical considerations and computational considerations. The theoretical material contained herein is included for two reasons: (1) Much of it is essential for the development and understanding of the computational methods; (2) the student should be exposed to a certain amount of fundamental mathematical theory so that he will be capable of understanding other and more advanced material involving linear algebraic systems. The latter is essential so that the student will have the background necessary to be able to learn of—or perhaps to develop—new methodology in this field. A student who does not possess such a background and who is not willing to continue learning beyond his school years may soon find himself outdated, particularly in this age of rapidly changing technology.

Thus, an important part of the text will involve theorems and their proofs. Therefore, before proceeding to the main topics of the text, we shall devote several pages to a discussion of theorem proving. This discussion includes proof by mathematical induction, the use of counterexamples, and "if and only if" theorems.

1-2 THEOREMS

To illustrate some of the concepts to be discussed in this section, we shall employ several well-known results of plane geometry:

Consider the triangle below, with sides a, b, and c, and angles A, B, and C.

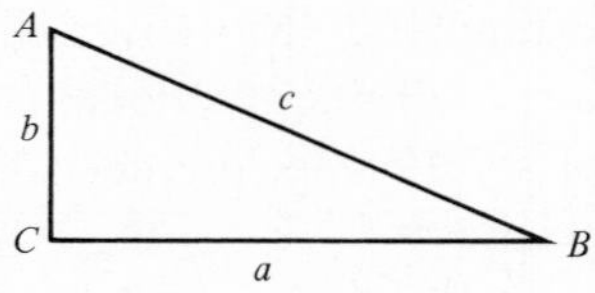

Theorem 1-1 Pythagoras's theorem If C is a right angle, then $c^2 = a^2 + b^2$.

Theorem 1-2 If $A = B$, then $a = b$.

Theorem 1-3 If $a = b$, then $A = B$.

Each of the three theorems above is of the form "If '*xxxx*,' then '*yyyy*.'" The '*xxxx*' part is often called the *hypothesis*, and the '*yyyy*' part the *conclusion*. Such a theorem may be stated equivalently as "'*xxxx*' implies '*yyyy*.'" If one wishes to apply such a theorem to establish the validity of '*yyyy*,' one must first show that '*xxxx*' is in fact true for the case in question. For example, if we had a triangle ABC and wished to use the equation $c^2 = a^2 + b^2$, we would first have to show that C was a right angle.

However, in order to *prove* the theorem "If '*xxxx*,' then '*yyyy*,'" we *assume* that '*xxxx*' and show through a sequence of logical arguments that this assumption leads necessarily to the conclusion that '*yyyy*.'

EXAMPLE 1-1

Theorem If ABC is a triangle with angles A, B, and C, and sides opposite these angles of lengths a, b, and c respectively, then $a^2 = b^2 + c^2 - 2bc\,(\cos A)$.

PROOF The hypothesis states that we have a triangle as shown below in the figure on the left. Since we don't have any information about

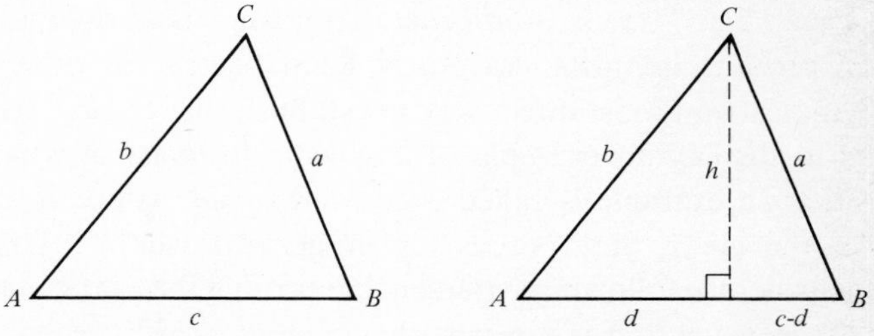

the angles A, B, and C, we cannot apply Pythagoras's theorem directly to triangle ABC. However, suppose we construct a line from C through and perpendicular to AB, intersecting AB at D. We now have two triangles ADC and CDB each with a right angle, as shown above in the figure on the right. Applying Pythagoras's theorem to each triangle yields:

Triangle ADC:

$$b^2 = h^2 + d^2 \qquad (1\text{-}1)$$

Triangle CDB:

$$a^2 = h^2 + (c - d)^2 \qquad (1\text{-}2)$$

Subtracting Eq. (1-1) from Eq. (1-2) yields

$$\begin{aligned} a^2 - b^2 &= h^2 + (c - d)^2 - h^2 - d^2 \\ &= h^2 + c^2 - 2cd + d^2 - h^2 - d^2 \\ &= c^2 - 2cd \end{aligned}$$

Adding b^2 to both sides of the above equation yields

$$a^2 = b^2 + c^2 - 2cd \qquad (1\text{-}3)$$

Now, we recall that, in triangle ADC, the cosine of angle A is equal to the side adjacent to A divided by the hypotenuse of ADC; i.e.,

$$\cos A = d/b \qquad (1\text{-}4)$$

Upon multiplying both sides of (1-4) by b, we obtain

$$d = b \cos A \qquad (1\text{-}5)$$

If we now substitute Eq. (1-5) into Eq. (1-3), we immediately obtain the desired result:

$$a^2 = b^2 + c^2 - 2cb\,(\cos A) \qquad \text{QED}\dagger$$

Frequently in the course of developing theorems we shall encounter situations in which we desire to show that some statement is *not* generally true. To say that "'*xxxx*' is, in general, not true" is to say that '*xxxx*' might be true in certain circumstances but will not be true in other circumstances. The simplest and most direct way to establish that "'*xxxx*' is, in general, not true" is to display an example of a specific instance in which '*xxxx*' is *not* true. Such an example is called a *counterexample* (since it is an example counter to the claim that "'*xxxx*' is in general true"). To illustrate this point, consider the following statement concerning the triangle ABC with sides a, b, and c: "In general, it is not true that $c^2 = a^2 + b^2$."

An example of a triangle in which $c^2 \neq a^2 + b^2$ is a simple matter to construct, e.g.,

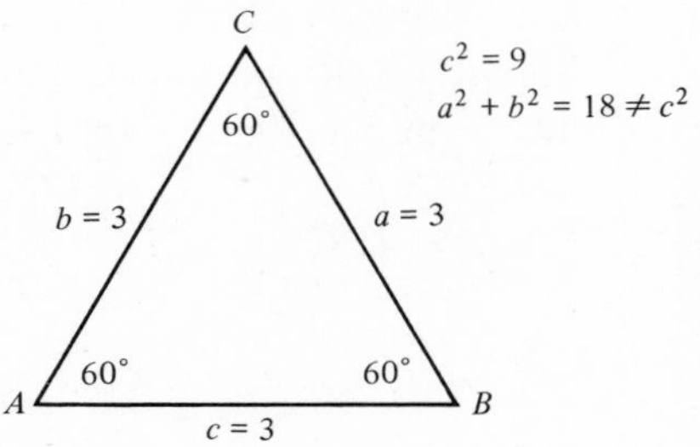

† QED is the abbreviation for the Latin *quod erat demonstrandum*, which translates as "which was to be demonstrated." QED has for centuries been the traditional designation for the conclusion of the proof of a theorem.

Let us now consider the relationship of Theorems 1-2 and 1-3. Theorem 1-2 states, "If '*xxxx*,' then '*yyyy*'" and Theorem 1-3 states, "If '*yyyy*,' then '*xxxx*'"; Theorem 1-3 is called the *converse* of Theorem 1-2. A theorem and its converse theorem are often combined into one theorem, in one of the following three forms:

Theorem: If '*xxxx*,' then '*yyyy*'; conversely, if '*yyyy*,' then '*xxxx*.'
Theorem: '*xxxx*' if and only if '*yyyy*.'
Theorem: A necessary and sufficient condition for '*xxxx*' is '*yyyy*.'

The reader should be certain he is convinced that the above three forms are completely equivalent. Perhaps a substitution of $A = B$ for '*xxxx*' and $a = b$ for '*yyyy*' will be of some help in clarifying this point.

Regarding the third form above, we note that "If '*xxxx*,' then '*yyyy*'" could also be expressed, "A sufficient condition for '*yyyy*' is that '*xxxx*,'" or "'*xxxx*' is sufficient for it to be true that '*yyyy*.'" The "necessary" part is just another way of stating the converse. We shall not employ this third form in this text, but the reader should familiarize himself with it, since he will undoubtedly encounter it elsewhere.

Since any theorem of one of these three forms is really two theorems, the most straightforward method of proof is to prove it as two separate theorems. Thus, the theorem "'*xxxx*' if and only if '*yyyy*'" would be proved as follows:

1 Assume '*xxxx*' and prove '*yyyy*.'
2 Assume '*yyyy*' and prove '*xxxx*.'

1-3 MATHEMATICAL INDUCTION

In this section we shall discuss a particular method of proof, called *mathematical induction* or just *induction*, which is extremely powerful for proving certain types of theorems.

We begin with an illustration:

Theorem 1-4 For any positive integer n, the sum of the integers from 1 to n is equal to $n(n + 1)/2$.

This theorem may be stated symbolically:

$$1 + 2 + 3 + \cdots + n = \frac{n(n+1)}{2} \tag{1-6}$$

or†

$$\sum_{k=1}^{n} k = \frac{n(n+1)}{2}$$

Since Theorem 1-4 states that Eq. (1-6) holds true for *any* positive integer n (i.e., *every* positive integer n), it is not sufficient to show that (1-6) is satisfied for just a few values of n. For example, clearly the theorem is true for $n = 1$ and $n = 2$. However, the statement "The sum of the integers from 1 to n is equal to $(2n - 1)$" is also true for $n = 1$ and $n = 2$, but is *not* true for all positive integers n (counterexample?). Thus, we must somehow prove the validity of Eq. (1-6) for all positive integers n. How can we accomplish this? Consider the following slightly different theorem:

Theorem SD (Slightly different) For any integer $n \geq 2$, if "the sum of the integers from 1 to $(n - 1)$ is equal to $(n - 1)(n)/2$," then "the sum of the integers from 1 to n is equal to $n(n + 1)/2$."

PROOF We restate the theorem symbolically:
If

$$\sum_{k=1}^{n-1} k = \frac{(n-1)(n)}{2} \tag{1-7}$$

then

$$\sum_{k=1}^{n} k = \frac{n(n+1)}{2} \tag{1-6}$$

Proceeding with the proof: If we add n to both sides of Eq. (1-7), then its left-hand side becomes $\sum_{k=1}^{n-1} k + n$, which is equal to $\sum_{k=1}^{n} k$; i.e., the sum of the integers 1 to $(n - 1)$, plus n, is equal to the sum of the integers from 1 to n. Hence,

† The symbol $\sum$ is the standard mathematical notation for a sum, and $\sum_{k=1}^{n} a_k$ is read, "the sum of the elements a_k, from $k = 1$ to $k = n$." Thus, $\sum_{k=1}^{n} a_k = a_1 + a_2 + \cdots + a_n$. The letter k above is called the *index of summation* and is essentially a "dummy variable." That is, we can replace k with any other letter and the result will be an equivalent expression:

$$\sum_{k=1}^{n} a_k \equiv \sum_{j=1}^{n} a_j \equiv a_1 + a_2 + \cdots + a_n$$

$$\sum_{k=1}^{n-1} k + n = \frac{(n-1)(n)}{2} + n$$

$$\sum_{k=1}^{n} k = \frac{(n-1)(n)}{2} + n$$

$$= \frac{(n-1)(n) + 2n}{2}$$

$$= \frac{n^2 - n + 2n}{2}$$

$$= \frac{n^2 + n}{2}$$

$$= \frac{n(n+1)}{2} \qquad \text{QED}$$

Let us denote by $S(n)$ the statement, "the sum of the integers from 1 to n is equal to $n(n+1)/2$." Then, we can write Theorem SD as "For any integer $n \geq 2$, if $S(n-1)$, then $S(n)$." It is clear that we have not yet proved Theorem 1-4, since we have not yet established that the hypothesis of Theorem SD $S(n-1)$ is true. In fact, we can also easily prove the following:

For any integer $n \geq 2$, if $\sum_{k=1}^{n-1} k = (n-1)(n)/2 + 10$, then $\sum_{k=1}^{n} k = n(n+1)/2 + 10$.

In this latter "theorem," of course, the hypothesis is never satisfied.

However, substituting successively the values $n = 2, 3, 4, \ldots$ into Theorem SD, we obtain the following:

$n = 2$: If $S(1)$, then $S(2)$.
$n = 3$: If $S(2)$, then $S(3)$.
$n = 4$: If $S(3)$, then $S(4)$.
$\vdots$

The above sequence indicates that if we can prove that $S(1)$ is true, then $S(n)$ will be true for $n = 2, 3, \ldots$. It is a trivial matter to establish $S(1)$: we merely substitute $n = 1$ into Eq. (1-6) and obtain an identity.

Thus, combining the proof of Theorem SD with the proof of $S(1)$ yields the proof of Theorem 1-4.

We have just illustrated the method of mathematical induction. The procedure may be summarized as follows:

Mathematical Induction

Given the theorem, "For all positive integers $n \geq a$, $S(n)$," where $S(n)$ is a statement (or equation) involving the integer n, the proof by induction is as follows:

1 Prove $S(a)$; typically, $a = 0$, 1, or 2.
2 Prove the SD theorem: If $S(n-1)$, then $S(n)$. In this proof, the hypothesis $S(n-1)$ is called the *induction hypothesis.*

We conclude this section with another example.

Theorem 1-5 For all positive integers $n \geq 1$, the sum of the squares of the integers from 1 to n is equal to $n(n+1)(2n+1)/6$, i.e.,

$$\sum_{k=1}^{n} k^2 = \frac{n(n+1)(2n+1)}{6} \tag{1-8}$$

PROOF We may consider $S(n)$ to be Eq. (1-8).
1. Then, with $a = 1$, we have $S(1)$:

$$\sum_{k=1}^{1} k^2 \overset{?}{=} \frac{1(1+1)(2 \cdot 1 + 1)}{6}$$
$$1 = 1$$

Thus, $S(1)$ is valid.
2. Now, we wish to prove, "If $S(n-1)$, then $S(n)$". $S(n-1)$ is

$$\sum_{k=1}^{n-1} k^2 = \frac{(n-1)(n)(2n-1)}{6} \tag{1-9}$$

Adding n^2 to both sides of Eq. (1-9)—the induction hypothesis—yields

$$\sum_{k=1}^{n-1} k^2 + n^2 = \frac{(n-1)(n)(2n-1)}{6} + n^2$$

$$\sum_{k=1}^{n} k^2 = \frac{2n^3 - 3n^2 + n}{6} + n^2$$

$$= \frac{2n^3 - 3n^2 + n + 6n^2}{6}$$

$$= \frac{2n^3 + 3n^2 + n}{6}$$

$$= \frac{n(2n^2 + 3n + 1)}{6}$$

$$= \frac{n(n+1)(2n+1)}{6}$$

QED

1-4 COMPLEX NUMBERS

We include in this chapter a brief introduction to complex numbers, as we will require some knowledge of this subject, primarily in Chaps. 9 and 10.

The reader may recall from elementary algebra that the quadratic equation

$$ax^2 + bx + c = 0 \qquad (1\text{-}10)$$

(in which a, b, and c are given real numbers) has two roots† x_1, x_2 provided $a \neq 0$, which are given by

$$\begin{aligned} x_1 &= \frac{-b + \sqrt{b^2 - 4ac}}{2a} \\ x_2 &= \frac{-b - \sqrt{b^2 - 4ac}}{2a} \end{aligned} \qquad (1\text{-}11)$$

There are several possibilities for the values of these roots:

(*a*) $b^2 - 4ac = 0$. Then $x_1 = x_2 = -b/2a$.
(*b*) $b^2 - 4ac > 0$. Then x_1 and x_2 are two distinct real numbers.
(*c*) $b^2 - 4ac < 0$. Then the expressions (1-11) for x_1 and x_2 contain the square root of a negative number. It is this case which requires the introduction of "complex numbers."

The square root of a negative number is called an *imaginary number*. Since $\sqrt{xy} = \sqrt{x}\sqrt{y}$ for any numbers x, y, we can always express the square root of a negative number $(-x)$ as

$$\sqrt{(-x)} = \sqrt{x}\sqrt{(-1)}$$

The symbol $\sqrt{(-1)}$ is usually represented by the lowercase letter i:

$$i \equiv \sqrt{(-1)}$$

Thus, any imaginary number will be of the form iy. A number of the form $x + iy$ is called a *complex number*; x is called its *real part* and y is called its *imaginary part*.

Below, we introduce some basic definitions and properties regarding complex numbers which will be of use to us later on.

If $z_1 = x_1 + iy_1$ and $z_2 = x_2 + iy_2$ are two complex numbers (x_1, x_2, y_1, y_2 are assumed to be real numbers), then the sum $(z_1 + z_2)$ is defined by:

$$z_1 + z_2 \equiv (x_1 + x_2) + i(y_1 + y_2) \qquad (1\text{-}12)$$

† A root is a value of x which satisfies the equation.

The product $z_1 z_2$ is defined by:

$$z_1 z_2 = (x_1 x_2 - y_1 y_2) + i(x_1 y_2 + x_2 y_1) \tag{1-13}$$

For $z_2 \neq 0$, the quotient† z_1/z_2 is defined by:

$$\frac{z_1}{z_2} = \left(\frac{x_1 x_2 + y_1 y_2}{x_2^2 + y_2^2}\right) + i\left(\frac{x_2 y_1 - x_1 y_2}{x_2^2 + y_2^2}\right) \tag{1-14}$$

These definitions follow our usual notions of addition, multiplication, and division as the reader may easily verify.

Graphically, we may consider the set of all complex numbers as forming a plane, since each complex number $z = x + iy$ has two components, x and y. This plane is usually called the "complex plane" and is usually graphed with the two rectangular axes x and iy. Thus, a complex number $z_1 = x_1 + iy_1$ is considered as a point in the complex plane and is graphed as illustrated below:

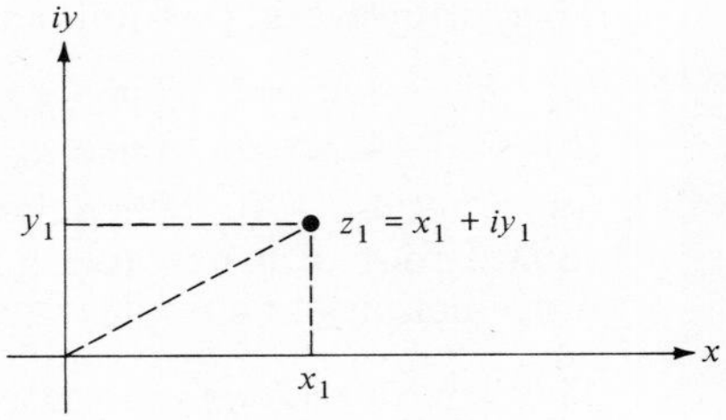

To indicate its correspondence with a point in the complex plane, $z_1 = x_1 + iy_1$ is often written as $z_1 = (x_1, y_1)$.

The *length*, or *modulus*, of a complex number z_1 is defined as its distance in the complex plane from the origin, and is denoted by $|z_1|$. Since $|z_1|$ is the hypotenuse of the right triangle formed by the three points (0, 0), $(x_1, 0)$, and (x_1, y_1), it is clear that

$$|z_1| = \sqrt{x_1^2 + y_1^2} \tag{1-15}$$

Note that if $y_1 = 0$ (i.e., $z_1 = x_1$ is a real number), then $|z_1|$ is the conventional absolute value of the real number x_1. We may occasionally interchange the terms *absolute value* and *modulus*.

The *conjugate* of the complex number $z_1 = x_1 + iy_1$ is denoted by $\bar{z}_1$ and is defined to be the complex number

$$\bar{z}_1 = x_1 - iy_1 \tag{1-16}$$

† $z_2 = 0$ if and only if $x_2 = y_2 = 0$ (by definition).

The following equations may easily be derived by elementary algebra and/or direct application of Eqs. (1-12) through (1-16):

For any complex numbers $z_1 = x_1 + iy_1$, $z_2 = x_2 + iy_2$, $z_3 = x_3 + iy_3$:

$$\frac{z_1}{z_2} = z_1\left(\frac{1}{z_2}\right) \qquad \text{if } z_2 \neq 0$$

$$\frac{1}{z_1 z_2} = \frac{1}{z_1}\frac{1}{z_2} \qquad \text{if } z_1 \neq 0,\ z_2 \neq 0$$

$$z_1(z_2 + z_3) = z_1 z_2 + z_1 z_3$$

$$(z_1 + z_2)z_3 = z_1 z_3 + z_2 z_3$$

$$\overline{(\bar{z}_1)} = z_1$$

$$\overline{(z_1 + z_2)} = \bar{z}_1 + \bar{z}_2$$

$$\overline{(z_1 z_2)} = \bar{z}_1 \bar{z}_2$$

$$\overline{(z_1/z_2)} = \bar{z}_1/\bar{z}_2 \qquad \text{if } z_2 \neq 0$$

$$|z_1| = |\bar{z}_1|$$

$$|z_1\bar{z}_1| = |z_1|^2 = |\bar{z}_1|^2 = z_1\bar{z}_1$$

$$|z_1||z_2| = |z_1 z_2|$$

$$|z_1| \geq |x_1|, \qquad |z_1| \geq |y_1|$$

The following theorems are also easily proved:

Theorem 1-6 $z_1 z_2 = 0$ if and only if either $z_1 = 0$ or $z_2 = 0$.

Theorem 1-7 $z_1 = \bar{z}_1$, if and only if z_1 is a real number (i.e., if $z_1 = x_1 + iy_1$, then $y_1 = 0$).

Theorem 1-8 $z_1 = -\bar{z}_1$, if and only if z_1 is a purely imaginary number (i.e., if $z_1 = x_1 + iy_1$, then $x_1 = 0$).

We leave all of the above as exercises for the interested reader. These results are included in this text only for future reference.

Theorem 1-9 Triangle inequality (*a*) $|z_1 + z_2| \leq |z_1| + |z_2|$
(*b*) For any n complex numbers $z_1, z_2, \ldots, z_n$:

$$|z_1 + z_2 + \cdots + z_n| \leq |z_1| + |z_2| + \cdots + |z_n|$$

PROOF (*a*)
$$\begin{aligned}|z_1 + z_2|^2 &= (z_1 + z_2)\overline{(z_1 + z_2)}\\
&= (z_1 + z_2)(\bar{z}_1 + \bar{z}_2)\\
&= (z_1\bar{z}_1 + z_2\bar{z}_2) + (z_1\bar{z}_2 + \bar{z}_1 z_2)\\
&= |z_1|^2 + |z_2|^2 + \{z_1\bar{z}_2 + \overline{(z_1\bar{z}_2)}\}\\
&= \{|z_1|^2 + |z_2|^2 + 2|z_1||z_2|\} - 2|z_1||z_2|\\
&\quad + \{z_1\bar{z}_2 + \overline{(z_1\bar{z}_2)}\}\\
&= (|z_1| + |z_2|)^2 - 2|z_1||z_2|\\
&\quad + \{z_1\bar{z}_2 + \overline{(z_1\bar{z}_2)}\}\end{aligned}$$

Hence

$$|z_1 + z_2|^2 - (|z_1| + |z_2|)^2 = -2|z_1||z_2| + \{z_1\bar{z}_2 + \overline{(z_1\bar{z}_2)}\} \tag{1-17}$$

Now, let $z_1\bar{z}_2 = x + iy$, where x, y are real numbers. Then, the right-hand side of (1-17) becomes:

$$\begin{aligned}-2|z_1||z_2| + \{x + x\} &= -2\{|z_1||\bar{z}_2| - x\}\\
&= -2\{|z_1\bar{z}_2| - x\}\\
&\leq 0 \qquad \text{since } |z_1\bar{z}_2| \geq x\end{aligned}$$

Thus, $|z_1 + z_2|^2 - (|z_1| + |z_2|)^2 \leq 0$

$$|z_1 + z_2|^2 \leq (|z_1| + |z_2|)^2$$

$$|z_1 + z_2| \leq |z_1| + |z_2|$$

QED

Part (*b*) may be proved by induction.

Throughout the text, whenever we wish to refer to a number without distinguishing between real numbers and complex numbers, we shall use the term *scalar*, which the reader may consider to be synonymous with "number."

2
MATRICES

2-1 INTRODUCTION AND BASIC DEFINITIONS

In Sec. 1-4 we briefly discussed complex numbers and presented a list of equations, many of which may have seemed to the reader to be somewhat trivial. However, this is because these equations, which together are known as "the laws of scalar algebra," have undoubtedly become so familiar to the reader that he now accepts them without question. In this chapter, we shall be concerned with quantities known as "matrices" (as opposed to "scalars"), and we shall desire to develop an analogous algebra for matrices.

In Table 2-1 we summarize the laws of scalar algebra. The lowercase letters *a*, *b*, and *c* refer to any scalar (real or complex number) unless otherwise specified.

In the remainder of this section and in Sec. 2-2 we shall define and develop the laws of matrix algebra.

We begin by defining: Any rectangular array of scalars (e.g., a table) is called a *matrix*. The scalars contained in a matrix are called its *elements*. Thus, the following are examples of matrices:

$$\begin{bmatrix} 1 & 0 & -3 \\ 2 & 1 & 0 \\ 7 & \frac{1}{2} & 1 \\ -4 & 6 & 8 \end{bmatrix} \qquad \begin{bmatrix} 4 & 1+2i \\ 0 & 6 \end{bmatrix} \qquad \begin{bmatrix} 1 \\ 3 \\ -2 \end{bmatrix} \qquad [1 \;\; 0 \;\; 2 \;\; 4] \qquad [-2]$$

The elements of a matrix will be enclosed by a pair of square brackets, as indicated above. An entire matrix will be denoted by a boldface capital letter. Thus, we may write

$$\mathbf{A} = \begin{bmatrix} 1 & 0 & -3 \\ 2 & 1 & 0 \\ 7 & \frac{1}{2} & 1 \\ -4 & 6 & 8 \end{bmatrix} \qquad \mathbf{B} = \begin{bmatrix} 4 & 1+2i \\ 0 & 6 \end{bmatrix}$$

Table 2-1 LAWS OF SCALAR ALGEBRA

A. Equivalence properties:
   1. For any two scalars a, b, either $a = b$ or $a \neq b$ (*determinative* property)
   2. For any scalar a, $a = a$ (*reflexive* property)
   3. If $a = b$, then $b = a$ (*symmetric* property)
   4. If $a = b$ and $b = c$, then $a = c$ (*transitive* property)

B. Addition of scalars:
   1. $a + (b + c) = (a + b) + c$ (*associative law of addition*)
   2. $a + b = b + a$ (*commutative law of addition*)
   3. If $a + c = b + c$, then $a = b$ (*cancellation law of addition*)
   4. $a + 0 = a$ (0 is called the *identity element for addition*)
   5. $a + (-a) = 0$ ($-a$ is called the *inverse of a with respect to addition*)

C. Multiplication of scalars:
   1. $a(bc) = (ab)c$ (*associative law of multiplication*)
   2. $ab = ba$ (*commutative law of multiplication*)
   3. If $ab = ac$ and $a \neq 0$, then $b = c$
      If $ab = cb$ and $b \neq 0$, then $a = c$ } (*cancellation law of multiplication*)
   4. $a \cdot 1 = 1 \cdot a = a$ (1 is called the *identity element for multiplication*)
   5. If $a \neq 0$, then $a \cdot a^{-1} = 1$ (a^{-1} is called the *inverse of a with respect to multiplication*)
      (i.e., for every nonzero scalar a, there exists an element a^{-1} such that $a \cdot a^{-1} = 1$.)

D. Distributive laws:
   1. $a(b + c) = ab + ac$
      $(a + b)c = ac + bc$

Some additional notational devices which we shall employ are:

1 The element in the ith row and jth column of the matrix **A** shall be denoted by a_{ij}. Thus, if **A** is a matrix with m rows and n columns, then

$$\mathbf{A} = \begin{bmatrix} a_{11} & a_{12} & \cdots & a_{1n} \\ a_{21} & a_{22} & \cdots & a_{2n} \\ \vdots & \vdots & & \vdots \\ a_{m1} & a_{m2} & \cdots & a_{mn} \end{bmatrix} \tag{2-1}$$

2 A matrix with m rows and n columns is called an *$m \times n$ matrix* (read "m by n") or a matrix *of order $m \times n$*. If $m = n$, the matrix is said to be *of order n*.

3 The matrix **A** of Eq. (2-1) may be written in abbreviated form as

$$\mathbf{A} = [a_{ij}]_{m \times n} \qquad \text{or} \qquad \mathbf{A} = [a_{ij}]$$

The latter form may be employed if the order of **A** is either clear from the context or unimportant.

4 A matrix with only one row or only one column is called a *vector* (*row vector* or *column vector*, respectively) and shall be denoted by a lowercase boldface letter. Thus,

$$\mathbf{a} = [1 \quad 3 \quad -2] \qquad \mathbf{b} = \begin{bmatrix} 0 \\ 1 \\ 6 \\ 0 \end{bmatrix} \qquad \mathbf{x} = \begin{bmatrix} x_1 \\ x_2 \\ x_3 \end{bmatrix}$$

As in the last example above, we may use only one subscript to denote the elements of a vector. The latter are also called the *components* of the vector.

We are now ready to develop the algebra of matrices. The first step is to define the meaning of *equality* so that we may write matrix equations:

Two matrices $\mathbf{A} = [a_{ij}]_{m \times n}$ and $\mathbf{B} = [b_{ij}]_{p \times q}$ are *equal*, written $\mathbf{A} = \mathbf{B}$, if and only if

1 $m = p$
2 $n = q$
3 $a_{ij} = b_{ij}$ for each pair of subscripts (i, j)

Thus, two matrices are equal if and only if they are of the same order and their corresponding elements are equal. With this definition, it is clear that the four equivalence properties of scalar algebra also apply to matrices (substituting the word *matrix* for *scalar* in Table 2-1 yields the corresponding equivalence relations for matrices).

Now, we define matrix addition:

Given two matrices $\mathbf{A} = [a_{ij}]_{m \times n}$ $\mathbf{B} = [b_{ij}]_{m \times n}$, of the same order, the sum $\mathbf{A} + \mathbf{B}$ is defined as the $m \times n$ matrix $\mathbf{C} = [c_{ij}]$, where $c_{ij} = a_{ij} + b_{ij}$, for each pair of subscripts (i, j).

Thus, two matrices may be together if and only if they are of the same order; otherwise, matrix addition is undefined.

With this definition, it is a simple matter to prove the following (which correspond to properties B1 through B5 of Table 2-1):

Theorem 2-1 If $\mathbf{A} = [a_{ij}]_{m \times n}$, $\mathbf{B} = [b_{ij}]_{m \times n}$, $\mathbf{C} = [c_{ij}]_{m \times n}$, then:

1 $\mathbf{A} + (\mathbf{B} + \mathbf{C}) = (\mathbf{A} + \mathbf{B}) + \mathbf{C}$

2 $\mathbf{A} + \mathbf{B} = \mathbf{B} + \mathbf{A}$

3 If $\mathbf{A} + \mathbf{C} = \mathbf{B} + \mathbf{C}$, then $\mathbf{A} = \mathbf{B}$.

4 $\mathbf{A} + \mathbf{O}_{m \times n} = \mathbf{A}$ (where $\mathbf{O}_{m \times n}$ is a matrix whose elements are all zero; such a matrix is called a *zero matrix*).

5 $\mathbf{A} + (-\mathbf{A}) = \mathbf{O}_{m \times n}$ [where $-\mathbf{A}$ is the $m \times n$ matrix whose elements are $(-a_{ij})$].

We leave the proof of this theorem as an exercise for the reader.

EXAMPLE 2-1† Suppose we are conducting a poll of voters within a state and we have compiled the data in Table 2-2.

This information is clearly representable as a 4×6 matrix; we may also divide the information into two 4×3 matrices, one containing the urban

Table 2-2 NUMBERS OF VOTERS (IN UNITS OF 10,000) FOR CANDIDATE 1, 2, 3†

	Urban areas			Rural areas		
Categories	1	2	3	1	2	3
I	2	1	3	3	2	1
II	3	1	2	4	3	1
III	3	1	1	4	2	2
IV	1	2	2	3	2	1

† Total number of voters = 50(10,000) = 500,000. The categories I, II, III, and IV represent a socioeconomic breakdown.

† This example has been borrowed by permission of one of the authors, from Cooper and Steinberg [2].

areas data and the other the rural area data. If we denote these two matrices by **A** and **B**, respectively, then

$$\mathbf{A} = \begin{bmatrix} 2 & 1 & 3 \\ 3 & 1 & 2 \\ 3 & 1 & 1 \\ 1 & 2 & 2 \end{bmatrix} \qquad \mathbf{B} = \begin{bmatrix} 3 & 2 & 1 \\ 4 & 3 & 1 \\ 4 & 2 & 2 \\ 3 & 2 & 1 \end{bmatrix}$$

Moreover, if we do not wish to distinguish between the urban areas data and the rural areas data, then these may be combined by performing the matrix addition

$$\mathbf{C} = \mathbf{A} + \mathbf{B} = \begin{bmatrix} 2 & 1 & 3 \\ 3 & 1 & 2 \\ 3 & 1 & 1 \\ 1 & 2 & 2 \end{bmatrix} + \begin{bmatrix} 3 & 2 & 1 \\ 4 & 3 & 1 \\ 4 & 2 & 2 \\ 3 & 2 & 1 \end{bmatrix} = \begin{bmatrix} 5 & 3 & 4 \\ 7 & 4 & 3 \\ 7 & 3 & 3 \\ 4 & 4 & 3 \end{bmatrix}$$

Instead of studying these data in terms of numbers of voters in each category, we may desire to convert the data into percentage form. To do this, we would multiply each element of the matrix **C** by the scalar $\alpha = \frac{1}{50} \times 100 = 2$, obtaining the matrix

$$\mathbf{P} = \begin{bmatrix} 10 & 6 & 8 \\ 14 & 8 & 6 \\ 14 & 6 & 6 \\ 8 & 8 & 6 \end{bmatrix}$$

////

This last operation—multiplying each element by a scalar—occurs so frequently that we shall officially define:

Given any matrix $\mathbf{A} = [a_{ij}]$ and any scalar α, the product of α and **A** is defined as the matrix obtained by multiplying each element of **A** by α; this product is written $\alpha\mathbf{A}$ or $\mathbf{A}\alpha$. If $\mathbf{B} = [b_{ij}] = \alpha\mathbf{A} = \mathbf{A}\alpha$, then $b_{ij} = \alpha a_{ij}$ for each pair of subscripts (i, j).

The reader should be able to easily verify:

$$\alpha(\beta\mathbf{A}) = (\alpha\beta)\mathbf{A} = (\beta\alpha)\mathbf{A} \tag{2-2}$$

for any matrix **A** and scalars α, β.

2-2 MATRIX MULTIPLICATION

In the previous section we observed that a matrix may be considered as a table of data. If we were to interchange the rows and the columns of such a table of data, we would essentially have the same table of data (with the column and row headings interchanged), but the corresponding matrix would not be equal, in general, to the original matrix. The matrix obtained from a given matrix $\mathbf{A}$ by interchanging the corresponding rows and columns of $\mathbf{A}$ is called the *transpose* of $\mathbf{A}$, and is denoted by $\mathbf{A}^T$ (or, in some texts, by $\mathbf{A}^t$ or $\mathbf{A}'$). If $\mathbf{A} = [a_{ij}]_{m \times n}$ and $\mathbf{A}^T = [b_{ij}]$, then $\mathbf{A}^T$ is of order $n \times m$ and $a_{ij} = b_{ji}$.

EXAMPLE 2-2 (*a*) If
$$\mathbf{A} = \begin{bmatrix} a_{11} & a_{12} & a_{13} \\ a_{21} & a_{22} & a_{23} \end{bmatrix}$$
then
$$\mathbf{A}^T = \begin{bmatrix} a_{11} & a_{21} \\ a_{12} & a_{22} \\ a_{13} & a_{23} \end{bmatrix}$$

(*b*) If
$$\mathbf{A} = \begin{bmatrix} 1 & 3 & 2 \\ 0 & -1 & 1 \\ 2 & 0 & 4 \\ -3 & 7 & 0 \end{bmatrix}$$
then
$$\mathbf{A}^T = \begin{bmatrix} 1 & 0 & 2 & -3 \\ 3 & -1 & 0 & 7 \\ 2 & 1 & 4 & 0 \end{bmatrix}$$

(*c*) If
$$\mathbf{a} = \begin{bmatrix} 1 \\ 2 \\ 0 \\ 1 \end{bmatrix}$$
then $\mathbf{a}^T = [1 \quad 2 \quad 0 \quad 1]$

Example (*c*) above indicates that the transpose of a column vector is a row vector. Hereinafter, we shall assume that all vectors are column vectors; a row vector will always be denoted as the transpose of a column vector.

We shall begin our discussion of matrix multiplication by defining the product of the row vector $\mathbf{a}^T = [a_1 \quad a_2 \quad \dots \quad a_n]$ and the column vector $\mathbf{b} = \begin{bmatrix} b_1 \\ b_2 \\ \vdots \\ b_n \end{bmatrix}$, written $\mathbf{a}^T\mathbf{b}$:

$$\begin{aligned} \mathbf{a}^T\mathbf{b} &= [a_1 \quad a_2 \quad \cdots \quad a_n] \begin{bmatrix} b_1 \\ b_2 \\ \vdots \\ b_n \end{bmatrix} \\ &= a_1 b_1 + a_2 b_2 + \cdots + a_n b_n \\ &= \sum_{k=1}^{n} a_k b_k \end{aligned} \tag{2-3}$$

The product $\mathbf{a}^T\mathbf{b}$ is called the *scalar product* of $\mathbf{a}$ with $\mathbf{b}$, since the result is a scalar quantity.

EXAMPLE 2-3 If
$$\mathbf{a} = \begin{bmatrix} a_1 \\ a_2 \end{bmatrix}$$
and
$$\mathbf{b} = \begin{bmatrix} b_1 \\ b_2 \end{bmatrix} \qquad \text{then} \qquad \mathbf{a}^T\mathbf{b} = [a_1 b_1 + a_2 b_2]$$

This result may be interpreted geometrically by considering $\begin{bmatrix} a_1 \\ a_2 \end{bmatrix}$ and $\begin{bmatrix} b_1 \\ b_2 \end{bmatrix}$ as points in the plane. Thus, the line segment from the origin to $\begin{bmatrix} a_1 \\ a_2 \end{bmatrix}$ and the line segment from the origin to $\begin{bmatrix} b_1 \\ b_2 \end{bmatrix}$ form an angle θ, as shown below:

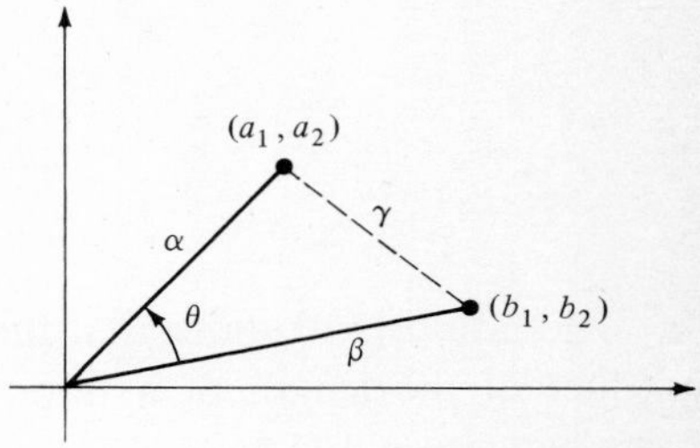

Moreover, by connecting the points $\begin{bmatrix} a_1 \\ a_2 \end{bmatrix}$ and $\begin{bmatrix} b_1 \\ b_2 \end{bmatrix}$, we obtain a triangle. By the theorem of Example 1-1,

$$\gamma^2 = \alpha^2 + \beta^2 - 2\alpha\beta \cos \theta \tag{2-4}$$

And we also know that

$$\begin{aligned} \alpha^2 &= a_1{}^2 + a_2{}^2 \\ \beta^2 &= b_1{}^2 + b_2{}^2 \\ \gamma^2 &= (a_1 - b_1)^2 + (a_2 - b_2)^2 \end{aligned} \tag{2-5}$$

Substitution of Eqs. (2-5) into (2-4) yields

$$(a_1 - b_1)^2 + (a_2 - b_2)^2 = (a_1{}^2 + a_2{}^2) + (b_1{}^2 + b_2{}^2)$$
$$-2\sqrt{a_1{}^2 + a_2{}^2}\,\sqrt{b_1{}^2 + b_2{}^2} \cos \theta \tag{2-6}$$

Upon rearranging and combining Eq. (2-6), we obtain

$$\cos \theta = \frac{a_1 b_1 + a_2 b_2}{\sqrt{a_1{}^2 + a_2{}^2}\,\sqrt{b_1{}^2 + b_2{}^2}} \tag{2-7}$$

Furthermore, the reader will observe that

$$\mathbf{a}^T\mathbf{a} = a_1{}^2 + a_2{}^2$$

$$\mathbf{b}^T\mathbf{b} = b_1{}^2 + b_2{}^2$$

Thus, the scalar product of a vector with itself is the square of the length of the vector. In general, the square root of the scalar product of any real-component vector $\mathbf{x}^T = [x_1 \quad x_2 \quad \cdots \quad x_n]$ with itself shall be called the *length* of $\mathbf{x}$ and shall be denoted by $\|\mathbf{x}\|$:

$$\|\mathbf{x}\| = \left[\sum_{k=1}^{n} x_k{}^2\right]^{1/2} \tag{2-8}$$

Hence, we may write Eq. (2-7) as follows:

$$\cos\theta = \frac{\mathbf{a}^T\mathbf{b}}{\|\mathbf{a}\|\,\|\mathbf{b}\|}$$

or

$$\mathbf{a}^T\mathbf{b} = \|\mathbf{a}\| \cdot \|\mathbf{b}\| \cos\theta \tag{2-9}$$

Equation (2-9) states that the scalar product of **a** and **b** is equal to the product of the length of **a** times the length of **b** times the cosine of the angle between **a** and **b**.

The scalar product of three-component vectors also satisfies Eq. (2-9), and, indeed, the "angle" between two n-component vectors is frequently defined by (2-9). ////

Consider now the system of simultaneous equations

$$\begin{aligned}
a_{11}x_1 + a_{12}x_2 + \cdots + a_{1n}x_n &= b_1 \\
a_{21}x_1 + a_{22}x_2 + \cdots + a_{2n}x_n &= b_2 \\
a_{31}x_1 + a_{32}x_2 + \cdots + a_{3n}x_n &= b_3 \\
\cdots\cdots\cdots\cdots\cdots\cdots\cdots\cdots\cdots \\
a_{m1}x_1 + a_{m2}x_2 + \cdots + a_{mn}x_n &= b_m
\end{aligned} \tag{2-10}$$

If we define

$$\mathbf{x} = \begin{bmatrix} x_1 \\ x_2 \\ \vdots \\ x_n \end{bmatrix} \qquad \mathbf{a}_1 = \begin{bmatrix} a_{11} \\ a_{12} \\ \vdots \\ a_{1n} \end{bmatrix} \qquad \mathbf{a}_2 = \begin{bmatrix} a_{21} \\ a_{22} \\ \vdots \\ a_{2n} \end{bmatrix} \qquad \cdots \qquad \mathbf{a}_m = \begin{bmatrix} a_{m1} \\ a_{m2} \\ \vdots \\ a_{mn} \end{bmatrix}$$

then we can express the system of Eqs. (2-10) as follows:

$$\begin{aligned} \mathbf{a}_1{}^T\mathbf{x} &= b_1 \\ \mathbf{a}_2{}^T\mathbf{x} &= b_2 \\ \vdots \quad & \quad \vdots \\ \mathbf{a}_m{}^T\mathbf{x} &= b_m \end{aligned} \tag{2-11}$$

Thus, the left-hand side of each equation in (2-10) is the scalar product of one row of the *coefficient* matrix

$$\mathbf{A} = \begin{bmatrix} a_{11} & a_{12} & \cdots & a_{1n} \\ a_{21} & a_{22} & \cdots & a_{2n} \\ \vdots & \vdots & & \vdots \\ a_{m1} & a_{m2} & \cdots & a_{mn} \end{bmatrix}$$

with the vector of variables $\mathbf{x}$.

The system of Eqs. (2-11) may also be written:

$$\begin{aligned} [a_{11} \quad a_{12} \quad \cdots \quad a_{1n}] \begin{bmatrix} x_1 \\ x_2 \\ \vdots \\ x_n \end{bmatrix} &= b_1 \\ [a_{21} \quad a_{22} \quad \cdots \quad a_{2n}] \begin{bmatrix} x_1 \\ x_2 \\ \vdots \\ x_n \end{bmatrix} &= b_2 \\ &\vdots \\ [a_{m1} \quad a_{m2} \quad \cdots \quad a_{mn}] \begin{bmatrix} x_1 \\ x_2 \\ \vdots \\ x_n \end{bmatrix} &= b_m \end{aligned} \tag{2-12}$$

The above, however, seems somewhat cumbersome and inefficient. Suppose instead we write:

$$\begin{bmatrix} [a_{11} \quad a_{12} \quad \cdots \quad a_{1n}] \\ [a_{21} \quad a_{22} \quad \cdots \quad a_{2n}] \\ \vdots \\ [a_{m1} \quad a_{m2} \quad \cdots \quad a_{mn}] \end{bmatrix} \begin{bmatrix} x_1 \\ x_2 \\ \vdots \\ x_n \end{bmatrix} = \begin{bmatrix} b_1 \\ b_2 \\ \vdots \\ b_m \end{bmatrix} \tag{2-13}$$

The expression (2-13) is not really an equation (yet), unless we *define* the indicated product on the left-hand side to be equivalent to the left-hand side of (2-12). Note that if the inner brackets on the left-hand side of (2-13) are deleted, we obtain:

$$\begin{bmatrix} a_{11} & a_{12} & \cdots & a_{1n} \\ a_{21} & a_{22} & \cdots & a_{2n} \\ \vdots & \vdots & & \vdots \\ a_{m1} & a_{m2} & \cdots & a_{mn} \end{bmatrix} \begin{bmatrix} x_1 \\ x_2 \\ \vdots \\ x_n \end{bmatrix} = \begin{bmatrix} b_1 \\ b_2 \\ \vdots \\ b_m \end{bmatrix}$$

or

$$\mathbf{Ax} = \mathbf{b} \tag{2-14}$$

(where $\mathbf{b} = [b_1 \quad b_2 \quad \cdots \quad b_m]^T$).

Thus, we have expressed the system of equations (2-10) by the matrix "equation" (2-14). However, in order for (2-14) to be a legitimate equation, we must define:

The product of the $m \times n$ matrix $\mathbf{A}$ and the n-component vector $\mathbf{x}$, written $\mathbf{Ax}$, is an m-component vector, whose ith component is the scalar product of the ith row of $\mathbf{A}$ and the vector $\mathbf{x}$ [as indicated in (2-12)]; i.e., the ith component is $\sum_{k=1}^{n} a_{ik} x_k$.

We have now defined the product of an $m \times n$ matrix by a one-column matrix. We now extend this definition to include the product of an $m \times n$ matrix by a p-column matrix:

If $\mathbf{A} = [a_{ij}]_{m \times n}$ and $\mathbf{X} = [x_{ij}]_{n \times p}$, then the product $\mathbf{AX}$ is the $m \times p$ matrix $\mathbf{B} = [b_{ij}]_{m \times p}$, where the jth column of $\mathbf{B}$ is defined as the product of $\mathbf{A}$ and the jth column of $\mathbf{X}$:

$$j\text{th column of } \mathbf{B} = \begin{bmatrix} b_{1j} \\ b_{2j} \\ \vdots \\ b_{mj} \end{bmatrix} = \begin{bmatrix} a_{11} & a_{12} & \cdots & a_{1n} \\ a_{21} & a_{22} & \cdots & a_{2n} \\ \vdots & \vdots & & \vdots \\ a_{m1} & a_{m2} & \cdots & a_{mn} \end{bmatrix} \begin{bmatrix} x_{1j} \\ x_{2j} \\ \vdots \\ x_{nj} \end{bmatrix} \tag{2-15}$$

From (2-15), it is clear that

$$b_{ij} = \sum_{k=1}^{n} a_{ik} x_{kj} = a_{i1} x_{1j} + a_{i2} x_{2j} + \cdots + a_{in} x_{nj} \tag{2-16}$$

We summarize:

If $\mathbf{A} = [a_{ij}]_{m \times n}$, $\mathbf{X} = [x_{ij}]_{n \times p}$, $\mathbf{B} = [b_{ij}]_{m \times p}$, and $\mathbf{AX} = \mathbf{B}$, where b_{ij} is defined by Eq. (2-16), then we have defined the product of two matrices $\mathbf{A}$, $\mathbf{X}$, provided that the number of columns of $\mathbf{A}$ is equal to the number of rows of $\mathbf{X}$ (so that the scalar products implicit in the definition are well-defined). Thus, whereas *addition* of two matrices is defined only if the two matrices are of the same order, the *multiplication* of two matrices is defined only if the condition stated in the previous sentence is satisfied.

The reader may be wondering at this point why we have not defined matrix multiplication in the more natural parallel to our definition of matrix addition (i.e., if $\mathbf{AX} = \mathbf{B}$, then elements of the product $\mathbf{B}$ are the products of the corresponding elements of $\mathbf{A}$ and $\mathbf{X}$: $b_{ij} = a_{ij}x_{ij}$). The reason we have not done so is simple: There is no particularly useful application for the latter. Our definition of matrix multiplication has been motivated by its usefulness, in terms of application.

EXAMPLE 2-4 If

$$\mathbf{A} = \begin{bmatrix} 1 & 8 & -2 & 0 \\ 3 & -5 & 6 & -9 \\ -4 & 0 & -7 & 2 \end{bmatrix} \qquad \mathbf{X} = \begin{bmatrix} 1 & 0 \\ 6 & -2 \\ 3 & 4 \\ 5 & 0 \end{bmatrix}$$

then

$$\mathbf{AX} = \begin{bmatrix} 1 & 8 & -2 & 0 \\ 3 & -5 & 6 & -9 \\ -4 & 0 & -7 & 2 \end{bmatrix} \begin{bmatrix} 1 & 0 \\ 6 & -2 \\ 3 & 4 \\ 5 & 0 \end{bmatrix}$$

$$= \begin{bmatrix} (1\cdot 1 + 8\cdot 6 + -2\cdot 3 + 0\cdot 5) & (1\cdot 0 + 8[-2] + -2\cdot 4 + 0\cdot 0) \\ (3\cdot 1 + -5\cdot 6 + 6\cdot 3 + -9\cdot 5) & (3\cdot 0 + -5[-2] + 6\cdot 4 + -9\cdot 0) \\ (-4\cdot 1 + 0\cdot 6 + -7\cdot 3 + 2\cdot 5) & (-4\cdot 0 + 0[-2] + -7\cdot 4 + 2\cdot 0) \end{bmatrix}$$

$$= \begin{bmatrix} 43 & -24 \\ -54 & 34 \\ -15 & -28 \end{bmatrix}$$

2-3 PROPERTIES OF MATRIX PRODUCTS

As we have noted in the previous section, our definition of matrix multiplication was chosen for its usefulness, not particularly for its simplicity. Unfortunately, as we shall see in this section, the consequences of our choice of definition are not altogether desirable.

Let us review this definition again, with a slight change in notation. Given two matrices $\mathbf{A} = [a_{ij}]_{m \times n}$, $\mathbf{B} = [b_{ij}]_{q \times p}$, the product $\mathbf{AB}$ is defined if and only if $n = q$. If $n = q$, then $\mathbf{A}$ and $\mathbf{B}$ are said to be *conformable* for multiplication, and, letting $\mathbf{AB} = \mathbf{C} = [c_{ij}]$, $\mathbf{C}$ will be an $m \times p$ matrix, with

$$c_{ij} = \sum_{k=1}^{n} a_{ik} b_{kj} \qquad \begin{array}{l} i = 1, 2, \ldots, m \\ j = 1, 2, \ldots, p \end{array}$$

In the product $\mathbf{AB}$, $\mathbf{B}$ is said to be *premultiplied* by $\mathbf{A}$, and $\mathbf{A}$ is said to be *postmultiplied* by $\mathbf{B}$.

Note that even if two matrices **A** and **B** are conformable for multiplication (i.e., **AB** exists), it is not in general true that **B** and **A** are conformable for multiplication. For example, if **A** is 2×3 and **B** is 3×5, then **AB** is defined and is a 2×5 matrix. However, **BA** is undefined. Thus, one rule of scalar algebra that does not extend to matrix algebra is the commutative law for multiplication:

Given two matrices **A** and **B**, it is *not* true in general that **AB** = **BA**.

Above we have seen that **AB** may exist while **BA** is undefined, or vice versa. Moreover, it is easy to see that both products will exist if and only if **A** is $m \times n$ and **B** is $n \times m$. In this case, **AB** is an $m \times m$ matrix and **BA** is an $n \times n$ matrix. Thus, both products will in general be of different order, let alone equal to each other. In fact, it is clear that **AB** and **BA** will both be of the same order if and only if both **A** and **B** are square matrices of the same order. However, it is unfortunately still true that **AB** and **BA** will in general be unequal even in this case.

EXAMPLE 2-5 If

$$\mathbf{A} = \begin{bmatrix} 1 & 0 & 2 \\ -1 & 1 & 3 \\ 2 & -4 & 0 \end{bmatrix} \qquad \mathbf{B} = \begin{bmatrix} 4 & 3 & 5 \\ 2 & 0 & 2 \\ -1 & 1 & 0 \end{bmatrix}$$

then

$$\mathbf{AB} = \begin{bmatrix} 1 & 0 & 2 \\ -1 & 1 & 3 \\ 2 & -4 & 0 \end{bmatrix} \begin{bmatrix} 4 & 3 & 5 \\ 2 & 0 & 2 \\ -1 & 1 & 0 \end{bmatrix} = \begin{bmatrix} 2 & 5 & 5 \\ -5 & 0 & -3 \\ 0 & 6 & 2 \end{bmatrix}$$

$$\mathbf{BA} = \begin{bmatrix} 4 & 3 & 5 \\ 2 & 0 & 2 \\ -1 & 1 & 0 \end{bmatrix} \begin{bmatrix} 1 & 0 & 2 \\ -1 & 1 & 3 \\ 2 & -4 & 0 \end{bmatrix} = \begin{bmatrix} 11 & -17 & 17 \\ 6 & -8 & 4 \\ -2 & 1 & 1 \end{bmatrix}$$

$$\mathbf{AB} \neq \mathbf{BA}$$

////

We now continue our assault on the properties of matrix multiplication by observing that the cancellation law does not hold, in general:

EXAMPLE 2-6 If

$$\mathbf{A} = \begin{bmatrix} 1 & -1 & 3 \\ 2 & 0 & 1 \\ 0 & 2 & -5 \end{bmatrix} \qquad \mathbf{B} = \begin{bmatrix} 1 & 0 & 1 \\ -1 & 1 & 3 \\ 2 & 4 & 0 \end{bmatrix}$$

$$\mathbf{C} = \begin{bmatrix} 1 & 1 & 0 \\ -1 & -4 & 8 \\ 2 & 2 & 2 \end{bmatrix}$$

Then

$$\mathbf{AB} = \begin{bmatrix} 1 & -1 & 3 \\ 2 & 0 & 1 \\ 0 & 2 & -5 \end{bmatrix} \begin{bmatrix} 1 & 0 & 1 \\ -1 & 1 & 3 \\ 2 & 4 & 0 \end{bmatrix} = \begin{bmatrix} 8 & 11 & -2 \\ 4 & 4 & 2 \\ -12 & -18 & 6 \end{bmatrix}$$

$$\mathbf{AC} = \begin{bmatrix} 1 & -1 & 3 \\ 2 & 0 & 1 \\ 0 & 2 & -5 \end{bmatrix} \begin{bmatrix} 1 & 1 & 0 \\ -1 & -4 & 8 \\ 2 & 2 & 2 \end{bmatrix} = \begin{bmatrix} 8 & 11 & -2 \\ 4 & 4 & 2 \\ -12 & -18 & 6 \end{bmatrix}$$

Hence, $\mathbf{AB} = \mathbf{AC}$ and $\mathbf{A} \neq \mathbf{O}_{3\times3}$, but clearly $\mathbf{B} \neq \mathbf{C}$. Thus, the cancellation law does not hold. ////

Moreover, a special case of the cancellation law of scalar algebra is: if $ab = 0$, then either $a = 0$ or $b = 0$ or both. The example below indicates that this law does not extend to matrix algebra either.

EXAMPLE 2-7 If

$$\mathbf{A} = \begin{bmatrix} 1 & -1 & 3 \\ 2 & 0 & 1 \\ 0 & 2 & -5 \end{bmatrix} \qquad \mathbf{B} = \begin{bmatrix} 1 & -2 & -1 \\ -5 & 10 & 5 \\ -2 & 4 & 2 \end{bmatrix}$$

Then

$$\mathbf{AB} = \begin{bmatrix} 1 & -1 & 3 \\ 2 & 0 & 1 \\ 0 & 2 & -5 \end{bmatrix} \begin{bmatrix} 1 & -2 & -1 \\ -5 & 10 & 5 \\ -2 & 4 & 2 \end{bmatrix} = \begin{bmatrix} 0 & 0 & 0 \\ 0 & 0 & 0 \\ 0 & 0 & 0 \end{bmatrix}$$

Thus, $\mathbf{AB} = \mathbf{O}_{3\times3}$, but $\mathbf{A} \neq \mathbf{O}_{3\times3}$ and $\mathbf{B} \neq \mathbf{O}_{3\times3}$.
Note also, that $\mathbf{BA} \neq \mathbf{O}_{3\times3}$:

$$\mathbf{BA} = \begin{bmatrix} 1 & -2 & -1 \\ -5 & 10 & 5 \\ -2 & 4 & 2 \end{bmatrix} \begin{bmatrix} 1 & -1 & 3 \\ 2 & 0 & 1 \\ 0 & 2 & -5 \end{bmatrix} = \begin{bmatrix} -3 & -3 & 6 \\ 15 & 5 & -30 \\ 6 & 6 & -12 \end{bmatrix}$$ ////

Thus far in this section we have discovered that matrix multiplication does not obey the commutative or cancellation laws, in general. Now, we shall at last turn to the positive side of the story, and investigate those laws which are applicable to matrix multiplication. We remark, first of all, that under certain circumstances, to be investigated in Sec. 2-4, the cancellation law can be applied to matrix equations. However, for the moment, we shall consider:

Theorem 2-2 Associative law If $\mathbf{A} = [a_{ij}]_{m\times n}$, $\mathbf{B} = [b_{ij}]_{n\times q}$, and $\mathbf{C} = [c_{ij}]_{q\times p}$, then

$$(\mathbf{AB})\mathbf{C} = \mathbf{A}(\mathbf{BC}) \qquad (2\text{-}17)$$

PROOF First, we observe that $\mathbf{A}_{m \times n}\mathbf{B}_{n \times q}$ is an $m \times q$ matrix, which when postmultiplied by $\mathbf{C}_{q \times p}$ yields an $m \times p$ matrix. Similarly, on the right-hand side of Eq. (2-17), $\mathbf{B}_{n \times q}\mathbf{C}_{q \times p}$ is an $n \times p$ matrix, which when premultiplied by $\mathbf{A}_{m \times n}$ also yields an $m \times p$ matrix. Thus, the matrix products on both sides of Eq. (2-17) yield matrices of the same order. It remains now to show that the corresponding elements of these matrices are equal.

To do so, we will first calculate $\mathbf{D} = [d_{ij}]_{m \times q}$:

$$\mathbf{D} = \mathbf{AB}$$

Thus,

$$d_{ij} = \sum_{k=1}^{n} a_{ik} b_{kj}$$

Then, we calculate $\mathbf{DC} = \mathbf{E} = [e_{ij}]_{m \times p}$:

$$e_{ij} = \sum_{r=1}^{q} d_{ir} c_{rj}$$

$$= \sum_{r=1}^{q} \left\{ \sum_{k=1}^{n} a_{ik} b_{kr} \right\} c_{rj}$$

$$e_{ij} = \sum_{r=1}^{q} \sum_{k=1}^{n} a_{ik} b_{kr} c_{rj} \tag{2-18}$$

and $\mathbf{E} = (\mathbf{AB})\mathbf{C}$

Now, letting $\mathbf{BC} = \mathbf{F} = [f_{ij}]_{n \times p}$, we find that

$$f_{ij} = \sum_{t=1}^{q} b_{it} c_{tj}$$

Thus, if $\mathbf{AF} = \mathbf{G} = [g_{ij}]_{m \times p}$, then

$$g_{ij} = \sum_{s=1}^{n} a_{is} f_{sj}$$

$$= \sum_{s=1}^{n} a_{is} \left\{ \sum_{t=1}^{q} b_{st} c_{tj} \right\}$$

$$= \sum_{s=1}^{n} \sum_{t=1}^{q} a_{is} b_{st} c_{tj}$$

$$g_{ij} = \sum_{t=1}^{q} \sum_{s=1}^{n} a_{is} b_{st} c_{tj} \tag{2-19}$$

In the last step above, we have interchanged the order of the two summations; this is permissible (in the case of finite sums) since it is equivalent to reordering the terms of the summation (i.e., adding the terms in a different order). Furthermore, the two indices of summation t, s may legitimately be replaced by any two other indices, such as r and k:

$$g_{ij} = \sum_{r=1}^{q} \sum_{k=1}^{n} a_{ik} b_{kr} c_{rj} \qquad (2\text{-}20)$$

Upon comparing Eqs. (2-18) and (2-20), it is evident that $e_{ij} = g_{ij}$ and hence that $\mathbf{E} = \mathbf{G}$. QED

Theorem 2-2 enables us to delete the parentheses in expressions involving matrix products. Thus, we may write

$$\mathbf{A(BC) = (AB)C = ABC}$$

We now state (and leave the proof as an exercise for the reader):

Theorem 2-3 Distributive laws If $\mathbf{A}$ and $\mathbf{B}$ are $m \times n$ matrices, $\mathbf{C}$ is $n \times p$, and $\mathbf{D}$ is $q \times m$, then

$$\mathbf{(A + B)C = AC + BC} \qquad (2\text{-}21)$$

$$\mathbf{D(A + B) = DA + DB} \qquad (2\text{-}22)$$

The proof of Theorem 2-2 (and of Theorem 2-3) involves computations and comparisons with the elements of matrices, resulting in an extremely tedious proof. Whenever possible, we shall endeavor—and we strongly urge the reader to do likewise—to employ manipulations involving only matrices and not their specific elements.

EXAMPLE 2-8

Theorem If $\mathbf{A}$ and $\mathbf{B}$ are of order $m \times n$, and if $\mathbf{C}$ and $\mathbf{D}$ are of order $n \times p$, then

$$\mathbf{(A + B)(C + D) = AC + AD + BC + BD}$$

PROOF Let $\mathbf{C + D = E}$. Then

$$\begin{aligned} \mathbf{(A + B)(C + D)} &= \mathbf{(A + B)E} \\ &= \mathbf{AE + BE} \qquad \text{by Eq. (2-21)} \\ &= \mathbf{A(C + D) + B(C + D)} \\ &= \mathbf{AC + AD + BC + BD} \qquad \text{by Eq. (2-22)} \end{aligned}$$

QED

We conclude this section with one additional, very important theorem, whose proof unfortunately involves the elements of matrices.

Theorem 2-4 If **A**, **B**, and **C** are matrices of order $m \times n$, $n \times p$, and $m \times n$, respectively,

$$(\alpha \mathbf{A})^T = \alpha(\mathbf{A}^T) \qquad \alpha \text{ any scalar} \tag{2-23}$$

$$(\mathbf{A}^T)^T = \mathbf{A} \tag{2-24}$$

$$(\mathbf{A} + \mathbf{C})^T = \mathbf{A}^T + \mathbf{C}^T \tag{2-25}$$

$$(\mathbf{AB})^T = \mathbf{B}^T\mathbf{A}^T \tag{2-26}$$

PROOF Equations (2-23), (2-24), and (2-25) are quite easy to prove, and we leave the details to the reader. To prove (2-26), let us first compute $(\mathbf{AB})^T = \mathbf{D} = [d_{ij}]$.

Note that $\mathbf{D}^T = ((\mathbf{AB})^T)^T = \mathbf{AB}$, by Eq. (2-24), and if $\mathbf{D}^T = [\hat{d}_{ij}]$, then

$$\hat{d}_{ij} = \sum_{k=1}^{n} a_{ik} b_{kj}$$

Moreover, since $\hat{d}_{ij} = d_{ji}$,

$$d_{ji} = \sum_{k=1}^{n} a_{ik} b_{kj}$$

or

$$d_{ij} = \sum_{k=1}^{n} a_{jk} b_{ki}$$

Now, letting $\mathbf{A}^T = [\hat{a}_{ij}]$, $\mathbf{B}^T = [\hat{b}_{ij}]$, and letting $\mathbf{B}^T\mathbf{A}^T = \mathbf{E} = [e_{ij}]$,

$$e_{ij} = \sum_{k=1}^{n} \hat{b}_{ik} \hat{a}_{kj}$$

But, again, $\hat{a}_{ij} = a_{ji}$ and $\hat{b}_{ij} = b_{ji}$, so that

$$e_{ij} = \sum_{k=1}^{n} b_{ki} a_{jk}$$

$$= \sum_{k=1}^{n} a_{jk} b_{ki}$$

Hence, $e_{ij} = d_{ij}$ QED

2-4 DIAGONAL MATRICES AND MATRIX INVERSES

In this section we shall be considering square ($n \times n$) matrices, unless otherwise noted.

Given any square matrix $\mathbf{A} = [a_{ij}]$, the elements $a_{11}, a_{22}, \ldots, a_{nn}$ are called the *diagonal elements* of $\mathbf{A}$. A matrix whose nondiagonal elements are all zero is called a *diagonal matrix*. A matrix all of whose elements above its diagonal are zero is called a *lower triangular matrix*; a matrix all of whose elements below its diagonal are zero is called an *upper triangular matrix*.

EXAMPLE 2-9

Diagonal matrices: $\begin{bmatrix} 1 & 0 & 0 \\ 0 & 2 & 0 \\ 0 & 0 & -1 \end{bmatrix}$ $\begin{bmatrix} 3 & 0 & 0 \\ 0 & 3 & 0 \\ 0 & 0 & 3 \end{bmatrix}$ $\begin{bmatrix} 2 & 0 & 0 & 0 \\ 0 & 0 & 0 & 0 \\ 0 & 0 & 4 & 0 \\ 0 & 0 & 0 & 7 \end{bmatrix}$

$\begin{bmatrix} 1 & 0 & 0 \\ 0 & 1 & 0 \\ 0 & 0 & 1 \end{bmatrix}$ $\begin{bmatrix} 0 & 0 & 0 \\ 0 & 0 & 0 \\ 0 & 0 & 0 \end{bmatrix}$

Lower triangular matrices: $\begin{bmatrix} 3 & 0 & 0 \\ -1 & 2 & 0 \\ 0 & 5 & 4 \end{bmatrix}$ $\begin{bmatrix} 2 & 0 & 0 & 0 \\ 3 & 5 & 0 & 0 \\ -1 & 0 & 0 & 0 \\ 0 & 1 & 3 & 4 \end{bmatrix}$ $\begin{bmatrix} 0 & 0 & 0 \\ 1 & 0 & 0 \\ 2 & 5 & 0 \end{bmatrix}$

Upper triangular matrices: $\begin{bmatrix} 3 & -1 & 0 \\ 0 & 2 & 5 \\ 0 & 0 & 4 \end{bmatrix}$ $\begin{bmatrix} 3 & 6 & 7 \\ 0 & 0 & 0 \\ 0 & 0 & 1 \end{bmatrix}$ $\mathbf{L}^T$, where $\mathbf{L}$ is any lower triangular matrix ////

A diagonal matrix whose diagonal elements are all equal to each other is called a *scalar matrix*. A scalar matrix acts like a scalar in matrix multiplication. That is, if $\mathbf{A} = [a_{ij}]_{n \times n}$ and if

$$\mathbf{S} = \begin{bmatrix} \alpha & 0 & \cdots & 0 \\ 0 & \alpha & \cdots & 0 \\ \vdots & \vdots & \ddots & \vdots \\ 0 & 0 & \cdots & \alpha \end{bmatrix}_{n \times n}$$

then $\mathbf{AS} = \mathbf{SA} = \alpha\mathbf{A}$. If $\alpha = 1$, we have a very special scalar matrix, called the *identity matrix*. An identity matrix of order n is denoted by $\mathbf{I}_n$; if the order is clear from the context, we may drop the subscript, denoting the identity matrix by $\mathbf{I}$.

If $\mathbf{A}$ is any $m \times n$ matrix, then it is obvious that

$$\mathbf{I}_m \mathbf{A} = \mathbf{A}\mathbf{I}_n = \mathbf{A} \qquad (2\text{-}27)$$

Thus, an identity matrix is the matrix counterpart of the scalar multiplicative identity 1 (see Table 2-1, property C4). There remains one property of scalar algebra whose matrix counterpart we have not yet considered, property C5: If $a \neq 0$, then $a \cdot a^{-1} = 1$ (a^{-1} is called the inverse of a with respect to multiplication). Therefore, we now define:

If $\mathbf{A}$ is an $m \times n$ matrix and if $\mathbf{A}_L$ is an $n \times m$ matrix such that

$$\mathbf{A}_L \mathbf{A} = \mathbf{I}_n \qquad (2\text{-}28)$$

then $\mathbf{A}_L$ is called a *left inverse* of $\mathbf{A}$. Similarly, if $\mathbf{A}_R$ is an $n \times m$ matrix such that

$$\mathbf{A}\mathbf{A}_R = \mathbf{I}_m$$

then $\mathbf{A}_R$ is called a *right inverse* of $\mathbf{A}$.

EXAMPLE 2-10 If $\mathbf{A} = \begin{bmatrix} 1 & 1 & 0 \\ -1 & 2 & 1 \end{bmatrix}$ and $\mathbf{A}_L = \begin{bmatrix} \alpha_1 & \beta_1 \\ \alpha_2 & \beta_2 \\ \alpha_3 & \beta_3 \end{bmatrix}$

then

$$\mathbf{A}_L \mathbf{A} = \begin{bmatrix} \alpha_1 & \beta_1 \\ \alpha_2 & \beta_2 \\ \alpha_3 & \beta_3 \end{bmatrix} \begin{bmatrix} 1 & 1 & 0 \\ -1 & 2 & 1 \end{bmatrix}$$

$$= \begin{bmatrix} \alpha_1 - \beta_1 & \alpha_1 + 2\beta_1 & \beta_1 \\ \alpha_2 - \beta_2 & \alpha_2 + 2\beta_2 & \beta_2 \\ \alpha_3 - \beta_3 & \alpha_3 + 2\beta_3 & \beta_3 \end{bmatrix} = \begin{bmatrix} 1 & 0 & 0 \\ 0 & 1 & 0 \\ 0 & 0 & 1 \end{bmatrix}$$

Equating corresponding elements leads to the nine equations:

$$\begin{array}{lll} \alpha_1 - \beta_1 = 1 & \alpha_2 - \beta_2 = 0 & \alpha_3 - \beta_3 = 0 \\ \alpha_1 + 2\beta_1 = 0 & \alpha_2 + 2\beta_2 = 1 & \alpha_3 + 2\beta_3 = 0 \\ \beta_1 = 0 & \beta_2 = 0 & \beta_3 = 1 \end{array}$$

We see that these equations have no solution (since, for example, $\beta_1 = 0$ implies that $\alpha_1 = 1$ *and* $\alpha_1 = 0$). Thus, $\mathbf{A}$ has no left inverse.

Let us see whether it possesses a right inverse:

We let

$$\mathbf{A}_R = \begin{bmatrix} \alpha_1 & \alpha_2 \\ \beta_1 & \beta_2 \\ \gamma_1 & \gamma_2 \end{bmatrix}$$

Then

$$\mathbf{A}\mathbf{A}_R = \begin{bmatrix} 1 & 1 & 0 \\ -1 & 2 & 1 \end{bmatrix} \begin{bmatrix} \alpha_1 & \alpha_2 \\ \beta_1 & \beta_2 \\ \gamma_1 & \gamma_2 \end{bmatrix}$$

$$= \begin{bmatrix} \alpha_1 + \beta_1 & \alpha_2 + \beta_2 \\ -\alpha_1 + 2\beta_1 + \gamma_1 & -\alpha_2 + 2\beta_2 + \gamma_2 \end{bmatrix} = \begin{bmatrix} 1 & 0 \\ 0 & 1 \end{bmatrix}$$

$$\alpha_1 + \beta_1 = 1 \qquad \alpha_2 + \beta_2 = 0$$

$$-\alpha_1 + 2\beta_1 + \gamma_1 = 0 \qquad -\alpha_2 + 2\beta_2 + \gamma_2 = 1$$

Eliminating α_1 and α_2 yields:

$$3\beta_1 + \gamma_1 = 1 \qquad 3\beta_2 + \gamma_2 = 1$$

Hence, any values of β_1, β_2, γ_1, γ_2, satisfying the above two equations will yield a right inverse. Therefore **A** has an infinite number of right inverses. For example, letting $\beta_1 = \beta_2 = 0$, we find that $\gamma_1 = \gamma_2 = 1$, $\alpha_1 = 1$, $\alpha_2 = 0$, and

$$\mathbf{A}_R = \begin{bmatrix} 1 & 0 \\ 0 & 0 \\ 1 & 1 \end{bmatrix}$$

Another right inverse is obtained by letting, for example, $\beta_1 = \beta_2 = 1$, which leads to $\alpha_1 = 0$, $\alpha_2 = -1$, $\gamma_1 = -2$, $\gamma_2 = -2$, yielding the right inverse $\begin{bmatrix} 0 & -1 \\ 1 & 1 \\ -2 & -2 \end{bmatrix}$ ////

EXAMPLE 2-11 If

$$\mathbf{A} = \begin{bmatrix} 1 & 1 \\ 0 & -1 \\ 1 & 4 \end{bmatrix}$$

then a procedure analogous to that employed in Example 2-10 will show that **A** does not have a right inverse but possesses an infinite number of left inverses, two of which are, for example,

$$\begin{bmatrix} 1 & 1 & 0 \\ -1 & 2 & 1 \end{bmatrix} \qquad \begin{bmatrix} 0 & 4 & 1 \\ 0 & -1 & 0 \end{bmatrix}$$ ////

EXAMPLE 2-12 If $\mathbf{A} = \begin{bmatrix} 1 & 2 \\ 0 & 1 \end{bmatrix}$ and $\mathbf{A}_R = \begin{bmatrix} \alpha & \beta \\ \gamma & \delta \end{bmatrix}$

$$\mathbf{A}\mathbf{A}_R = \begin{bmatrix} 1 & 2 \\ 0 & 1 \end{bmatrix} \begin{bmatrix} \alpha & \beta \\ \gamma & \delta \end{bmatrix} = \begin{bmatrix} \alpha + 2\gamma & \beta + 2\delta \\ \gamma & \delta \end{bmatrix} = \begin{bmatrix} 1 & 0 \\ 0 & 1 \end{bmatrix}$$

or $\gamma = 0$, $\delta = 1$, $\alpha = 1$, $\beta = -2$. Hence, **A** has a unique right inverse,

$$\mathbf{A}_R = \begin{bmatrix} 1 & -2 \\ 0 & 1 \end{bmatrix}$$

Moreover, if $\mathbf{A}_L = \begin{bmatrix} a & b \\ c & d \end{bmatrix}$

then
$$\mathbf{A}_L\mathbf{A} = \begin{bmatrix} a & b \\ c & d \end{bmatrix}\begin{bmatrix} 1 & 2 \\ 0 & 1 \end{bmatrix} = \begin{bmatrix} a & 2a+b \\ c & 2c+d \end{bmatrix} = \begin{bmatrix} 1 & 0 \\ 0 & 1 \end{bmatrix}$$

from which we see that $a = 1$, $c = 0$, $b = -2$, $d = 1$; hence, **A** has a unique left inverse,

$$\mathbf{A}_L = \begin{bmatrix} 1 & -2 \\ 0 & 1 \end{bmatrix}$$

Observe that $\mathbf{A}_R = \mathbf{A}_L$. We shall prove below that if any matrix has both a left and a right inverse, then these are both unique and identical. But first, we note that not every matrix has an inverse. The reader can easily verify that the matrix $\mathbf{A} = \begin{bmatrix} 1 & 1 \\ 1 & 1 \end{bmatrix}$ has neither a right nor a left inverse.

////

Theorem 2-5 If a matrix **A**, of order $m \times n$ has both a left inverse $\mathbf{A}_L$ and a right matrix $\mathbf{A}_R$, then:

1 $\mathbf{A}_L = \mathbf{A}_R$
2 $\mathbf{A}_L$ is unique.

PROOF By definition,

$$\mathbf{A}_L\mathbf{A} = \mathbf{I}_n \qquad (2\text{-}29)$$

$$\mathbf{A}\mathbf{A}_R = \mathbf{I}_m \qquad (2\text{-}30)$$

Postmultiplying both sides of (2-29) by $\mathbf{A}_R$ yields

$$(\mathbf{A}_L\mathbf{A})\mathbf{A}_R = \mathbf{I}_n\mathbf{A}_R$$

or

$$\mathbf{A}_L(\mathbf{A}\mathbf{A}_R) = \mathbf{A}_R \qquad (2\text{-}31)$$

By Eq. (2-30), we may substitute $\mathbf{A}\mathbf{A}_R = \mathbf{I}_m$ for the quantity in parentheses in (2-31), obtaining

$$\mathbf{A}_L(\mathbf{I}_m) = \mathbf{A}_R$$
$$\mathbf{A}_L = \mathbf{A}_R$$

Moreover, if $\mathbf{A}$ were to possess another left inverse, say $\hat{\mathbf{A}}_L$, then we could apply the same sequence of steps as above to show that $\hat{\mathbf{A}}_L = \mathbf{A}_R$. Thus, since $\mathbf{A}_L = \mathbf{A}_R$ and $\hat{\mathbf{A}}_L = \mathbf{A}_R$, it must be true that $\hat{\mathbf{A}}_L = \mathbf{A}_L$. Hence $\mathbf{A}_L$ (and of course also $\mathbf{A}_R$) is unique. QED

The following results will be developed in later chapters:

1 If $\mathbf{A}$ is $m \times n$, $m \neq n$, then $\mathbf{A}$ possesses either a right inverse or a left inverse, *but not both.* See Examples 2-10 and 2-11. (Discussed in Chap. 6).
2 If $\mathbf{A}$ is square, then it possesses either both a right and a left inverse (and hence these are equal and unique by Theorem 2-5) or no inverse at all (Chap. 4).

If a square matrix $\mathbf{A}$ possesses a right and a left inverse, then this common (and unique) inverse is called *the inverse of* $\mathbf{A}$, and is denoted by $\mathbf{A}^{-1}$.

Observe that, if $\mathbf{A}^{-1}$ exists, then the cancellation laws are valid:

$$\begin{aligned} \mathbf{A}^{-1}(\mathbf{AB}) &= \mathbf{A}^{-1}(\mathbf{AC}) \\ (\mathbf{A}^{-1}\mathbf{A})\mathbf{B} &= (\mathbf{A}^{-1}\mathbf{A})\mathbf{C} \\ \mathbf{IB} &= \mathbf{IC} \\ \mathbf{B} &= \mathbf{C} \end{aligned}$$

Also, if $\mathbf{AB} = \mathbf{O}$ and $\mathbf{A}^{-1}$ exists, then

$$\begin{aligned} \mathbf{A}^{-1}(\mathbf{AB}) &= \mathbf{A}^{-1}\mathbf{O} \\ (\mathbf{A}^{-1}\mathbf{A})\mathbf{B} &= \mathbf{O} \\ \mathbf{B} &= \mathbf{O} \end{aligned}$$

We now prove some very useful results regarding inverses of square matrices:

Theorem 2-6 Let $\mathbf{A}$ and $\mathbf{B}$ be square matrices of order n. If $\mathbf{A}^{-1}$ and $\mathbf{B}^{-1}$ exist, then:

1 $(\mathbf{A}^{-1})^{-1}$ exists.

2 $(\mathbf{A}^{-1})^{-1} = \mathbf{A}$ (2-32)

3 $(\mathbf{AB})^{-1}$ exists.

4 $(\mathbf{AB})^{-1} = \mathbf{B}^{-1}\mathbf{A}^{-1}$ (2-33)

5 $(\mathbf{A}^T)^{-1}$ exists.

6 $(\mathbf{A}^T)^{-1} = (\mathbf{A}^{-1})^T$ (2-34)

PROOF Suppose **C** is the inverse of **A**. Then by definition,

$$\mathbf{AC} = \mathbf{CA} = \mathbf{I}_n \tag{2-35}$$

But the above expression also may be interpreted as "**A** is the inverse of **C**." Since $\mathbf{A}^{-1}$ is unique, $\mathbf{C} = \mathbf{A}^{-1}$ and we have just stated that **A** is the inverse of $\mathbf{A}^{-1}$, which is precisely what Eq. (2-32) states. Moreover, this clearly implies that $(\mathbf{A}^{-1})^{-1}$ exists, as it is equal to **A**.

To prove (2-33), we merely observe that

$$(\mathbf{AB})(\mathbf{B}^{-1}\mathbf{A}^{-1}) = \mathbf{I}_n$$
$$(\mathbf{B}^{-1}\mathbf{A}^{-1})(\mathbf{AB}) = \mathbf{I}_n$$

Hence, by definition, $(\mathbf{B}^{-1}\mathbf{A}^{-1})$ is the inverse of **AB** (and, therefore, $(\mathbf{AB})^{-1}$ exists).

Finally, to prove that $(\mathbf{A}^T)^{-1} = (\mathbf{A}^{-1})^T$ (and therefore exists), we let $(\mathbf{A}^{-1})^T = \mathbf{C}$ and premultiply both sides by $\mathbf{A}^T$:

$$\mathbf{A}^T(\mathbf{A}^{-1})^T = \mathbf{A}^T\mathbf{C} \tag{2-36}$$

By Theorem 2-4 [Eq. (2-26)], the left-hand side of (2-36) is equal to $(\mathbf{A}^{-1}\mathbf{A})^T$. Hence

$$\begin{aligned} \mathbf{A}^T\mathbf{C} &= (\mathbf{A}^{-1}\mathbf{A})^T \\ &= \mathbf{I}_n{}^T \\ &= \mathbf{I}_n \end{aligned} \tag{2-37}$$

Equation (2-37) implies that "**C** is the inverse of $\mathbf{A}^T$": $\mathbf{C} = (\mathbf{A}^T)^{-1}$ But $\mathbf{C} = (\mathbf{A}^{-1})^T$, by definition. QED

2-5 SYMMETRIC AND HERMITIAN MATRICES

In Sec. 2-2 we defined the transpose of a matrix **A**, denoted by $\mathbf{A}^T$, as the matrix obtained from **A** by interchanging the corresponding rows and columns of **A**.

If $\mathbf{A} = \mathbf{A}^T$, then **A** is called a *symmetric matrix*. Thus, if $\mathbf{A} = [a_{ij}]$ is symmetric, then $a_{ij} = a_{ji}$ and **A** is square. Below are examples of symmetric matrices:

$$\begin{bmatrix} 1 & 0 & -2 & 3 \\ 0 & 5 & 1 & 2 \\ -2 & 1 & 4 & -7 \\ 3 & 2 & -7 & 0 \end{bmatrix} \qquad \begin{bmatrix} 0 & 2 & 3 \\ 2 & 1 & 0 \\ 3 & 0 & 8 \end{bmatrix} \qquad \begin{bmatrix} 3 & 0 & 0 \\ 0 & -2 & 0 \\ 0 & 0 & 6 \end{bmatrix}$$

Real symmetric matrices (those whose elements are all real numbers) play a very important role in the applications of matrix algebra. For example, if the element a_{ij} of a matrix **A** represents the distance between location i and location j, then the resulting matrix will be symmetric.† In statistical problems in which more than one random variable are involved, two matrices which are frequently relevant are the covariance matrix and the correlation matrix. Both of these matrices are always symmetric.

EXAMPLE 2-13

The Method of Least Squares

The admissions office of a small—but prestigious—midwestern university wishes to establish some criteria for determining which of its multitudes of applicants it should admit to its freshman class for the coming year. The office has compiled a set of data of previous students, with the high school rank and college rank of each student, and has decided to try and use high school rank as a "predictor" of college rank. Thus, letting

$$x = \text{high school rank}$$

$$y = \text{college rank}$$

the admissions office desires to find constants a, b such that

$$y = ax + b \qquad (2\text{-}38)$$

In this manner, given an applicant's high school rank x, the admissions office can then use Eq. (2-38) to predict this applicant's college rank (as a measure of his chances for success in the school).

Suppose the admission office has m sets of data, (x_i, y_i), $i = 1, 2, \ldots, m$. Plotting such data will yield a graph as shown on page 36. Thus, the admissions office desires to find the values of a and b so that the straight line $y = ax + b$ "best" fits the data. Substitution of the data into Eq. (2-34) yields m equations:

$$y_i = (ax_i + b) + \varepsilon_i \qquad i = 1, 2, \ldots, m \qquad (2\text{-}39)$$

The ε_i in (2-39) are needed because, as is illustrated on page 36, the data will not, in general, all lie on the same line (since two points determine a straight line, it would be most accidental if all m data points happened to lie on the same line).

† Unless the "distance" is direction-oriented, in which case $a_{ij} = -a_{ji}$. Such matrices are called *skew-symmetric*. If **A** is skew-symmetric, then $\mathbf{A} = -\mathbf{A}^T$.

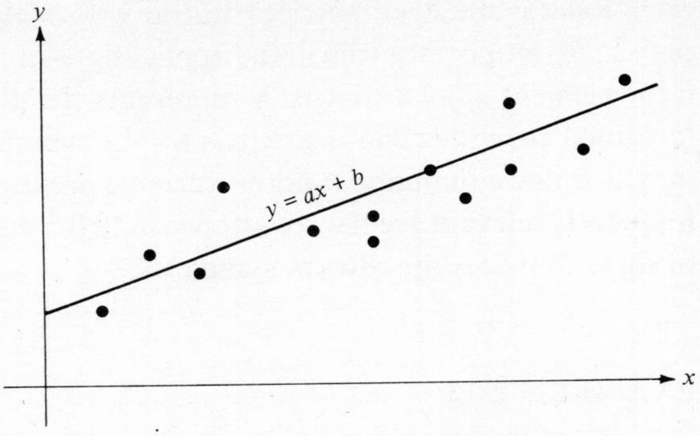

The classical least-squares approach to this problem is to find the values of a and b such that the sum of squares of the "errors" ε_i is a minimum. Hence, we seek to minimize

$$\begin{aligned} L &= \sum_{i=1}^{m} \varepsilon_i^2 \\ &= \sum_{i=1}^{m} (y_i - ax_i - b)^2 \end{aligned}$$

From calculus,† we know that the values of a and b for which L is a minimum are the solutions to the following system of simultaneous equations:

$$\begin{bmatrix} \sum_{i=1}^{m} x_i^2 & \sum_{i=1}^{m} x_i \\ \sum_{i=1}^{m} x_i & m \end{bmatrix} \begin{bmatrix} a \\ b \end{bmatrix} = \begin{bmatrix} \sum_{i=1}^{m} x_i y_i \\ \sum_{i=1}^{m} y_i \end{bmatrix} \tag{2-40}$$

Observe that the coefficient matrix above is symmetric. Moreover, suppose we write the m equations $y_i = ax_i + b$ in matrix notation:

$$\text{Let } \mathbf{y} = \begin{bmatrix} y_1 \\ y_2 \\ \vdots \\ y_m \end{bmatrix} \qquad \mathbf{x} = \begin{bmatrix} x_1 & 1 \\ x_2 & 1 \\ \vdots & \vdots \\ x_m & 1 \end{bmatrix} \qquad \mathbf{a} = \begin{bmatrix} a \\ b \end{bmatrix}$$

Then $\mathbf{Xa} = \mathbf{y}$ (2-41)

Observe also that if we premultiply both sides of Eq. (2-41) by $\mathbf{X}^T$, we obtain

$$(\mathbf{X}^T\mathbf{X})\mathbf{a} = \mathbf{X}^T\mathbf{y} \tag{2-42}$$

† For the reader who is unfamiliar with calculus, we note that this is our only reference to this subject in this text.

$$\left(\begin{bmatrix} x_1 & x_2 & \cdots & x_m \\ 1 & 1 & \cdots & 1 \end{bmatrix}\begin{bmatrix} x_1 & 1 \\ x_2 & 1 \\ \vdots & \vdots \\ x_m & 1 \end{bmatrix}\right)\begin{bmatrix} a \\ b \end{bmatrix} = \begin{bmatrix} x_1 & x_2 & \cdots & x_m \\ 1 & 1 & \cdots & 1 \end{bmatrix}\begin{bmatrix} y_1 \\ y_2 \\ \vdots \\ y_m \end{bmatrix}$$

$$\begin{bmatrix} \sum_{i=1}^{m} x_i^2 & \sum_{i=1}^{m} x_i \\ \sum_{i=1}^{m} x_i & m \end{bmatrix}\begin{bmatrix} a \\ b \end{bmatrix} = \begin{bmatrix} \sum_{i=1}^{m} x_i y_i \\ \sum_{i=1}^{m} y_i \end{bmatrix}$$

Thus, Eqs. (2-40) and (2-42) are identical.

Suppose our admissions office decides it wants to throw several more factors into its formula for predicting college rank (such as Scholastic Aptitude Verbal and Mathematics Test scores). If there are n such factors $x_1, x_2, \ldots, x_n$, then the predictor formula will be of the form

$$y_i = \sum_{j=1}^{n} a_j x_{ij} + \varepsilon_i \qquad i = 1, 2, \ldots, m \tag{2-43}$$

where x_{ij} denotes the ith data value of the jth factor x_j, and we wish to find the values of $a_1, a_2, \ldots, a_n$ which minimize $L = \sum_{i=1}^{m} \varepsilon_i^2$. (In the previous case, $a_1 = a$, $a_2 = b$, and $x_{i2} = 1$, $i = 1, 2, \ldots, m$.) In matrix notation, the m equations

$$y_i = \sum_{j=1}^{n} a_j x_{ij} \qquad i = 1, 2, \ldots, m$$

may be written

$$\mathbf{Xa} = \mathbf{y} \tag{2-44}$$

where $\mathbf{X} = [x_{ij}]_{m \times n}$, $\mathbf{a} = [a_1 \quad a_2 \quad \ldots \quad a_n]^T$, $\mathbf{y} = [y_1 \quad y_2 \quad \ldots \quad y_m]^T$. Again, it can be shown† that the values of $a_1, a_2, \ldots, a_n$ which minimize $L = \sum_{i=1}^{m} \varepsilon_i^2$ are found by solving

$$(\mathbf{X}^T\mathbf{X})\mathbf{a} = \mathbf{X}^T\mathbf{y}$$

By direct application of Theorem 2-4, we can show that $\mathbf{X}^T\mathbf{X}$ is *always* a symmetric matrix. Letting $\mathbf{A} = \mathbf{X}^T\mathbf{X}$, we wish to show that $\mathbf{A} = \mathbf{A}^T$. But

$$\begin{aligned} \mathbf{A}^T &= (\mathbf{X}^T\mathbf{X})^T \\ &= \mathbf{X}^T(\mathbf{X}^T)^T && \text{by Eq. (2-26)} \\ &= \mathbf{X}^T\mathbf{X} && \text{by Eq. (2-24)} \\ &= \mathbf{A} \end{aligned}$$

////

† See, for example, Noble [4].

As we shall learn in Chap. 9, real symmetric matrices have certain special properties, which unfortunately do not generalize to symmetric matrices with complex elements. However, there is another generalization to certain matrices with complex elements in which these special properties of real symmetric matrices are also valid. To define these matrices, we first define:

The *transpose conjugate* of a matrix **A**, denoted by $\mathbf{A}^*$ (or in some texts by $\mathbf{A}^H$) is the matrix obtained from **A** by taking the transpose of the matrix whose elements are the complex conjugates of those of **A**. In other words, if $\mathbf{A} = [a_{ij}]_{m \times n}$ and $\mathbf{A}^* = [a_{ij}^*]$, then $a_{ij}^* = \bar{a}_{ji}$ and $\mathbf{A}^*$ is of order $n \times m$. (In some texts, $\mathbf{A}^*$ is called the *tranjugate* of **A** or *adjoint* of **A**).

If $\mathbf{A} = \mathbf{A}^*$, then **A** is called a *Hermitian matrix*. Thus, **A** is Hermitian if $a_{ij} = \bar{a}_{ji}$. Note that a real symmetric matrix is a special case of a Hermitian matrix.

The results below follow directly from the definition of $\mathbf{A}^*$ and Theorems 2-4 and 2-6:

Theorem 2-7 If **A**, **B**, and **C** are matrices of order $m \times n$, $n \times p$, and $m \times n$, respectively, then

$$(\alpha\mathbf{A})^* = \bar{\alpha}\mathbf{A}^* \qquad \text{any scalar} \tag{2-45}$$

$$(\mathbf{A}^*)^* = \mathbf{A} \tag{2-46}$$

$$(\mathbf{A} + \mathbf{C})^* = \mathbf{A}^* + \mathbf{C}^* \tag{2-47}$$

$$(\mathbf{AB})^* = \mathbf{B}^*\mathbf{A}^* \tag{2-48}$$

$$(\mathbf{A}^{-1})^* = (\mathbf{A}^*)^{-1} \tag{2-49}$$

While we are on the subject of appropriate generalizations from real matrices to complex matrices, we note that the proper generalization of the scalar product of two real component vectors is:

The *inner product* of the vectors $\mathbf{a} = [\alpha_1 \quad \alpha_2 \quad \cdots \quad \alpha_n]^T$ and $\mathbf{b} = [\beta_1 \quad \beta_2 \quad \cdots \quad \beta_n]^T$, where the α_j and β_j are complex numbers, is defined by

$$\mathbf{a}^*\mathbf{b} = \sum_{j=1}^{n} \bar{\alpha}_j \beta_j \tag{2-50}$$

Note that the inner product of any vector with itself is always a non-negative real number:

$$\mathbf{a}^*\mathbf{a} = \sum_{j=1}^{n} \bar{\alpha}_j \alpha_j = \sum_{j=1}^{n} |\alpha_j|^2 \geq 0 \tag{2-51}$$

Moreover, $\mathbf{a}^*\mathbf{a} = 0$ if and only if every $\alpha_j = 0$, $j = 1, 2, \ldots, n$.

2-6 MATRIX PARTITIONING

We have seen (e.g., in Example 2-8 and the proof of Theorem 2-6) that matrix notation and manipulations with matrices (as opposed to their elements) can greatly facilitate the development of theoretical results. We shall now extend this concept further by defining an additional notational convenience known as matrix partitioning.

It is frequently desirable to subdivide (or partition) the elements of a matrix into groups of elements each group of which is itself a matrix (i.e., a rectangular array). Each such group is called a *submatrix* of the original matrix. The partitioning itself is often indicated by dashed lines. For example, consider the matrix

$$\mathbf{A} = \begin{bmatrix} 2 & 4 & 0 & -3 & 7 \\ 1 & 0 & 0 & 5 & 7 \\ -4 & -2 & 0 & 4 & 3 \\ 1 & -1 & 2 & 8 & 5 \end{bmatrix}$$

One possible partitioned form is

$$\mathbf{A} = \left[\begin{array}{cc:ccc} 2 & 4 & 0 & -3 & 7 \\ 1 & 0 & 0 & 5 & 7 \\ \hdashline -4 & -2 & 0 & 4 & 3 \\ 1 & -1 & 2 & 8 & 5 \end{array}\right]$$

The submatrices of **A** would then be

$$\mathbf{A}_{11} = \begin{bmatrix} 2 & 4 \\ 1 & 0 \end{bmatrix} \qquad \mathbf{A}_{12} = \begin{bmatrix} 0 & -3 & 7 \\ 0 & 5 & 7 \end{bmatrix}$$

$$\mathbf{A}_{21} = \begin{bmatrix} -4 & -2 \\ 1 & -1 \end{bmatrix} \qquad \mathbf{A}_{22} = \begin{bmatrix} 0 & 4 & 3 \\ 2 & 8 & 5 \end{bmatrix}$$

and we could then write

$$\mathbf{A} = \begin{bmatrix} \mathbf{A}_{11} & \mathbf{A}_{12} \\ \mathbf{A}_{21} & \mathbf{A}_{22} \end{bmatrix} \tag{2-52}$$

Equation (2-52) suggests that we may wish to treat the submatrices as *elements* of **A**. Frequently, this is precisely what is desired. Suppose, for example, that we wished to perform the matrix multiplication **AB**, where **A** and **B** are of orders $m \times n$ and $n \times p$, respectively, and suppose further that **A** and **B** have been partitioned as follows:

$$\mathbf{A} = \begin{bmatrix} \mathbf{A}_{11} & \mathbf{A}_{12} \\ \mathbf{A}_{21} & \mathbf{A}_{22} \end{bmatrix} \qquad \mathbf{B} = \begin{bmatrix} \mathbf{B}_{11} & \mathbf{B}_{12} \\ \mathbf{B}_{21} & \mathbf{B}_{22} \end{bmatrix}$$

Then, symbolically, we can write:

$$\mathbf{AB} = \begin{bmatrix} \mathbf{A}_{11} & \mathbf{A}_{12} \\ \mathbf{A}_{21} & \mathbf{A}_{22} \end{bmatrix} \begin{bmatrix} \mathbf{B}_{11} & \mathbf{B}_{12} \\ \mathbf{B}_{21} & \mathbf{B}_{22} \end{bmatrix}$$

$$= \begin{bmatrix} (\mathbf{A}_{11}\mathbf{B}_{11} + \mathbf{A}_{12}\mathbf{B}_{21}) & (\mathbf{A}_{11}\mathbf{B}_{12} + \mathbf{A}_{12}\mathbf{B}_{22}) \\ (\mathbf{A}_{21}\mathbf{B}_{11} + \mathbf{A}_{22}\mathbf{B}_{21}) & (\mathbf{A}_{21}\mathbf{B}_{12} + \mathbf{A}_{22}\mathbf{B}_{22}) \end{bmatrix} \tag{2-53}$$

In order for (2-53) to be a legitimate equation, each of the indicated matrix operations on the right-hand side must be defined. Thus, each of the matrix products $\mathbf{A}_{11}\mathbf{B}_{11}$, $\mathbf{A}_{12}\mathbf{B}_{21}$, $\mathbf{A}_{11}\mathbf{B}_{12}$, $\mathbf{A}_{12}\mathbf{B}_{22}$, etc., must be defined and each of the pairs of matrices $\mathbf{A}_{11}\mathbf{B}_{11}$ and $\mathbf{A}_{12}\mathbf{B}_{21}$, $\mathbf{A}_{11}\mathbf{B}_{12}$ and $\mathbf{A}_{12}\mathbf{B}_{22}$, etc., must be of the same order (so that the indicated matrix additions are defined).

We can satisfy these conditions by partitioning **A** and **B** so that

$$\begin{array}{ll} \mathbf{A}_{11} \text{ is } r \times s & \mathbf{A}_{12} \text{ is } r \times (n-s) \\ \mathbf{A}_{21} \text{ is } (m-r) \times s & \mathbf{A}_{22} \text{ is } (m-r) \times (n-s) \\ \mathbf{B}_{11} \text{ is } s \times q & \mathbf{B}_{12} \text{ is } s \times (p-q) \\ \mathbf{B}_{21} \text{ is } (n-s) \times q & \mathbf{B}_{22} \text{ is } (n-s) \times (p-q) \end{array} \tag{2-54}$$

EXAMPLE 2-14 Partition **B** so that we can compute **AB**, if

$$\mathbf{A} = \left[\begin{array}{cc:ccc} 2 & 4 & 0 & -3 & 7 \\ 1 & 0 & 0 & 5 & 7 \\ \hdashline -4 & -2 & 0 & 4 & 3 \end{array}\right] \qquad \mathbf{B} = \begin{bmatrix} 0 & -1 & 2 & 5 \\ 1 & 0 & 0 & 8 \\ 2 & -3 & 0 & 4 \\ 5 & -2 & 3 & 0 \\ 1 & -1 & 1 & 1 \end{bmatrix}$$

and if **A** is partitioned as indicated.

Thus, $m = 3$, $n = 5$, $p = 4$, and

$$\mathbf{A}_{11} = \begin{bmatrix} 2 & 4 \\ 1 & 0 \end{bmatrix}_{2\times 2} \qquad \mathbf{A}_{12} = \begin{bmatrix} 0 & -3 & 7 \\ 0 & 5 & 7 \end{bmatrix}_{2\times 3}$$

$$\mathbf{A}_{21} = [-4 \quad -2]_{1\times 2} \qquad \mathbf{A}_{22} = [0 \quad 4 \quad 3]_{1\times 3}$$

According to (2-54), then, $r = s = 2$. Therefore, if

$$\mathbf{B} = \begin{bmatrix} \mathbf{B}_{11} & \mathbf{B}_{12} \\ \mathbf{B}_{21} & \mathbf{B}_{22} \end{bmatrix}$$

we must partition the rows of $\mathbf{B}$ so that $\mathbf{B}_{11}$ has $s = 2$ rows and $\mathbf{B}_{21}$ has $(n - s) = 3$ rows. We may partition the columns arbitrarily; e.g., let $q = 1$. We then have

$$\mathbf{B}_{11} = \begin{bmatrix} 0 \\ 1 \end{bmatrix} \qquad \mathbf{B}_{12} = \begin{bmatrix} -1 & 2 & 5 \\ 0 & 0 & 8 \end{bmatrix}$$

$$\mathbf{B}_{21} = \begin{bmatrix} 2 \\ 5 \\ 1 \end{bmatrix} \qquad \mathbf{B}_{22} = \begin{bmatrix} -3 & 0 & 4 \\ -2 & 3 & 0 \\ -1 & 1 & 1 \end{bmatrix}$$

and

$$\mathbf{AB} = \begin{bmatrix} \mathbf{A}_{11} & \mathbf{A}_{12} \\ \mathbf{A}_{21} & \mathbf{A}_{22} \end{bmatrix} \begin{bmatrix} \mathbf{B}_{11} & \mathbf{B}_{12} \\ \mathbf{B}_{21} & \mathbf{B}_{22} \end{bmatrix} = \begin{bmatrix} (\mathbf{A}_{11}\mathbf{B}_{11} + \mathbf{A}_{12}\mathbf{B}_{21}) & (\mathbf{A}_{11}\mathbf{B}_{12} + \mathbf{A}_{12}\mathbf{B}_{22}) \\ (\mathbf{A}_{21}\mathbf{B}_{11} + \mathbf{A}_{22}\mathbf{B}_{21}) & (\mathbf{A}_{21}\mathbf{B}_{12} + \mathbf{A}_{22}\mathbf{B}_{22}) \end{bmatrix}$$

$$= \begin{bmatrix} \left(\begin{bmatrix} 2 & 4 \\ 1 & 0 \end{bmatrix} \begin{bmatrix} 0 \\ 1 \end{bmatrix} + \begin{bmatrix} 0 & -3 & 7 \\ 0 & 5 & 7 \end{bmatrix} \begin{bmatrix} 2 \\ 5 \\ 1 \end{bmatrix} \right) & \left(\begin{bmatrix} 2 & 4 \\ 1 & 0 \end{bmatrix} \begin{bmatrix} -1 & 2 & 5 \\ 0 & 0 & 8 \end{bmatrix} + \begin{bmatrix} 0 & -3 & 7 \\ 0 & 5 & 7 \end{bmatrix} \begin{bmatrix} -3 & 0 & 4 \\ -2 & 3 & 0 \\ -1 & 1 & 1 \end{bmatrix} \right) \\ \left([-4 \quad -2] \begin{bmatrix} 0 \\ 1 \end{bmatrix} + [0 \quad 4 \quad 3] \begin{bmatrix} 2 \\ 5 \\ 1 \end{bmatrix} \right) & \left([-4 \quad -2] \begin{bmatrix} -1 & 2 & 5 \\ 0 & 0 & 8 \end{bmatrix} + [0 \quad 4 \quad 3] \begin{bmatrix} -3 & 0 & 4 \\ -2 & 3 & 0 \\ -1 & 1 & 1 \end{bmatrix} \right) \end{bmatrix}$$

$$= \begin{bmatrix} \left(\begin{bmatrix} 4 \\ 0 \end{bmatrix} + \begin{bmatrix} -8 \\ 32 \end{bmatrix} \right) & \left(\begin{bmatrix} -2 & 4 & 42 \\ -1 & 2 & 5 \end{bmatrix} + \begin{bmatrix} -1 & -2 & 7 \\ -17 & 22 & 7 \end{bmatrix} \right) \\ \left([-2] + [23] \right) & \left([4 \quad -8 \quad -36] + [-11 \quad 15 \quad 3] \right) \end{bmatrix}$$

$$= \begin{bmatrix} \begin{bmatrix} -4 \\ 32 \end{bmatrix} & \begin{bmatrix} -3 & 2 & 49 \\ -18 & 24 & 12 \end{bmatrix} \\ [\ 21] & [-7 \quad 7 \quad -33] \end{bmatrix}$$

$$= \begin{bmatrix} -4 & -3 & 2 & 49 \\ 32 & -18 & 24 & 12 \\ 21 & -7 & 7 & -33 \end{bmatrix}$$

Other valid ways of partitioning $\mathbf{B}$ such that $\mathbf{AB}$ can be computed as above include:

$$\left[\begin{array}{cc|cc} 0 & -1 & 2 & 5 \\ 1 & 0 & 0 & 8 \\ \hline 2 & -3 & 0 & 4 \\ 5 & -2 & 3 & 0 \\ 1 & -1 & 1 & 1 \end{array}\right] \qquad \left[\begin{array}{ccc|c} 0 & -1 & 2 & 5 \\ 1 & 0 & 0 & 8 \\ \hline 2 & -3 & 0 & 4 \\ 5 & -2 & 3 & 0 \\ 1 & -1 & 1 & 1 \end{array}\right] \qquad \left[\begin{array}{c|c|c|c} 0 & -1 & 2 & 5 \\ 1 & 0 & 0 & 8 \\ \hline 2 & -3 & 0 & 4 \\ 5 & -2 & 3 & 0 \\ 1 & -1 & 1 & 1 \end{array}\right] \qquad ////$$

The last form above indicates that we are not restricted to partitioning a matrix into four submatrices. Two particular forms for partitioning a matrix which are frequently quite useful are partitioning an $m \times n$ matrix into n one-column submatrices or into m one-row submatrices :

If $\mathbf{a}_j = j$th column of a matrix $\mathbf{A}$, then

$$\mathbf{A} = [\mathbf{a}_1 \quad \mathbf{a}_2 \quad \cdots \quad \mathbf{a}_n]$$

Similarly, if we define ${\mathbf{a}_i}^T$ to be the ith row of $\mathbf{A}$, then

$$\mathbf{A} = \begin{bmatrix} {\mathbf{a}_1}^T \\ {\mathbf{a}_2}^T \\ \vdots \\ {\mathbf{a}_m}^T \end{bmatrix}$$

The reader may recall that we actually employed matrix partitioning—without calling it that—in our discussion of the definition of matrix multiplication [see Eq. (2-13)]. In fact, if we wish to multiply the two matrices $\mathbf{A}$, $\mathbf{B}$,

$$\mathbf{A} = \begin{bmatrix} {\mathbf{a}_1}^T \\ {\mathbf{a}_2}^T \\ \vdots \\ {\mathbf{a}_m}^T \end{bmatrix} \qquad \mathbf{B} = [\mathbf{b}_1 \quad \mathbf{b}_2 \quad \cdots \quad \mathbf{b}_p]$$

(assuming that $\mathbf{A}$ is $m \times n$ and $\mathbf{B}$ is $n \times p$), then

$$\mathbf{AB} = \begin{bmatrix} {\mathbf{a}_1}^T \\ {\mathbf{a}_2}^T \\ \vdots \\ {\mathbf{a}_m}^T \end{bmatrix} [\mathbf{b}_1 \quad \mathbf{b}_2 \quad \cdots \quad \mathbf{b}_p]$$

$$= \begin{bmatrix} {\mathbf{a}_1}^T\mathbf{b}_1 & {\mathbf{a}_1}^T\mathbf{b}_2 & \cdots & {\mathbf{a}_1}^T\mathbf{b}_p \\ {\mathbf{a}_2}^T\mathbf{b}_1 & {\mathbf{a}_2}^T\mathbf{b}_2 & \cdots & {\mathbf{a}_2}^T\mathbf{b}_p \\ \vdots & \vdots & & \vdots \\ {\mathbf{a}_m}^T\mathbf{b}_1 & {\mathbf{a}_m}^T\mathbf{b}_2 & \cdots & {\mathbf{a}_m}^T\mathbf{b}_p \end{bmatrix}$$

Thus, the element in the ith row and jth column of the product is just the scalar product of the ith row of $\mathbf{A}$ with the jth column of $\mathbf{B}$: ${\mathbf{a}_i}^T\mathbf{b}_j$.

REFERENCES

1. BRONSON, RICHARD: "Matrix Methods: An Introduction," Academic, New York, 1969.
2. COOPER, LEON, and DAVID STEINBERG: "Introduction to Methods of Optimization," Saunders, Philadelphia, 1970.
3. HOHN, FRANZ E.: "Elementary Matrix Algebra," 2d ed., Macmillan, New York, 1964.
4. NOBLE, BEN: "Applied Linear Algebra," Prentice-Hall, Englewood Cliffs, N.J., 1969.

EXERCISES

The matrices below pertain to Exercises 1 to 20. In each exercise, perform the indicated operation or state that the operation is undefined:

$$\mathbf{A} = \begin{bmatrix} 1 & 0 & 3 \\ 0 & -1 & 5 \\ 7 & 3 & 2 \end{bmatrix} \quad \mathbf{B} = \begin{bmatrix} 0 & 0 & 1 \\ 1 & 0 & -2 \\ 7 & -5 & 4 \\ 3 & -3 & 0 \end{bmatrix} \quad \mathbf{C} = \begin{bmatrix} 1 & 8 & 0 & 0 \\ -4 & 0 & 6 & 1 \\ 0 & -1 & 0 & -2 \end{bmatrix}$$

$$\mathbf{D} = \begin{bmatrix} 2 & 0 & 0 \\ 0 & -1 & 0 \\ 0 & 0 & 3 \end{bmatrix} \quad \mathbf{v} = \begin{bmatrix} 1+i \\ 1-i \\ 3 \end{bmatrix} \quad \mathbf{w} = \begin{bmatrix} 3i \\ 1+2i \\ 0 \end{bmatrix} \quad \mathbf{x} = \begin{bmatrix} 1 \\ 0 \\ -2 \\ 4 \end{bmatrix} \quad \mathbf{y} = \begin{bmatrix} 0 \\ 1 \\ 2 \\ 1 \end{bmatrix}$$

1 $\mathbf{AB}$

2 $\mathbf{BA}$

3 $\mathbf{AC}$

4 $\mathbf{BC}$

5 $\mathbf{CB}$

6 (*a*) $\mathbf{AB}^T$
(*b*) $\mathbf{BA}^T$
(*c*) $(\mathbf{BA}^T)^T$

7 (*a*) $\mathbf{AD}$
(*b*) $\mathbf{DA}$

8 (*a*) $\mathbf{xy}$
(*b*) $\mathbf{x}^T\mathbf{y}$
(*c*) $\mathbf{y}^T\mathbf{x}$
(*d*) $\mathbf{xy}^T$
(*e*) $\mathbf{yx}^T$

9 (*a*) $\mathbf{Cx}$
(*b*) $\mathbf{xC}$
(*c*) $\mathbf{xC}^T$

10 $\mathbf{A}^T\mathbf{B}^T$

11 (*a*) $\mathbf{B} + \mathbf{C}$
(*b*) $\mathbf{B} + \mathbf{C}^T$
(*c*) $\mathbf{B}^T + \mathbf{C}$
(*d*) $(\mathbf{B}^T + \mathbf{C})^T$

12 (*a*) $\mathbf{AA}$
(*b*) $\mathbf{A}^T\mathbf{A}$
(*c*) $\mathbf{AA}^T$

13 (*a*) $\mathbf{v}^*\mathbf{w}$
(*b*) $\mathbf{w}^*\mathbf{v}$

14 (*a*) $\mathbf{v}^*\mathbf{v}$
(*b*) $\mathbf{v}^T\mathbf{v}$
(*c*) $\mathbf{w}^*\mathbf{w}$
(*d*) $\mathbf{w}^T\mathbf{w}$

15 (*a*) $(\mathbf{BA})\mathbf{C}$
(*b*) $\mathbf{B}(\mathbf{AC})$
(*c*) $(\mathbf{CB})\mathbf{A}$
(*d*) $\mathbf{ABC}$

16 (*a*) $\mathbf{Dv}$
(*b*) $\mathbf{v}^*\mathbf{D}^*$
(*c*) $(\mathbf{Dv})^*$

17 Let $\alpha = 2 + i$
(*a*) $\alpha\mathbf{w}$
(*b*) $(\alpha\mathbf{w})^*$
(*c*) $(\alpha\mathbf{v})^*\mathbf{w}$
(*d*) $\alpha(\mathbf{v}^*\mathbf{w})$

18 (*a*) $(\mathbf{A} + \mathbf{D})\mathbf{C}$
(*b*) $\mathbf{B}(\mathbf{A} + \mathbf{D})$

19 $\mathbf{D}(\mathbf{A} + \mathbf{CB})\mathbf{A}$

20 (*a*) $\mathbf{y}^T\mathbf{CBx}$
(*b*) $\mathbf{y}^T\mathbf{BAx}$

21 Prove Theorem 2-1.

22 Use Theorem 2-3 and mathematical induction to prove: If $\mathbf{A}_1, \mathbf{A}_2, \ldots, \mathbf{A}_k$ are $m \times n$ matrices, then for every positive integer k,

$$(\mathbf{A}_1 + \mathbf{A}_2 + \cdots + \mathbf{A}_k)\mathbf{C} = \sum_{j=1}^{k} \mathbf{A}_j\mathbf{C}$$

$$\mathbf{D}(\mathbf{A}_1 + \mathbf{A}_2 + \cdots + \mathbf{A}_k) = \sum_{j=1}^{k} \mathbf{D}\mathbf{A}_j$$

where $\mathbf{C}$ is any $n \times p$ matrix and $\mathbf{D}$ is any $q \times m$ matrix.

23 Prove Theorem 2-4.

24 Use Theorem 2-4 and induction to prove: If $\mathbf{A}_1, \mathbf{A}_2, \ldots, \mathbf{A}_k$ are matrices such that the product $\mathbf{A}_1\mathbf{A}_2 \cdots \mathbf{A}_k$ is defined, then, for every positive integer k,

$$(\mathbf{A}_1\mathbf{A}_2 \cdots \mathbf{A}_k)^T = \mathbf{A}_k{}^T\mathbf{A}_{k-1}^T \cdots \mathbf{A}_2{}^T\mathbf{A}_1{}^T$$

25 For each of the matrices below, compute a left and a right inverse or show that such doesn't exist:

(a) $\mathbf{A} = \begin{bmatrix} 3 & 1 \\ 5 & 2 \end{bmatrix}$ (b) $\mathbf{B} = \begin{bmatrix} 2 & 1 & 1 \\ 1 & 2 & 1 \end{bmatrix}$ (c) $\mathbf{C} = \begin{bmatrix} 1 & 0 \\ 0 & 2 \\ 3 & 1 \end{bmatrix}$

26 If

$$\mathbf{A} = \begin{bmatrix} 3 & 1 \\ 5 & 2 \end{bmatrix} \quad \text{and} \quad \mathbf{B} = \begin{bmatrix} 4 & -3 \\ 5 & -4 \end{bmatrix}$$

compute:

(a) $\mathbf{A}^{-1}$ (b) $\mathbf{B}^{-1}$ (c) $(\mathbf{AB})^{-1}$ (d) $(\mathbf{BA})^{-1}$ (e) $(\mathbf{A}^T)^{-1}$

27 Prove that, if $\mathbf{A}$ is any $m \times n$ matrix, $\mathbf{A}^T\mathbf{A}$ and $\mathbf{AA}^T$ are both symmetric.

28 Show by counterexample that, in general, $\mathbf{AA}^T \neq \mathbf{A}^T\mathbf{A}$.

29 Show that if $\mathbf{A}$ is $m \times n$ and $\mathbf{B}$ is a symmetric matrix of order m, then $\mathbf{A}^T\mathbf{BA}$ is symmetric of order n.

30 Use Theorem 2-6 and induction to prove: If $\mathbf{A}_1, \mathbf{A}_2, \ldots, \mathbf{A}_k$ are square matrices whose inverses all exist, then

$$(\mathbf{A}_1\mathbf{A}_2 \cdots \mathbf{A}_k)^{-1} = \mathbf{A}_k^{-1}\mathbf{A}_{k-1}^{-1} \cdots \mathbf{A}_2^{-1}\mathbf{A}_1^{-1}$$

for every positive integer k.

31 Show that for any square matrix $\mathbf{A}$ and scalar matrix $\mathbf{D}$ of the same order, $\mathbf{AD} = \mathbf{DA}$.

32 Show by counterexample that if $\mathbf{A}$ is a square matrix and $\mathbf{D}$ is a diagonal matrix of the same order, then in general $\mathbf{AD} \neq \mathbf{DA}$.

33 The product of a square matrix $\mathbf{A}$ with itself is usually denoted by $\mathbf{AA} \equiv \mathbf{A}^2$; the product of $\mathbf{A}$ with itself k times is denoted by $\mathbf{A}^k$.

(a) Show by counterexample that in general

$$(\mathbf{A} + \mathbf{B})^2 \neq \mathbf{A}^2 + 2\mathbf{AB} + \mathbf{B}^2$$
$$(\mathbf{A} - \mathbf{B})^2 \neq \mathbf{A}^2 - \mathbf{B}^2$$

(b) Expand $(\mathbf{A} + \mathbf{B})^3$.

34 Show by counterexample that if $\mathbf{A}$ and $\mathbf{B}$ are both symmetric matrices of the same order, in general $\mathbf{AB}$ is not symmetric.

35 Prove that if $\mathbf{A}$ and $\mathbf{B}$ are both symmetric and of the same order, and if $\mathbf{AB} = \mathbf{BA}$, then $\mathbf{AB}$ is symmetric.

36 If

$$\mathbf{A} = \left[\begin{array}{cc|c|cc} 3 & 1 & 4 & 3 & 2 \\ 1 & 0 & 5 & -1 & 0 \\ \hline 0 & 0 & 2 & 0 & -2 \\ -3 & 1 & 1 & 0 & 0 \end{array}\right] \qquad \mathbf{B} = \begin{bmatrix} 1 & 0 & 3 & -2 & 5 & 0 \\ 0 & 1 & 2 & 0 & 1 & -3 \\ 0 & 0 & 1 & 0 & -1 & 0 \\ 2 & 2 & -2 & 1 & 3 & 0 \\ 0 & 0 & 0 & 0 & 1 & 0 \end{bmatrix}$$

partition $\mathbf{B}$ and calculate $\mathbf{AB}$ in partitioned form (with $\mathbf{A}$ partitioned as shown).

37 If $\mathbf{A}$ is an $m \times n$ matrix and $\mathbf{B}$ is an $n \times p$ matrix whose columns are $\mathbf{b}_1, \mathbf{b}_2, \ldots, \mathbf{b}_p$, prove that

$$\mathbf{AB} = [\mathbf{Ab}_1 \quad \mathbf{Ab}_2 \quad \cdots \quad \mathbf{Ab}_p]$$

38 Let $\mathbf{A}$ and $\mathbf{B}$ be square matrices of orders m and n, respectively, and let $\mathbf{C}$ be any $n \times m$ matrix. Show that, if $\mathbf{A}^{-1}$ and $\mathbf{B}^{-1}$ exist, then

$$\left[\begin{array}{c|c} \mathbf{A} & \mathbf{O} \\ \hline \mathbf{C} & \mathbf{B} \end{array}\right]^{-1} = \left[\begin{array}{c|c} \mathbf{A}^{-1} & \mathbf{O} \\ \hline \mathbf{T} & \mathbf{B}^{-1} \end{array}\right]$$

where $\mathbf{T} = -\mathbf{B}^{-1}\mathbf{C}\mathbf{A}^{-1}$.

39 Prove Theorem 2-7.

40 Let $\mathbf{A}$ and $\mathbf{B}$ be nth-order matrices and assume that $\mathbf{A}^{-1}$ exists. Show that in general, $\mathbf{A}^{-1}\mathbf{B} \neq \mathbf{B}\mathbf{A}^{-1}$ and hence the scalar arithmetic operation "division" is not defined for matrices; i.e., the notation "$\mathbf{A}/\mathbf{B}$," corresponding to $a/b = ab^{-1} = b^{-1}a$, is not well-defined.

3

DETERMINANTS AND ELEMENTARY MATRICES

3-1 INTRODUCTION

In this chapter and the next our discussion will concentrate on the subject of square matrices. We shall begin by introducing *determinants*, which are scalars associated with square matrices. Determinants play an important role in the study of properties of solutions to systems of simultaneous algebraic equations (the main topic of Chap. 4). After spending some time considering properties of determinants and their calculation, we shall study briefly a very special set of matrices, called *elementary matrices*.

The definition of a determinant may, unfortunately, seem rather artificial and contrived, at first. It is hoped, however, that the reader will gain some appreciation for the versatility of the concept of determinants by the conclusion of this chapter.

We observed in Chap. 2 that a matrix is not a number (or scalar) but rather an array of numbers. Thus, a matrix has no "value," per se. We will now define for each square matrix a number (scalar) which we shall call its *determinant.* Given a matrix **A**, the determinant of **A** will be denoted either by det (**A**) or by $|\mathbf{A}|$.†

First, let us consider the determinant of the 1×1 matrix $\mathbf{A} = [a_{11}]$. Since such matrices can be treated essentially as scalars, we shall define

$$\det([a_{11}]) = a_{11} \tag{3-1}$$

That is, the determinant of the matrix consisting of the one element a_{11} is just a_{11}. Now, we shall build upon the above definition by defining the determinant of a 2×2 matrix in terms of determinants of certain 1×1 matrices, and then by defining the determinant of a 3×3 matrix in terms of determinants of 2×2 matrices, etc. In this manner we will thus define the determinant of an $n \times n$ matrix in terms of determinants of certain $(n-1) \times (n-1)$ matrices.

In order to facilitate the discussion as to the nature of these "certain $(n-1) \times (n-1)$ matrices," we pause to define:

The determinant of the $(n-1) \times (n-1)$ matrix obtained from a matrix $\mathbf{A} = [a_{ij}]$ by deleting the ith row and the jth column is called the *minor* of the element a_{ij}, and is denoted by m_{ij}.

The number A_{ij},

$$A_{ij} = (-1)^{i+j} m_{ij} \tag{3-2}$$

is called the *cofactor* of a_{ij}.

EXAMPLE 3-1 If

$$\mathbf{A} = \begin{bmatrix} 3 & 1 & 0 \\ -1 & 2 & 4 \\ 5 & 6 & -2 \end{bmatrix}$$

then the minor of $a_{11} = 3$ is

$$m_{11} = \det\left(\begin{bmatrix} 2 & 4 \\ 6 & -2 \end{bmatrix}\right)$$

the minor of $a_{12} = 1$ is

$$m_{12} = \det\left(\begin{bmatrix} -1 & 4 \\ 5 & -2 \end{bmatrix}\right)$$

etc. The cofactor of a_{11} is

$$(-1)^{1+1} m_{11} = (-1)^2 \det\left(\begin{bmatrix} 2 & 4 \\ 6 & -2 \end{bmatrix}\right)$$

† The vertical lines employed to denote the determinant of a matrix should not be confused with the absolute value or modulus notations, which are used only with *scalars*.

the cofactor of a_{12} is

$$(-1)^{1+2}m_{12} = (-1)^3 \det\left(\begin{bmatrix} -1 & 4 \\ 5 & -2 \end{bmatrix}\right)$$

etc. ////

EXAMPLE 3-2 The reader may have previously encountered the concept of determinants for 2 × 2 (and perhaps for 3 × 3) matrices, in which case he may recall the formula

$$\det\left(\begin{bmatrix} a_{11} & a_{12} \\ a_{21} & a_{22} \end{bmatrix}\right) = a_{11}a_{22} - a_{12}a_{21} \tag{3-3}$$

Note that the cofactor of a_{11} is $A_{11} = (-1)^{1+1} \det([a_{22}]) = a_{22}$, and the cofactor of a_{12} is $A_{12} = (-1)^{1+2} \det([a_{21}]) = -a_{21}$. Thus, if we substitute $A_{11} = a_{22}$, $A_{12} = -a_{21}$ into Eq. (3-3), we obtain

$$\det\left(\begin{bmatrix} a_{11} & a_{12} \\ a_{21} & a_{22} \end{bmatrix}\right) = a_{11}A_{11} + a_{12}A_{12} \tag{3-4}$$

////

Equation (3-4) suggests one possible way of extending our definition of determinants to higher-order matrices. From the right-hand side of Eq. (3-4), we see that (at least according to the definition of the 2 × 2 determinant assumed in Example 3-2) the determinant of $\mathbf{A} = \begin{bmatrix} a_{11} & a_{12} \\ a_{21} & a_{22} \end{bmatrix}$ is the sum of the first element in row 1 multiplied by its cofactor and the second element in row 1 multiplied by its cofactor. For an $n \times n$ matrix $\mathbf{A} = [a_{ij}]$, we might therefore write

$$\begin{aligned} \det(\mathbf{A}) &= a_{11}A_{11} + a_{12}A_{12} + a_{13}A_{13} + \cdots + a_{1n}A_{1n} \\ &= \sum_{j=1}^{n} a_{1j}A_{1j} \end{aligned} \tag{3-5}$$

Indeed, this is in fact exactly what we shall do: The determinant of the $n \times n$ matrix $\mathbf{A} = [a_{ij}]$ is *defined* by Eq. (3-5). The right-hand side of Eq. (3-5) indicates that det (**A**) is the sum of each element of the first row of **A** multiplied by its corresponding cofactor. Since each cofactor is merely plus or minus a determinant of an $(n-1) \times (n-1)$ matrix, we have defined the determinant of the $n \times n$ matrix **A** in terms of the determinants of n $(n-1) \times (n-1)$ matrices.

EXAMPLE 3-3 If
$$\mathbf{A} = \begin{bmatrix} 3 & 1 & 0 \\ -1 & 2 & 4 \\ 5 & 6 & -2 \end{bmatrix}$$
then
$$\begin{aligned}\det(\mathbf{A}) &= a_{11}A_{11} + a_{12}A_{12} + a_{13}A_{13} \\ &= 3(-1)^{1+1}\det\left(\begin{bmatrix} 2 & 4 \\ 6 & -2 \end{bmatrix}\right) + 1(-1)^{1+2}\det\left(\begin{bmatrix} -1 & 4 \\ 5 & -2 \end{bmatrix}\right) \\ &\quad + 0(-1)^{1+3}\det\left(\begin{bmatrix} -1 & 2 \\ 5 & 6 \end{bmatrix}\right) \\ &= 3(-4-24) + -1(2-20) + 0(-6-10) \\ &= -66\end{aligned}$$
////

EXAMPLE 3-4 If
$$\mathbf{A} = \begin{bmatrix} a_{11} & a_{12} & a_{13} \\ a_{21} & a_{22} & a_{23} \\ a_{31} & a_{32} & a_{33} \end{bmatrix}$$
then
$$\begin{aligned}\det(\mathbf{A}) &= a_{11}A_{11} + a_{12}A_{12} + a_{13}A_{13} \\ &= a_{11}(-1)^{1+1}\det\left(\begin{bmatrix} a_{22} & a_{23} \\ a_{32} & a_{33} \end{bmatrix}\right) \\ &\quad + a_{12}(-1)^{1+2}\det\left(\begin{bmatrix} a_{21} & a_{23} \\ a_{31} & a_{33} \end{bmatrix}\right) \\ &\quad + a_{13}(-1)^{1+3}\det\left(\begin{bmatrix} a_{21} & a_{22} \\ a_{31} & a_{32} \end{bmatrix}\right) \\ &= a_{11}(a_{22}a_{33} - a_{23}a_{32}) - a_{12}(a_{21}a_{33} - a_{23}a_{31}) \\ &\quad + a_{13}(a_{21}a_{32} - a_{22}a_{31}) \\ &= a_{11}a_{22}a_{33} - a_{11}a_{23}a_{32} - a_{12}a_{21}a_{33} + a_{12}a_{23}a_{31} \\ &\quad + a_{13}a_{21}a_{32} - a_{13}a_{22}a_{31} \qquad (3\text{-}6)\end{aligned}$$
////

Note that each term in the right-hand side of Eq. (3-6) contains three factors: exactly one element from each of the three rows of **A** and exactly one element from each of the three columns of **A**, with no two elements from the same row or the same column. Since there are $3! = 6$ possible ways of choosing three elements in the above manner, we see that all six possible terms of the above type appear in the expression for det (**A**) in Eq. (3-6). Some of these terms are preceded by a plus sign, others by a minus sign.

In the general $n \times n$ case, det (**A**) is also the sum of all possible terms containing n elements, one from each row, one from each column, with no two elements from the same row or the same column, with some terms preceded by a plus sign, others by a minus sign. In fact, in some texts (e.g., Hohn [1]), determinants are defined in this manner, along with the rule for determining the appropriate sign for each term. The reader should keep in mind that both definitions are equivalent and lead to the same set of properties (to be discussed below and in the next section) for determinants.

Observe now that rearranging the terms of the right-hand side of Eq. (3-6) leads to the following equivalent expressions for det (**A**):

$$\begin{aligned}(a)\quad \det(\mathbf{A}) &= a_{11}(a_{22}a_{33} - a_{23}a_{32}) - a_{21}(a_{12}a_{33} - a_{13}a_{32}) \\ &\quad + a_{31}(a_{12}a_{23} - a_{13}a_{22}) \\ &= a_{11}(-1)^{1+1}\det\left(\begin{bmatrix} a_{22} & a_{23} \\ a_{32} & a_{33}\end{bmatrix}\right) \\ &\quad + a_{21}(-1)^{1+2}\det\left(\begin{bmatrix} a_{12} & a_{13} \\ a_{32} & a_{33}\end{bmatrix}\right) \\ &\quad + a_{31}(-1)^{1+3}\det\left(\begin{bmatrix} a_{12} & a_{13} \\ a_{22} & a_{23}\end{bmatrix}\right) \\ &= a_{11}A_{11} + a_{21}A_{21} + a_{31}A_{31}\end{aligned}$$

$$\begin{aligned}(b)\quad \det(\mathbf{A}) &= -a_{21}(a_{12}a_{33} - a_{13}a_{32}) + a_{22}(a_{11}a_{33} - a_{13}a_{31}) \\ &\quad - a_{23}(a_{11}a_{32} - a_{12}a_{31}) \\ &= a_{21}(-1)^{2+1}\det\left(\begin{bmatrix} a_{12} & a_{13} \\ a_{32} & a_{33}\end{bmatrix}\right) \\ &\quad + a_{22}(-1)^{2+2}\det\left(\begin{bmatrix} a_{11} & a_{13} \\ a_{31} & a_{33}\end{bmatrix}\right) \\ &\quad + a_{23}(-1)^{2+3}\det\left(\begin{bmatrix} a_{11} & a_{12} \\ a_{31} & a_{32}\end{bmatrix}\right) \\ &= a_{21}A_{21} + a_{22}A_{22} + a_{23}A_{23}\end{aligned}$$

$$\begin{aligned}(c)\quad \det(\mathbf{A}) &= a_{31}(a_{12}a_{23} - a_{13}a_{22}) - a_{32}(a_{11}a_{23} - a_{13}a_{21}) \\ &\quad + a_{33}(a_{11}a_{22} - a_{12}a_{21}) \\ &= a_{31}(-1)^{3+1}\det\left(\begin{bmatrix} a_{12} & a_{13} \\ a_{22} & a_{23}\end{bmatrix}\right) \\ &\quad + a_{32}(-1)^{3+2}\det\left(\begin{bmatrix} a_{11} & a_{13} \\ a_{21} & a_{23}\end{bmatrix}\right) \\ &\quad + a_{33}(-1)^{3+3}\det\left(\begin{bmatrix} a_{11} & a_{12} \\ a_{21} & a_{22}\end{bmatrix}\right) \\ &= a_{31}A_{31} + a_{32}A_{32} + a_{33}A_{33}\end{aligned}$$

Thus, we see that det (**A**) can be expressed in terms of the elements (and their corresponding cofactors) of the first row of **A**, the second row of **A**, the third row of **A**, or the first *column* of **A**. Similarly, we can easily see that det (**A**) can be expressed in terms of the elements and corresponding cofactors of the other two columns of **A**.

In the general $n \times n$ case, we find† that

$$\det(\mathbf{A}) = \sum_{k=1}^{n} a_{ik} A_{ik} \tag{3-7}$$

$$\det(\mathbf{A}) = \sum_{k=1}^{n} a_{kj} A_{kj} \tag{3-8}$$

Equation (3-7) states that det (**A**) is equal to the sum of the products of the elements of the ith row (for *any* row i) of **A** times their respective cofactors; Eq. (3-8) states that det (**A**) is equal to the sum of the products of the elements of the jth column (for *any* column j) of **A** times their respective cofactors.

The right-hand side of Eq. (3-7) is called the cofactor expansion of det (**A**) by the ith row; similarly, the right-hand side of Eq. (3-8) is called the cofactor expansion of det (**A**) by the jth column.

EXAMPLE 3-5 If

$$\mathbf{A} = \begin{bmatrix} 3 & 1 & 0 \\ -1 & 2 & 4 \\ 5 & 6 & -2 \end{bmatrix}$$

then the expansion of det (**A**) by the first row has been calculated in Example 3-3; the expansion of det (**A**) by the second row is

$$\begin{aligned}
\det(\mathbf{A}) &= a_{21}A_{21} + a_{22}A_{22} + a_{23}A_{23} \\
&= (-1)(-1)^{2+1} \det\left(\begin{bmatrix} 1 & 0 \\ 6 & -2 \end{bmatrix}\right) + 2(-1)^{2+2} \det\left(\begin{bmatrix} 3 & 0 \\ 5 & -2 \end{bmatrix}\right) \\
&\quad + 4(-1)^{2+3} \det\left(\begin{bmatrix} 3 & 1 \\ 5 & 6 \end{bmatrix}\right) \\
&= (-1)(-1)(-2-0) + 2(1)(-6-0) + 4(-1)(18-5) \\
&= -2 - 12 - 52 \\
&= -66
\end{aligned}$$

† We have not actually proved the validity of Eqs. (3-7) and (3-8), since the proof is tedious and relatively uninstructive. For a rigorous proof, the interested reader can consult Noble [2].

The expansion of det (**A**) by the third column is

$$\begin{aligned}\det(\mathbf{A}) &= a_{13}A_{13} + a_{23}A_{23} + a_{33}A_{33}\\ &= 0(-1)^{1+3}\det\left(\begin{bmatrix}-1 & 2\\ 5 & 6\end{bmatrix}\right) + 4(-1)^{2+3}\det\left(\begin{bmatrix}3 & 1\\ 5 & 6\end{bmatrix}\right)\\ &\quad + (-2)(-1)^{3+3}\det\left(\begin{bmatrix}3 & 1\\ -1 & 2\end{bmatrix}\right)\\ &= 0(1)(-6-10) + 4(-1)(18-5) + (-2)(1)(6+1)\\ &= 0 - 52 - 14\\ &= -66\end{aligned}$$

////

3-2 PROPERTIES OF DETERMINANTS

We shall now proceed to develop some basic properties of determinants. The first result which we can immediately obtain by inspecting Eqs. (3-7) and (3-8) is that, since both row and column cofactor expansions yield det (**A**), the determinant of the transpose of any square matrix **A** is equal to the determinant of **A**. If

$$\mathbf{A} = \begin{bmatrix} a_{11} & a_{12} & \cdots & a_{1n}\\ a_{21} & a_{22} & \cdots & a_{2n}\\ \vdots & \vdots & & \vdots\\ a_{n1} & a_{n2} & \cdots & a_{nn}\end{bmatrix} \qquad \mathbf{A}^T = \begin{bmatrix} a_{11} & a_{21} & \cdots & a_{n1}\\ a_{12} & a_{22} & \cdots & a_{n2}\\ \vdots & \vdots & & \vdots\\ a_{1n} & a_{2n} & \cdots & a_{nn}\end{bmatrix}$$

and the cofactor expansion of det (**A**) by row 1 is

$$\det(\mathbf{A}) = \sum_{j=1}^{n} a_{1j}A_{1j}$$

which is also the cofactor expansion of det (**A**) by the first column of $\mathbf{A}^T$. We summarize in:

Theorem 3-1 For any square matrix **A**, det (**A**) = det ($\mathbf{A}^T$).

Below are three additional results which are also immediate consequences of Eqs. (3-7) and (3-8):

Theorem 3-2 If any row (or column) of a square matrix **A** consists of all zero elements, then det (**A**) = 0.

PROOF Express det (**A**) in terms of the cofactor expression by the row (or column) which contains all zero elements. Then each term in the summation will consist of a zero multiplied by a cofactor. QED

Theorem 3-3 If each element of any row (or column) of a square matrix **A** is multiplied by a constant α, the determinant of the resulting matrix is $\alpha \det(\mathbf{A})$.

PROOF Suppose each element of row i is multiplied by α, yielding a new matrix whose ith row consists of the elements $\alpha a_{i1}, \alpha a_{i2}, \ldots, \alpha a_{in}$. The cofactor expansion of the determinant of this new matrix by its ith row is then equal to $\sum_{j=1}^{n} (\alpha a_{ij}) A_{ij} = \alpha(\sum_{j=1}^{n} a_{ij} A_{ij}) = \alpha \det(\mathbf{A})$.

QED

Theorem 3-4 (*a*) If two matrices $\mathbf{A} = [a_{ij}]$, $\mathbf{B} = [b_{ij}]$ are identical except for the elements of the kth row (for any row k), then $\det(\mathbf{A}) + \det(\mathbf{B}) = \det(\mathbf{C})$, where $\mathbf{C} = [c_{ij}]$ and

$$c_{ij} = \begin{cases} a_{ij} = b_{ij} & i \neq k \\ a_{kj} + b_{kj} & i = k \end{cases}$$

(*b*) If **A** and **B** are identical except for the elements of the pth column, then $\det(\mathbf{A}) + \det(\mathbf{B}) = \det(\mathbf{C})$, where $\mathbf{C} = [c_{ij}]$ and

$$c_{ij} = \begin{cases} a_{ij} = b_{ij} & j \neq p \\ a_{ip} + b_{ip} & j = p \end{cases}$$

PROOF (*a*) The cofactor expansion of $\det(\mathbf{C})$ in terms of the kth row is

$$\det(\mathbf{C}) = \sum_{j=1}^{n} c_{kj} C_{kj}$$

Since C_{kj} is $(-1)^{k+j}$ times the determinant obtained by deleting the kth row and jth column of **C**, it is obvious that $C_{kj} = A_{kj} = B_{kj}$, because **A**, **B**, and **C** differ only in the kth row. Moreover, $c_{kj} = a_{kj} + b_{kj}$. Hence,

$$\begin{aligned} \det(\mathbf{C}) &= \sum_{j=1}^{n} (a_{kj} + b_{kj}) A_{kj} \\ &= \sum_{j=1}^{n} a_{kj} A_{kj} + \sum_{j=1}^{n} b_{kj} A_{kj} \\ &= \sum_{j=1}^{n} a_{kj} A_{kj} + \sum_{j=1}^{n} b_{kj} B_{kj} \\ &= \det(\mathbf{A}) + \det(\mathbf{B}) \end{aligned}$$

Part (*b*) is proved in a similar manner. QED

EXAMPLE 3-6 Let

$$\mathbf{A} = \begin{bmatrix} 1 & -3 & 2 \\ -8 & 0 & 9 \\ 4 & 5 & -7 \end{bmatrix} \qquad \mathbf{B} = \begin{bmatrix} -2 & -3 & 1 \\ -8 & 0 & 9 \\ 4 & 5 & -7 \end{bmatrix} \qquad \mathbf{C} = \begin{bmatrix} -1 & -6 & 3 \\ -8 & 0 & 9 \\ 4 & 5 & -7 \end{bmatrix}$$

Then the elements of the first row of **C** are equal to the sums of the corresponding elements of the first row of **A** and the first row of **B**. Thus, Theorem 3-4*a* states that det (**A**) + det (**B**) = det (**C**). Let us verify this result:

$$\det(\mathbf{C}) = \begin{vmatrix} -1 & -6 & 3 \\ -8 & 0 & 9 \\ 4 & 5 & -7 \end{vmatrix}$$

$$= (-1)(-1)^{1+1}\begin{vmatrix} 0 & 9 \\ 5 & -7 \end{vmatrix} + (-6)(-1)^{1+2}\begin{vmatrix} -8 & 9 \\ 4 & -7 \end{vmatrix} + (3)(-1)^{1+3}\begin{vmatrix} -8 & 0 \\ 4 & 5 \end{vmatrix}$$

$$= -1(0 - 45) + (6)(56 - 36) + 3(-40 - 0)$$

$$= 45 + 120 - 120$$

$$= 45$$

$$\det(\mathbf{A}) = \begin{vmatrix} 1 & -3 & 2 \\ -8 & 0 & 9 \\ 4 & 5 & -7 \end{vmatrix}$$

$$= 1(-1)^{1+1}\begin{vmatrix} 0 & 9 \\ 5 & -7 \end{vmatrix} + (-3)(-1)^{1+2}\begin{vmatrix} -8 & 9 \\ 4 & -7 \end{vmatrix} + (2)(-1)^{1+3}\begin{vmatrix} -8 & 0 \\ 4 & 5 \end{vmatrix}$$

$$= (1)(0 - 45) + (3)(56 - 36) + (2)(-40 - 0)$$

$$= -45 + 60 - 80$$

$$= -65$$

$$\det(\mathbf{B}) = \begin{vmatrix} -2 & -3 & 1 \\ -8 & 0 & 9 \\ 4 & 5 & -7 \end{vmatrix}$$

$$= -2(-1)^{1+1}\begin{vmatrix} 0 & 9 \\ 5 & -7 \end{vmatrix} + (-3)(-1)^{1+2}\begin{vmatrix} -8 & 9 \\ 4 & -7 \end{vmatrix} + (1)(-1)^{1+3}\begin{vmatrix} -8 & 0 \\ 4 & 5 \end{vmatrix}$$

$$= (-2)(0 - 45) + (3)(56 - 36) + (1)(-40 - 0)$$
$$= 90 + 60 - 40$$
$$= 110$$

Thus,

$$\det(\mathbf{A}) + \det(\mathbf{B}) = -65 + 110$$
$$= 45 = \det(\mathbf{C}) \qquad ////$$

Theorem 3-4 shows the relationship of the determinants of matrices which are identical except for the elements of one row (or column). Let us now consider the relationship of the determinants of two matrices which are also almost identical. In particular, suppose that, given a matrix **A**, we form a matrix **B** by interchanging the first two rows of **A**, and set all other rows of **B** equal to the corresponding rows of **A**. Thus, row 1 of **A** is row 2 of **B**, row 2 of **A** is row 1 of **B**. Symbolically,

$$a_{1j} = b_{2j} \qquad j = 1, 2, \ldots, n$$
$$a_{2j} = b_{1j} \qquad j = 1, 2, \ldots, n$$
$$a_{ij} = b_{ij} \qquad i = 3, 4, \ldots, n \quad \text{and } j = 1, 2, \ldots, n$$

Computing det (**A**) by the cofactor expansion of the first row yields, by definition of det (**A**),

$$\det(\mathbf{A}) = \sum_{j=1}^{n} a_{1j} A_{1j} \tag{3-9}$$

Computing det (**B**) by the cofactor expansion of the *second* row yields:

$$\det(\mathbf{B}) = \sum_{j=1}^{n} b_{2j} B_{2j} \tag{3-10}$$

But observe that

$$B_{2j} = -A_{1j} \qquad j = 1, 2, \ldots, n$$

since

$$B_{2j} = (-1)^{2+j} \begin{vmatrix} b_{11} & b_{12} & \cdots & b_{1,j-1} & b_{1,j+1} & \cdots & b_{1n} \\ b_{31} & b_{32} & \cdots & b_{3,j-1} & b_{3,j+1} & \cdots & b_{3n} \\ \vdots & \vdots & & \vdots & \vdots & & \vdots \\ b_{n1} & b_{n2} & \cdots & b_{n,j-1} & b_{n,j+1} & \cdots & b_{nn} \end{vmatrix}$$

$$= (-1)^{2+j} \begin{vmatrix} a_{21} & a_{22} & \cdots & a_{2,j-1} & a_{2,j+1} & \cdots & a_{2n} \\ a_{31} & a_{32} & \cdots & a_{3,j-1} & a_{3,j+1} & \cdots & a_{3n} \\ \vdots & \vdots & & \vdots & \vdots & & \vdots \\ a_{n1} & a_{n2} & \cdots & a_{n,j-1} & a_{n,j+1} & \cdots & a_{nn} \end{vmatrix}$$

$= (-1)^{2+j}$ times the determinant of the matrix obtained by deleting the first row and the jth column of **A**

$= (-1)^{2+j} m_{1j}$ by definition of m_{1j}

$= (-1)^{2+j}(-1)^{1+j} A_{1j}$ by definition of A_{1j} (Eq. 3-2)

$= (-1)^{3+2j} A_{1j}$

$= (-1)^3(-1)^{2j} A_{1j}$

$= -A_{1j}$

Substituting $B_{2j} = -A_{1j}$ and $b_{2j} = a_{1j}$ into Eq. (3-10) yields

$$\det(\mathbf{B}) = \sum_{j=1}^{n} a_{1j}(-A_{1j})$$

$$= -\sum_{j=1}^{n} a_{1j} A_{1j}$$

$$= -\det(\mathbf{A}) \qquad \text{by Eq. (3-9)}$$

Thus, we have shown that interchanging the first two rows of a matrix yields a new matrix whose determinant is the negative of the determinant of the original matrix. We shall now show that interchanging *any two* rows of a matrix yields a matrix whose determinant is the negative of the determinant of the original matrix. First, let us prove:

Theorem 3-5 If a matrix $\mathbf{B} = [b_{ij}]$ is formed from a matrix $\mathbf{A} = [a_{ij}]$ by interchanging the first and kth rows of **A** ($k = 2, 3, \ldots, n$), then $\det(\mathbf{B}) = -\det(\mathbf{A})$.

PROOF We shall prove this theorem by induction on k. The $k = 2$ case has already been proved above. Thus, we now seek to prove:

If interchanging the first and $(k-1)$st rows of a matrix yields a matrix whose determinant is the negative of the determinant of the original matrix, then interchanging the first and kth rows of a matrix also yields a matrix whose determinant is the negative of the determinant of the original matrix, for $k = 3, 4, \ldots, n$.

If

$$b_{1j} = a_{kj}$$

$$b_{kj} = a_{1j}$$

$$b_{ij} = a_{ij} \qquad i \neq k, \quad i \neq 1$$

then we can write

$$\det(\mathbf{B}) = \sum_{j=1}^{n} b_{2j} B_{2j} \qquad (3\text{-}11)$$

But B_{2j} is $(-1)^{2+j}$ times the determinant of the matrix obtained from **B** by deleting its second row and jth column. Call this matrix **C**. Then,

$$B_{2j} = (-1)^{2+j} \det(\mathbf{C}) \qquad (3\text{-}12)$$

Observe that the $(k-1)$st row of **C** is $b_{k1}, b_{k2}, \ldots, b_{k,\,j-1}, b_{k,\,j+1}, b_{kn}$. In fact if we interchange the first and $(k-1)$st rows of **C**, we obtain a matrix identical with the matrix obtained by deleting the second row and jth column of **A**. Thus, by the induction hypothesis,

$$\det(\mathbf{C}) = -(-1)^{2+j} A_{2j} \qquad (3\text{-}13)$$

Substitution of $b_{2j} = a_{2j}$, along with Eqs. (3-12) and (3-13), into Eq. (3-11) yields the desired result:

$$\begin{aligned}
\det(\mathbf{B}) &= \sum_{j=1}^{n} b_{2j} B_{2j} \\
&= \sum_{j=1}^{n} a_{2j}(-1)^{2+j}[-(-1)^{2+j} A_{2j}] \\
&= -\sum_{j=1}^{n} (-1)^{4+2j} a_{2j} A_{2j} \\
&= -\sum_{j=1}^{n} a_{2j} A_{2j} \\
&= -\det(\mathbf{A})
\end{aligned}$$

QED

Now it is a simple matter to prove the result we have been after, which is:

Theorem 3-6 Interchanging any two rows (or any two columns) of a matrix yields a matrix whose determinant is the negative of the original matrix.

PROOF Suppose we wish to interchange the pth and qth rows of a matrix **A**, yielding a new matrix **B**. The cases $p = 1$ and $q = 1$ are covered by Theorem 3-5, and so we shall assume that $p \neq 1$, $q \neq 1$.

If we first interchange the first and pth rows of **A**, we obtain a new matrix $\mathbf{C}_1$ whose determinant is $-\det(\mathbf{A})$, by Theorem 3-5. If we then interchange the first and qth rows of $\mathbf{C}_1$, we obtain a new matrix $\mathbf{C}_2$ whose determinant is $-\det(\mathbf{C}_1) = +\det(\mathbf{A})$. Moreover, the first row of $\mathbf{C}_2$ is equal to the qth row of $\mathbf{A}$, the pth row of $\mathbf{C}_2$ is equal to the first row of $\mathbf{A}$, and the qth row of $\mathbf{C}_2$ is equal to the pth row of $\mathbf{A}$. Every other row of $\mathbf{C}_2$ is equal to the corresponding row of $\mathbf{A}$. Thus, if we now interchange the first and pth rows of $\mathbf{C}_2$, we obtain a matrix $\mathbf{C}_3$ which is identical to the matrix $\mathbf{B}$ obtained from $\mathbf{A}$ by interchanging the pth and qth rows of $\mathbf{A}$. And obviously, $\det(\mathbf{C}_3) = -\det(\mathbf{C}_2) = +\det(\mathbf{C}_1) = -\det(\mathbf{A})$. Schematically, we summarize the sequence of interchanges below:

ROWS OF A

Row	A	C_1	C_2	C_3
1	1	p	q	1
⋮	⋮	⋮	⋮	⋮
p	p	1	1	q
⋮	⋮	⋮	⋮	⋮
q	q	q	p	p

The proof for interchanging of columns results from direct application of Theorem 3-1:

$$\det(\mathbf{A}) = \det(\mathbf{A}^T) \qquad \text{QED}$$

Theorem 3-6 is one of the most important results of this chapter. It leads directly to two additional results, the second of which will be the key to developing an efficient computational method for calculating determinants.

Theorem 3-7 If a matrix $\mathbf{A}$ contains two rows (or two columns) which are identical, then $\det(\mathbf{A}) = 0$.

PROOF Suppose the pth and qth rows of $\mathbf{A}$ are identical. Then, by Theorem 3-6, interchanging these two rows yields a matrix $\mathbf{B}$ whose determinant is $-\det(\mathbf{A})$. But since we have interchanged two identical rows, $\mathbf{B} = \mathbf{A}$ and so $\det(\mathbf{B}) = \det(\mathbf{A})$. Combining

$$\det(\mathbf{B}) = -\det(\mathbf{A})$$

$$\det(\mathbf{B}) = \det(\mathbf{A})$$

yields $\det(\mathbf{B}) = \det(\mathbf{A}) = 0$ QED

Theorem 3-8 If a matrix $\mathbf{B}$ is identical with a matrix $\mathbf{A}$, except that the kth row (or column) of $\mathbf{B}$ is formed by adding α times the pth row (or column) of **A** to the kth row (or column) of **A** ($p \neq k$), then det (**B**) = det (**A**), where α is any scalar.

PROOF The theorem states that

$$b_{ij} = a_{ij} \qquad i \neq k$$
$$b_{kj} = a_{kj} + \alpha a_{pj}$$

Let us write the cofactor expansion of det (**B**) by the kth row:

$$\det(\mathbf{B}) = \sum_{j=1}^{n} b_{kj} B_{kj}$$
$$= \sum_{j=1}^{n} (a_{kj} + \alpha a_{pj}) B_{kj} \qquad (3\text{-}14)$$

But, since B_{kj} equals $(-1)^{k+j}$ times the determinant of the matrix obtained from **B** by deleting the kth row and jth column, and since **A** and **B** are identical except in their kth rows, it is clear that $B_{kj} = A_{kj}$. Hence, Eq. (3-14) becomes

$$\det(\mathbf{B}) = \sum_{j=1}^{n} (a_{kj} + \alpha a_{pj}) A_{kj}$$
$$= \sum_{j=1}^{n} a_{kj} A_{kj} + \alpha \sum_{j=1}^{n} a_{pj} A_{kj} \qquad (3\text{-}15)$$

Now, the first summation on the right-hand side of Eq. (3-15) is the cofactor expansion of det (**A**) by the kth row, and is hence equal to det (**A**); the second summation (ignoring the α) can be thought of as the cofactor expansion by the kth row of a matrix whose pth and kth rows are identical. Thus, by Theorem 3-7, this expression is equal to zero, and we have

$$\det(\mathbf{B}) = \det(\mathbf{A}) + \alpha \cdot 0 = \det(\mathbf{A}) \qquad \text{QED}$$

EXAMPLE 3-7 Let

$$\mathbf{A} = \begin{bmatrix} 1 & -3 & 2 \\ -8 & 0 & 9 \\ 4 & 5 & -7 \end{bmatrix}$$

Suppose we form **B** by adding 3 times row 1 of **A** to row 3. Thus,

$$\mathbf{B} = \begin{bmatrix} 1 & -3 & 2 \\ -8 & 0 & 9 \\ 7 & -4 & -1 \end{bmatrix}$$

and

$$\det(\mathbf{B}) = 1(-1)^{1+1}\begin{vmatrix} 0 & 9 \\ -4 & -1 \end{vmatrix} + (-3)(-1)^{1+2}\begin{vmatrix} -8 & 9 \\ 7 & -1 \end{vmatrix}$$
$$+ 2(-1)^{1+3}\begin{vmatrix} -8 & 0 \\ 7 & -4 \end{vmatrix}$$
$$= (1)(0 + 36) + (3)(8 - 63) + (2)(+32 - 0)$$
$$= +36 - 165 + 64$$
$$= -65$$
$$= \det(\mathbf{A}) \quad \text{as computed in Example 3-6}$$

////

Theorem 3-8 may be stated equivalently in the following way: Adding a multiple of one row (or column) to another row (or column) yields a new matrix whose determinant is the same as that of the original matrix.

EXAMPLE 3-8 Let

$$\mathbf{A} = \begin{bmatrix} 1 & -2 & 4 \\ -1 & 3 & -1 \\ 2 & -4 & 1 \end{bmatrix}$$

Then, if we add 1 times the first row to the second row of **A**, we obtain a new matrix **B**,

$$\mathbf{B} = \begin{bmatrix} 1 & -2 & 4 \\ 0 & 1 & 3 \\ 2 & -4 & 1 \end{bmatrix}$$

and det (**B**) = det (**A**). If we now add -2 times the first row of **B** to the third row of **B**, we obtain a new matrix **C**,

$$\mathbf{C} = \begin{bmatrix} 1 & -2 & 4 \\ 0 & 1 & 3 \\ 0 & 0 & -7 \end{bmatrix}$$

and det (**C**) = det (**B**) = det (**A**). But det (**C**) is much easier to calculate than det (**A**): The cofactor expansion of det (**C**) by the first column is

$$\det(\mathbf{C}) = 1(-1)^{1+1}\begin{vmatrix} 1 & 3 \\ 0 & -7 \end{vmatrix} + 0(-1)^{1+2}\begin{vmatrix} -2 & 4 \\ 0 & -7 \end{vmatrix} + 0(-1)^{1+3}\begin{vmatrix} -2 & 4 \\ 1 & 3 \end{vmatrix}$$
$$= \begin{vmatrix} 1 & 3 \\ 0 & -7 \end{vmatrix}$$
$$= -7 = \det(\mathbf{A})$$

Compare the above calculation with the cofactor expansion of det (**A**) by *any* row or column of **A**. ////

This example may suggest to the reader a possible method for calculating determinants of large matrices, where a general cofactor expansion would be too cumbersome. This will be the subject of the next section.

3-3 COMPUTATION OF DETERMINANTS

For many of the different kinds of computational problems treated in this text (e.g., finding determinants, solving systems of equations) we will be confronted with several methods of solution. In comparing various methods for solving the same type of problem there are numerous factors one must consider. Perhaps the primary factor is that of the relative efficiency of the methods under consideration.

One criterion commonly used in matrix-algebra-type problems for determining the efficiency of a particular method is to count the number of arithmetic operations the method requires to solve the problem. The rationale for this criterion is that for large problems—where efficiency is most important—the problem is most likely going to be solved on a digital computer, and on most such machines the speed of execution of arithmetic operations is slower than that of other types of programming instructions. Moreover, most such methods consist almost entirely (in terms of percentage of time required to solve a problem) of arithmetic operations.

Hence when discussing a particular method, we shall frequently calculate the number of arithmetic operations required by the method. Furthermore, since on some digital computers the addition or subtraction time is much faster than the multiply or divide time, we shall also distinguish between these two types of arithmetic operations.

Let us now calculate the number of arithmetic operations required to compute the determinant of an $n \times n$ matrix using Eq. (3-5):

$$\det(\mathbf{A}) = \sum_{j=1}^{n} a_{1j} A_{1j} \qquad (3\text{-}5)$$

Each A_{1j} is [ignoring the $(-1)^{1+j}$ factor] the determinant of an $(n-1) \times (n-1)$ matrix and thus may be expanded into the sum of $(n-1)$ terms similar to those in Eq. (3-5); for *each* of these $(n-1)$ terms, one of the factors is the determinant of an $(n-2) \times (n-2)$ matrix. Thus, so far, we have that det (**A**) is the sum of $n(n-1)$ terms, each containing three factors:

two of these are elements of **A** and the third is the determinant of an $(n-2) \times (n-2)$ matrix. Expanding each of these determinants by cofactors yields an expression for det (**A**) involving the sum of $n(n-1)(n-2)$ terms, each containing four factors: three elements of **A** and the determinant of an $(n-3) \times (n-3)$ matrix. Continuing in this manner, we eventually find that det (**A**) is the sum† of $n!$ terms each containing n factors, each factor being an element of **A**. Thus, if we were actually calculating det (**A**) by repeatedly applying cofactor expansions, we would require $(n-1)$ multiplications for each term, or a total of $n!(n-1)$ multiplications, and $(n!-1)$ additions. A 5×5 matrix, then, would require 480 multiplications and 119 additions; a 10×10 matrix would require over 38 million multiplications and over 4 million additions!

Fortunately, with a judicious application of the theorems developed in the last section we can drastically reduce the amount of computation required to calculate a determinant. For example, as we shall see, the determinant of a 10×10 matrix can be calculated with less than 340 multiplications and 300 additions.

The key to the method is the following:

Theorem 3-9 The determinant of an upper triangular matrix is equal to the product of its diagonal elements.

PROOF Recall from Chap. 2 that an upper triangular matrix has the form

$$\mathbf{A} = \begin{bmatrix} a_{11} & a_{12} & a_{13} & \cdots & a_{1n} \\ 0 & a_{22} & a_{23} & \cdots & a_{2n} \\ 0 & 0 & a_{33} & \cdots & a_{3n} \\ \vdots & \vdots & \vdots & \ddots & \vdots \\ 0 & 0 & 0 & \cdots & a_{nn} \end{bmatrix}$$

that is, **A** is upper triangular if all the elements below the diagonal are zero. Thus, if **A** is upper triangular, the cofactor expansion of det (**A**) by the first column is

$$\det(\mathbf{A}) = a_{11}(-1)^{1+1}A_{11} + 0(-1)^{2+1}A_{21} + \cdots + 0(-1)^{n+1}A_{n1}$$

$$= a_{11}\begin{vmatrix} a_{22} & a_{23} & \cdots & a_{2n} \\ 0 & a_{33} & \cdots & a_{3n} \\ \vdots & \vdots & \ddots & \vdots \\ 0 & 0 & \cdots & a_{nn} \end{vmatrix}$$

† We are ignoring the calculation of the sign for each term.

$$= a_{11}\left\{ a_{22}(-1)^{1+1} \begin{vmatrix} a_{33} & a_{34} & \cdots & a_{3n} \\ 0 & a_{44} & \cdots & a_{4n} \\ \vdots & \vdots & \ddots & \vdots \\ 0 & 0 & \cdots & a_{nn} \end{vmatrix} \right\}$$

$$= a_{11}a_{22}\left\{ a_{33}(-1)^{1+1} \begin{vmatrix} a_{44} & a_{45} & \cdots & a_{4n} \\ 0 & a_{55} & \cdots & a_{5n} \\ \vdots & \vdots & \ddots & \vdots \\ 0 & 0 & \cdots & a_{nn} \end{vmatrix} \right\}$$

$$\vdots$$

$$= a_{11}a_{22}a_{33} \cdots a_{nn}$$

In each step above, we have expanded the determinant in the right-hand side by its first column, which contains at most one nonzero element.

QED

Now, of course, not every matrix is upper triangular, and so the question is: How can we take advantage of Theorem 3-9 to calculate the determinant of *any* matrix?

The answer, at least partially, is by applying Theorem 3-8 methodically to form an upper triangular matrix from the given matrix. Theorem 3-8 states that adding a multiple of one row of a matrix **A** to another row yields a new matrix whose determinant equals det (**A**). Note that we can repeatedly add multiples of one row of **A** to another row and the determinant of the resulting matrices will all be equal to det (**A**).

Since our objective is to form an upper triangular matrix, suppose we employ the following procedure:

Step 1. Add multiples of row 1 to each of the remaining $(n-1)$ rows, choosing the appropriate multiple so that each element in column 1, except a_{11}, is zero.

Step 2. Next, add multiples of row 2 to each of the $(n-2)$ rows 3, 4, ..., n, so that each element in column 2 below the diagonal is zero.

Step k. For $k = 3, 4, \ldots, n-1$: Add multiples of row k to each of the remaining $(n-k)$ rows $k+1$, $k+2$, ..., n, so that each element in column k below the diagonal is zero.

Step n. The resulting matrix will be upper triangular. Hence, the desired determinant is the product of the diagonal elements of this matrix.

EXAMPLE 3-9 Let

$$\mathbf{A} = \begin{bmatrix} 1 & 3 & 2 & 1 & 0 \\ -2 & -4 & 0 & 1 & 3 \\ 3 & 0 & -9 & 1 & 2 \\ 1 & 2 & -1 & 2 & 5 \\ -1 & 2 & 3 & 4 & 7 \end{bmatrix}$$

Performing step 1 above yields the following matrix (where the numbers in parentheses on the right are the multiples of row 1 added to the corresponding row):

$$\begin{bmatrix} 1 & 3 & 2 & 1 & 0 \\ 0 & 2 & 4 & 3 & 3 \\ 0 & -9 & -15 & -2 & 2 \\ 0 & -1 & -3 & 1 & 5 \\ 0 & 5 & 5 & 5 & 7 \end{bmatrix} \quad \begin{matrix} \\ (2) \\ (-3) \\ (-1) \\ (1) \end{matrix}$$

Performing step 2 on the above matrix yields:

$$\begin{bmatrix} 1 & 3 & 2 & 1 & 0 \\ 0 & 2 & 4 & 3 & 3 \\ 0 & 0 & 3 & \frac{23}{2} & \frac{31}{2} \\ 0 & 0 & -1 & \frac{5}{2} & \frac{13}{2} \\ 0 & 0 & -5 & -\frac{5}{2} & -\frac{1}{2} \end{bmatrix} \quad \begin{matrix} \\ \\ (\frac{9}{2}) \\ (\frac{1}{2}) \\ (-\frac{5}{2}) \end{matrix}$$

Performing step 3 on this latter matrix yields:

$$\begin{bmatrix} 1 & 3 & 2 & 1 & 0 \\ 0 & 2 & 4 & 3 & 3 \\ 0 & 0 & 3 & \frac{23}{2} & \frac{31}{2} \\ 0 & 0 & 0 & \frac{19}{3} & \frac{35}{3} \\ 0 & 0 & 0 & \frac{50}{3} & \frac{76}{3} \end{bmatrix} \quad \begin{matrix} \\ \\ \\ (\frac{1}{3}) \\ (\frac{5}{3}) \end{matrix}$$

Finally, we perform step 4 by adding $-\frac{50}{19}$ times row 4 to row 5, and obtaining the following upper triangular matrix:

$$\begin{bmatrix} 1 & 3 & 2 & 1 & 0 \\ 0 & 2 & 4 & 3 & 3 \\ 0 & 0 & 3 & \frac{23}{2} & \frac{31}{2} \\ 0 & 0 & 0 & \frac{19}{3} & \frac{35}{3} \\ 0 & 0 & 0 & 0 & -\frac{102}{19} \end{bmatrix}$$

Thus,

$$\det(\mathbf{A}) = (1)(2)(3)(\tfrac{19}{3})(-\tfrac{102}{19}) = -204 \qquad ////$$

Let us now count the number of arithmetic operations necessary to implement step 1 through step n. First, we make several observations from the calculations of Example 3-9:

1 At the kth step ($k = 1, 2, \ldots, n-1$) the elements in rows $1, 2, \ldots, k$ remain unchanged, and hence no arithmetic operations are performed on these elements.

2 At the kth step ($k = 1, 2, \ldots, n-1$) the elements in columns $1, 2, \ldots, k-1$ remain unchanged; the elements in column k, rows $k+1, k+2, \ldots, n$, are *set* to zero. We shall not count this latter procedure as an arithmetic operation. Hence, no arithmetic operations are performed on any of the elements in rows 1 through k or on the elements in columns 1 through k.

In view of the above remarks, let us count the number of arithmetic operations performed at the kth step. The matrix to be operated on at the kth step is of the form:

$$\hat{\mathbf{A}} = \left[\begin{array}{cccccc|ccc} a_{11} & a_{12} & a_{13} & \cdots & a_{1k} & & a_{1,k+1} & \cdots & a_{1n} \\ 0 & \hat{a}_{22} & \hat{a}_{23} & \cdots & \hat{a}_{2k} & & \hat{a}_{2,k+1} & \cdots & \hat{a}_{2n} \\ 0 & 0 & \hat{a}_{33} & \cdots & \hat{a}_{3k} & & \hat{a}_{3,k+1} & \cdots & \hat{a}_{3n} \\ \vdots & \vdots & \vdots & \ddots & \vdots & & \vdots & & \vdots \\ 0 & 0 & 0 & \cdots & \hat{a}_{kk} & & \hat{a}_{k,k+1} & \cdots & \hat{a}_{kn} \\ \hline 0 & 0 & 0 & \cdots & \hat{a}_{k+1,k} & & \hat{a}_{k+1,k+1} & \cdots & \hat{a}_{k+1,n} \\ \vdots & \vdots & \vdots & & \vdots & & & \ddots & \vdots \\ 0 & 0 & 0 & \cdots & \hat{a}_{nk} & & \hat{a}_{n,k+1} & \cdots & \hat{a}_{nn} \end{array}\right] \tag{3-16}$$

The caret (ˆ) over the elements denotes that those elements are no longer the original elements of **A**.

All the arithmetic operations occur in the lower right submatrix, which is $(n-k) \times (n-k)$. To each of these $(n-k)^2$ elements, we add one number (which is a multiple of an element from row k). Thus, there are $(n-k)^2$ additions at the kth step.

For each of the $(n-k)$ rows, we must also determine the "appropriate" multiple of row k. Since the objective is to make the elements in column k equal to zero, these multiples are:

$$m_p = -\frac{\hat{a}_{pk}}{\hat{a}_{kk}} \qquad p = k+1, k+2, \ldots, n \tag{3-17}$$

where m_p denotes the multiple of row k to be added to row p. Computing these multiples therefore requires $(n-k)$ divisions. Furthermore, each of these $(n-k)$ multiples m_p is multiplied by the last $(n-k)$ elements of row k [and the results are added to the last $(n-k)$ rows, as discussed in the previous paragraph]. Then there are $(n-k)^2$ multiplications and $(n-k)$ divisions at the kth step. We shall combine these and say that there are $[(n-k)^2 + (n-k)]$ multiplications at the kth step.

Therefore, the total number of additions and multiplications required to perform steps 1 through $(n - 1)$ are:

$$n_a = \sum_{k=1}^{n-1} (n-k)^2 \qquad \text{additions}$$

$$n_m = \sum_{k=1}^{n-1} [(n-k)^2 + (n-k)] \qquad \text{multiplications}$$

We can evaluate the above quantities by applying the following two results, which were derived in Chap. 1:

$$\sum_{k=1}^{m} k = \frac{m(m+1)}{2}$$

$$\sum_{k=1}^{m} k^2 = \frac{m(m+1)(2m+1)}{6}$$

Thus,

$$\begin{aligned}
n_a &= \sum_{k=1}^{n-1} (n-k)^2 \\
&= \sum_{k=1}^{n-1} (n^2 - 2nk + k^2) \\
&= \sum_{k=1}^{n-1} n^2 - 2n \sum_{k=1}^{n-1} k + \sum_{k=1}^{n-1} k^2 \\
&= n^2(n-1) - 2n \frac{(n-1)(n)}{2} + \frac{(n-1)(n)(2n-1)}{6} \\
&= n^3 - n^2 - n^3 + n^2 + \frac{n^3}{3} - \frac{n^2}{2} + \frac{n}{6} \\
&= \left(\frac{n^3}{3} - \frac{n^2}{2} + \frac{n}{6}\right) \qquad \text{additions}
\end{aligned}$$

$$\begin{aligned}
n_m &= \sum_{k=1}^{n-1} [(n-k)^2 + (n-k)] \\
&= \sum_{k=1}^{n-1} (n-k)^2 + \sum_{k=1}^{n-1} (n-k) \\
&= n_a + \sum_{k=1}^{n-1} n - \sum_{k=1}^{n-1} k \\
&= \frac{n^3}{3} - \frac{n^2}{2} + \frac{n}{6} + n^2 - n - \frac{n^2}{2} + \frac{n}{2} \\
&= \left(\frac{n^3}{3} - \frac{n}{3}\right) \qquad \text{multiplications}
\end{aligned}$$

Now, since step n of our method requires $(n - 1)$ additional multiplications (multiplying the n diagonal elements together), the method described in this section for computing determinants requires a grand total of

$$\left(\frac{n^3}{3} - \frac{n^2}{2} + \frac{n}{6}\right) \qquad \text{additions}$$

and

$$\left(\frac{n^3}{3} + \frac{2n}{3} - 1\right) \qquad \text{multiplications}$$

to compute the determinant of an $n \times n$ matrix.

At this point we must consider one small detail which we have thus far ignored: Observe that in Eq. (3-17) we have divided by $\hat{a}_{kk}$, implicitly assuming, of course, that $\hat{a}_{kk} \neq 0$. However, clearly, situations will arise when at some step k in the calculation of a determinant, we may reach a point at which $\hat{a}_{kk} = 0$. When this situation occurs, we must modify the procedure slightly. We do so by making use of Theorem 3-6 (interchanging two rows reverses the sign of the determinant).

EXAMPLE 3-10 If

$$\mathbf{A} = \begin{bmatrix} 1 & -2 & 3 & 0 \\ -2 & 4 & -4 & 1 \\ -3 & 5 & 1 & 1 \\ 1 & -3 & 0 & 2 \end{bmatrix}$$

Then,

$$\det(\mathbf{A}) = \begin{vmatrix} 1 & -2 & 3 & 0 \\ -2 & 4 & -4 & 1 \\ -3 & 5 & 1 & 1 \\ 1 & -3 & 0 & 2 \end{vmatrix}$$

$$= \begin{vmatrix} 1 & -2 & 3 & 0 \\ 0 & 0 & 2 & 1 \\ 0 & 1 & 10 & 1 \\ 0 & -1 & -3 & 2 \end{vmatrix} \qquad \text{applying step 1}$$

Now, we cannot apply step 2 directly, since $\hat{a}_{22} = 0$. However, if we first interchange rows 2 and 3 and then apply step 2 to the resulting matrix, we obtain:

$$\det(\mathbf{A}) = \begin{vmatrix} 1 & -2 & 3 & 0 \\ 0 & 0 & 2 & 1 \\ 0 & 1 & 10 & 1 \\ 0 & -1 & -3 & 2 \end{vmatrix}$$

$$= (-1)\begin{vmatrix} 1 & -2 & 3 & 0 \\ 0 & 1 & 10 & 1 \\ 0 & 0 & 2 & 1 \\ 0 & -1 & -3 & 2 \end{vmatrix} \qquad \text{interchanging rows 2 and 3}$$

$$= (-1)\begin{vmatrix} 1 & -2 & 3 & 0 \\ 0 & 1 & 10 & 1 \\ 0 & 0 & 2 & 1 \\ 0 & 0 & 7 & 3 \end{vmatrix} \qquad \text{applying step 2}$$

Now, we can apply step 3 to the above and the result is

$$\det(\mathbf{A}) = (-1)\begin{vmatrix} 1 & -2 & 3 & 0 \\ 0 & 1 & 10 & 1 \\ 0 & 0 & 2 & 1 \\ 0 & 0 & 0 & -\frac{1}{2} \end{vmatrix}$$

$$= (-1)(1)(1)(2)(-\tfrac{1}{2})$$

$$= 1$$

////

Recall from our earlier discussion that at the kth step, we are really only working with the $(n-k) \times (n-k)$ submatrix which corresponds to rows $k+1, k+2, \ldots, n$ and columns $k+1, k+2, \ldots, n$ of $\hat{\mathbf{A}}$, as indicated in Eq. (3-16). In order to preserve this fact, when interchanging two rows at the kth step we will restrict ourself to interchanges involving row k and one of the rows $k+1, k+2, \ldots, n$ (performing such an interchange only when necessary, i.e., when $\hat{a}_{kk} = 0$). In interchanging row k with row q, we can choose any row q, $q = k+1, k+2, \ldots, n$, such that $\hat{a}_{qk} \neq 0$.

It is of course possible that no such row q exists. That is, it is possible to reach a situation at step k such that

$$\hat{a}_{kk} = 0$$

$$\hat{a}_{qk} = 0 \qquad q = k+1, k+2, \ldots, n$$

In such a case, the determinant of $\mathbf{A}$ can be shown to be zero [and hence no further calculations are necessary to find $\det(\mathbf{A})$]. The reader is asked to prove this fact in the Exercises.

EXAMPLE 3-11 If

$$\mathbf{A} = \begin{bmatrix} 1 & 1 & 2 & -1 \\ 2 & 3 & 7 & 0 \\ -1 & 0 & 1 & 4 \\ 1 & 1 & 2 & 0 \end{bmatrix}$$

Then,

$$\det(\mathbf{A}) = \begin{vmatrix} 1 & 1 & 2 & -1 \\ 2 & 3 & 7 & 0 \\ -1 & 0 & 1 & 4 \\ 1 & 1 & 2 & 0 \end{vmatrix}$$

$$= \begin{vmatrix} 1 & 1 & 2 & -1 \\ 0 & 1 & 3 & 2 \\ 0 & 1 & 3 & 3 \\ 0 & 0 & 0 & 1 \end{vmatrix} \qquad \text{applying step 1}$$

$$= \begin{vmatrix} 1 & 1 & 2 & -1 \\ 0 & 1 & 3 & 2 \\ 0 & 0 & 0 & 1 \\ 0 & 0 & 0 & 1 \end{vmatrix} \qquad \text{applying step 2}$$

Now, we have arrived at a point at which we cannot implement step 3, since $\hat{a}_{33} = 0$, and $\hat{a}_{43}$ also equals zero. However, we note that since rows 3 and 4 are identical, det $(\mathbf{A}) = 0$ by Theorem 3-7. ////

EXAMPLE 3-12 If

$$\mathbf{A} = \begin{bmatrix} 1 & 1 & 2 & -1 & 0 \\ 3 & 4 & 9 & -1 & 1 \\ 1 & 2 & 5 & 0 & 2 \\ 2 & 3 & 7 & 1 & -1 \\ 1 & 1 & 2 & 1 & 1 \end{bmatrix}$$

Then,

$$\det(\mathbf{A}) = \begin{vmatrix} 1 & 1 & 2 & -1 & 0 \\ 3 & 4 & 9 & -1 & 1 \\ 1 & 2 & 5 & 0 & 2 \\ 2 & 3 & 7 & 1 & -1 \\ 1 & 1 & 2 & 1 & 1 \end{vmatrix}$$

$$= \begin{vmatrix} 1 & 1 & 2 & -1 & 0 \\ 0 & 1 & 3 & 2 & 1 \\ 0 & 1 & 3 & 1 & 2 \\ 0 & 1 & 3 & 3 & -1 \\ 0 & 0 & 0 & 2 & 1 \end{vmatrix} \qquad \text{applying step 1}$$

$$= \begin{vmatrix} 1 & 1 & 2 & -1 & 0 \\ 0 & 1 & 3 & 2 & 1 \\ 0 & 0 & 0 & -1 & 1 \\ 0 & 0 & 0 & 1 & -2 \\ 0 & 0 & 0 & 2 & 1 \end{vmatrix} \qquad \text{applying step 2}$$

Now, we see that it is impossible to perform step 3; however, if we add -2 times row 4 to row 5, we will obtain

$$\det(\mathbf{A}) = \begin{vmatrix} 1 & 1 & 2 & -1 & 0 \\ 0 & 1 & 3 & 2 & 1 \\ 0 & 0 & 0 & -1 & 1 \\ 0 & 0 & 0 & 1 & -2 \\ 0 & 0 & 0 & 0 & 5 \end{vmatrix}$$

which is upper triangular with a zero along its diagonal, and hence $\det(\mathbf{A}) = 0$. ////

We have now completely described the method for computing determinants, which applies Theorem 3-8 (adding multiples of one row to another row) and Theorem 3-6 (interchanging two rows) to obtain an upper triangular matrix whose determinant is the product of its diagonals, by Theorem 3-9. This method is the most efficient (in terms of the number of arithmetic operations required) known method for computing the determinant of a general $n \times n$ matrix.

When using this method on a digital computer (or even a desk calculator), it is desirable for the elements of $\mathbf{A}$ to be of approximately the same order of magnitude, to reduce roundoff errors as much as possible. When necessary, this result may be achieved by application of Theorem 3-3 [multiplying one row of a matrix $\mathbf{A}$ by a constant α yields a matrix whose determinant is $\alpha \cdot \det(\mathbf{A})$]. A more complete discussion of roundoff problems will be presented in Chap. 5.

The three types of row (and column) operations contained in Theorems 3-3, 3-6, and 3-8 are so important that we shall devote the entire next section to a discussion of them.

3-4 ELEMENTARY MATRICES

The major result of this section will be to show that the determinant of the product of two square matrices $\mathbf{A}$, $\mathbf{B}$, is equal to the product of their determinants; i.e., $\det(\mathbf{AB}) = \det(\mathbf{A})\det(\mathbf{B})$. However, in order to prove this result, we must first consider some preliminary concepts and results.

We have been dealing with three basic types of row operations in our discussion of the computation of determinants:

(*a*) Adding a multiple of one row to another row
(*b*) Interchanging two rows
(*c*) Multiplying a row by a nonzero constant (scalar)

These three types of operations also play an important role in some methods for solving systems of linear algebraic equations, as we shall see in Chaps. 4 and 6. These operations are often called *elementary row operations*, and the corresponding column operations are similarly called *elementary column operations*.

Corresponding to each elementary operation we shall define an *elementary matrix*, which is the matrix obtained when an elementary operation is performed on an identity matrix.† In particular, we define:

(*a*) Let $\mathbf{E}_{kq}(\alpha)$ be the elementary matrix corresponding to the elementary operation "add α times row k to row q."
(*b*) Let $\mathbf{E}_{kq}$ be the elementary matrix corresponding to the elementary operation "interchange rows k and q."
(*c*) Let $\mathbf{E}_k(\alpha)$ be the elementary matrix corresponding to the elementary operation "multiply row k by the (nonzero) scalar α."
(*d*) Similarly, let $\mathbf{F}_{kq}(\alpha)$, $\mathbf{F}_{kq}$, and $\mathbf{F}_k(\alpha)$ be the elementary matrices corresponding to elementary column operations of types (*a*), (*b*), and (*c*), respectively.

EXAMPLE 3-13 If

$$\mathbf{I} = \begin{bmatrix} 1 & 0 & 0 & 0 \\ 0 & 1 & 0 & 0 \\ 0 & 0 & 1 & 0 \\ 0 & 0 & 0 & 1 \end{bmatrix}$$

then

$$\mathbf{E}_{1,3}(2) = \begin{bmatrix} 1 & 0 & 0 & 0 \\ 0 & 1 & 0 & 0 \\ 2 & 0 & 1 & 0 \\ 0 & 0 & 0 & 1 \end{bmatrix} \qquad \mathbf{E}_{2,3}(\alpha) = \begin{bmatrix} 1 & 0 & 0 & 0 \\ 0 & 1 & 0 & 0 \\ 0 & \alpha & 1 & 0 \\ 0 & 0 & 0 & 1 \end{bmatrix}$$

$$\mathbf{E}_{2,3} = \begin{bmatrix} 1 & 0 & 0 & 0 \\ 0 & 0 & 1 & 0 \\ 0 & 1 & 0 & 0 \\ 0 & 0 & 0 & 1 \end{bmatrix} \qquad \mathbf{E}_{1,3} = \begin{bmatrix} 0 & 0 & 1 & 0 \\ 0 & 1 & 0 & 0 \\ 1 & 0 & 0 & 0 \\ 0 & 0 & 0 & 1 \end{bmatrix}$$

$$\mathbf{E}_2(-3) = \begin{bmatrix} 1 & 0 & 0 & 0 \\ 0 & -3 & 0 & 0 \\ 0 & 0 & 1 & 0 \\ 0 & 0 & 0 & 1 \end{bmatrix} \qquad \mathbf{E}_4(\alpha) = \begin{bmatrix} 1 & 0 & 0 & 0 \\ 0 & 1 & 0 & 0 \\ 0 & 0 & 1 & 0 \\ 0 & 0 & 0 & \alpha \end{bmatrix}$$

† The *size* of the elementary matrix of course will depend on the size of the identity matrix we start with. This will depend on the particular application we are considering.

The reader should be able to verify readily that

$$\mathbf{E}_{1,3}(2) = \mathbf{F}_{3,1}(2) \qquad \mathbf{E}_{2,3}(\alpha) = \mathbf{F}_{3,2}(\alpha)$$
$$\mathbf{E}_{2,3} = \mathbf{F}_{2,3} \qquad \mathbf{E}_{1,3} = \mathbf{F}_{1,3}$$
$$\mathbf{E}_2(-3) = \mathbf{F}_2(-3) \qquad \mathbf{E}_4(\alpha) = \mathbf{F}_4(\alpha) \qquad ////$$

These latter results indicate that, in general, $\mathbf{E}_{kq}(\alpha) = \mathbf{F}_{qk}(\alpha)$, $\mathbf{E}_{kq} = \mathbf{F}_{kq}$, and $\mathbf{E}_k(\alpha) = \mathbf{F}_k(\alpha)$. In the Exercises the reader will be asked to prove the validity of this generalization.

EXAMPLE 3-14

(*a*)

$$\mathbf{E}_{1,3}(2) = \begin{bmatrix} 1 & 0 & 0 & 0 \\ 0 & 1 & 0 & 0 \\ 2 & 0 & 1 & 0 \\ 0 & 0 & 0 & 1 \end{bmatrix} \qquad \mathbf{E}_{1,3}(-2) = \begin{bmatrix} 1 & 0 & 0 & 0 \\ 0 & 1 & 0 & 0 \\ -2 & 0 & 1 & 0 \\ 0 & 0 & 0 & 1 \end{bmatrix}$$

and $[\mathbf{E}_{1,3}(2)][\mathbf{E}_{1,3}(-2)] = [\mathbf{E}_{1,3}(-2)][\mathbf{E}_{1,3}(2)] = \mathbf{I}_4$

Thus, $\mathbf{E}_{1,3}(-2) = [\mathbf{E}_{1,3}(2)]^{-1}$

(*b*)

$$\mathbf{E}_{2,3} = \begin{bmatrix} 1 & 0 & 0 & 0 \\ 0 & 0 & 1 & 0 \\ 0 & 1 & 0 & 0 \\ 0 & 0 & 0 & 1 \end{bmatrix} \qquad \mathbf{E}_{3,2} = \begin{bmatrix} 1 & 0 & 0 & 0 \\ 0 & 0 & 1 & 0 \\ 0 & 1 & 0 & 0 \\ 0 & 0 & 0 & 1 \end{bmatrix}$$

and $\mathbf{E}_{2,3}\,\mathbf{E}_{2,3} = \mathbf{I}_4 = \mathbf{E}_{3,2}\,\mathbf{E}_{3,2}$

Thus, $\mathbf{E}_{2,3} = \mathbf{E}_{3,2}$ and $\mathbf{E}_{2,3} = \mathbf{E}_{2,3}{}^{-1}$

(*c*)

$$\mathbf{E}_4(\alpha) = \begin{bmatrix} 1 & 0 & 0 & 0 \\ 0 & 1 & 0 & 0 \\ 0 & 0 & 1 & 0 \\ 0 & 0 & 0 & \alpha \end{bmatrix} \qquad \mathbf{E}_4\left(\frac{1}{\alpha}\right) = \begin{bmatrix} 1 & 0 & 0 & 0 \\ 0 & 1 & 0 & 0 \\ 0 & 0 & 1 & 0 \\ 0 & 0 & 0 & 1/\alpha \end{bmatrix}$$

and $[\mathbf{E}_4(\alpha)]\left[\mathbf{E}_4\left(\frac{1}{\alpha}\right)\right] = \left[\mathbf{E}_4\left(\frac{1}{\alpha}\right)\mathbf{E}_4(\alpha)\right] = \mathbf{I}_4$

Thus, $\mathbf{E}_4\left(\frac{1}{\alpha}\right) = [\mathbf{E}_4(\alpha)]^{-1}$ ////

The above results indicate that the inverse of an elementary matrix is also an elementary matrix of the same type.

We state this result formally:

Theorem 3-10 The inverse of an elementary matrix is also an elementary matrix of the same type.

We leave the proof of this theorem as an exercise for the reader. Another easily proved but very useful result is contained in:

Theorem 3-11 (*a*) $\det(\mathbf{E}_{kq}(\alpha)) = 1$
(*b*) $\det(\mathbf{E}_{kq}) = -1$
(*c*) $\det(\mathbf{E}_k(\alpha)) = \alpha$

PROOF We need only to note that each of the above elementary matrices has been formed by performing exactly *one* elementary operation on an identity matrix, whose determinant is unity (since **I** is upper triangular with all 1's along its diagonal). We then apply Theorems 3-8, 3-6, and 3-3 to prove (*a*), (*b*), and (*c*), respectively. QED

We shall now illustrate the most important relationship between elementary operations and elementary matrices.

EXAMPLE 3-15 Let

$$\mathbf{A} = \begin{bmatrix} 1 & 3 & 2 \\ 2 & 5 & 3 \\ 1 & 4 & 5 \end{bmatrix}$$

Suppose we add -2 times row 1 to row 2, yielding the matrix

$$\mathbf{B} = \begin{bmatrix} 1 & 3 & 2 \\ 0 & -1 & -1 \\ 1 & 4 & 5 \end{bmatrix}$$

Now, suppose instead of performing the above elementary row operation on **A**, we premultiply **A** by the elementary matrix corresponding to this elementary row operation; i.e., the elementary matrix

$$\mathbf{E}_{1,2}(-2) = \begin{bmatrix} 1 & 0 & 0 \\ -2 & 1 & 0 \\ 0 & 0 & 1 \end{bmatrix}$$

The result of this product is

$$[\mathbf{E}_{1,2}(-2)]\mathbf{A} = \begin{bmatrix} 1 & 0 & 0 \\ -2 & 1 & 0 \\ 0 & 0 & 1 \end{bmatrix} \begin{bmatrix} 1 & 3 & 2 \\ 2 & 5 & 3 \\ 1 & 4 & 5 \end{bmatrix}$$

$$= \begin{bmatrix} 1 & 3 & 2 \\ 0 & -1 & -1 \\ 1 & 4 & 5 \end{bmatrix}$$

$$= \mathbf{B}$$

Thus, the result of premultiplying $\mathbf{A}$ by the elementary matrix $\mathbf{E}_{1,2}(-2)$ is the matrix we obtained by performing the corresponding elementary row operation on $\mathbf{A}$.

Similarly, suppose we interchange rows 2 and 3 of $\mathbf{A}$, yielding

$$\mathbf{C} = \begin{bmatrix} 1 & 3 & 2 \\ 1 & 4 & 5 \\ 2 & 5 & 3 \end{bmatrix}$$

and also premultiply $\mathbf{A}$ by the elementary matrix $\mathbf{E}_{2,3}$:

$$\mathbf{E}_{2,3}\mathbf{A} = \begin{bmatrix} 1 & 0 & 0 \\ 0 & 0 & 1 \\ 0 & 1 & 0 \end{bmatrix} \begin{bmatrix} 1 & 3 & 2 \\ 2 & 5 & 3 \\ 1 & 4 & 5 \end{bmatrix}$$

$$= \begin{bmatrix} 1 & 3 & 2 \\ 1 & 4 & 5 \\ 2 & 5 & 3 \end{bmatrix}$$

$$= \mathbf{C}$$

////

We may generalize the results indicated in the above example to obtain the following theorem, the proof of which is relatively simple and is left as an exercise for the reader.

Theorem 3-12 (*a*) The matrix obtained by performing an elementary *row* operation on a given matrix $\mathbf{A}$ is equal to the product of $\mathbf{A}$ *premultiplied* by the corresponding elementary matrix.

(*b*) The matrix obtained by performing an elementary *column* operation on a given matrix $\mathbf{A}$ is equal to the product of $\mathbf{A}$ *post-multiplied* by the corresponding elementary matrix.

We now observe that, by performing a sequence of elementary row operations on any square matrix $\mathbf{A}$, we can obtain an upper triangular matrix, or else det (**A**) = 0. This is, in fact, precisely what the method described in the previous section accomplishes. In view of Theorem 3-12*a*, then, we may write, provided that det (**A**) ≠ 0,

$$\mathbf{T} = \mathbf{E}_m \mathbf{E}_{m-1} \cdots \mathbf{E}_1 \mathbf{A} \tag{3-18}$$

where **T** is upper triangular and $\mathbf{E}_1, \mathbf{E}_2, \ldots, \mathbf{E}_m$† are elementary matrices of any type—(*a*), (*b*), or (*c*).

Moreover, if det (**A**) ≠ 0, then clearly the diagonal elements of **T** are all nonzero. Thus, if we let

$$\mathbf{T} = \begin{bmatrix} t_{11} & t_{12} & \cdots & t_{1n} \\ 0 & t_{22} & \cdots & t_{2n} \\ \vdots & \vdots & & \vdots \\ 0 & 0 & \cdots & t_{nn} \end{bmatrix}$$

then it should be clear that we can perform a sequence of elementary row operations on **T** in such a way that we eventually form a diagonal matrix. Specifically, the procedure is: First add appropriate multiples of the *last* row of **T** to each of the first $(n-1)$ rows, so that the elements in the last column of these $(n-1)$ rows become zero. Secondly, add appropriate multiples of row $(n-1)$ of **T** to each of the first $(n-2)$ rows of **T** so that the elements in column $(n-1)$ of these rows will become zero (the elements in column n will automatically remain at zero). Continuing in this manner, we will eventually reduce **T** to a diagonal matrix. No interchange of rows will be necessary, since each $t_{ii} \neq 0$, $i = 1, 2, \ldots, n$ and none of these diagonal elements is altered by the above procedure.

Since each of the elementary operations employed in this procedure can be expressed as an elementary matrix, we can write

$$\mathbf{D} = \mathbf{E}_r \mathbf{E}_{r-1} \cdots \mathbf{E}_{m+1} \mathbf{T} \tag{3-19}$$

where

$$\mathbf{D} = \begin{bmatrix} t_{11} & 0 & \cdots & 0 \\ 0 & t_{22} & \cdots & 0 \\ \vdots & \vdots & \ddots & \vdots \\ 0 & 0 & \cdots & t_{nn} \end{bmatrix}$$

and, as in Eq. (3-18), the $\mathbf{E}_j$'s represent any elementary matrices. Substitution of Eq. (3-18) into Eq. (3-19) yields:

$$\begin{aligned} \mathbf{D} &= \mathbf{E}_r \mathbf{E}_{r-1} \cdots \mathbf{E}_{m+1}(\mathbf{E}_m \mathbf{E}_{m-1} \cdots \mathbf{E}_1 \mathbf{A}) \\ &= (\mathbf{E}_r \mathbf{E}_{r-1} \cdots \mathbf{E}_1)\mathbf{A} \end{aligned} \tag{3-20}$$

† The subscripts on these elementary matrices indicate the order in which the corresponding elementary row operations are performed.

Now, by Theorem 3-10, the inverse of each of the $\mathbf{E}_j$'s in Eq. (3-20) is also an elementary matrix. Thus, we may write

$$\begin{aligned} \mathbf{A} &= \mathbf{E}_1^{-1}\mathbf{E}_2^{-1} \cdots \mathbf{E}_r^{-1}\mathbf{D} \\ &= \hat{\mathbf{E}}_1\hat{\mathbf{E}}_2 \cdots \hat{\mathbf{E}}_r\mathbf{D} \end{aligned} \tag{3-21}$$

where $\hat{\mathbf{E}}_j$ denotes the elementary matrix which is the inverse of $\mathbf{E}_j$, $j = 1, 2, \ldots, r$.

Finally, we observe that $\mathbf{D}$ itself is the product of n elementary matrices of type (c):

$$\mathbf{D} = \mathbf{E}_1(t_{11})\mathbf{E}_2(t_{22}) \cdots \mathbf{E}_n(t_{nn})$$

For notational convenience, let us denote

$$\mathbf{E}_j(t_{jj}) = \hat{\mathbf{E}}_{r+j} \qquad j = 1, 2, \ldots, n$$

Thus, we have that

$$\mathbf{D} = \hat{\mathbf{E}}_{r+1}\hat{\mathbf{E}}_{r+2} \cdots \hat{\mathbf{E}}_{r+n}$$

and substitution of this result into Eq. (3-21) yields

$$\mathbf{A} = (\hat{\mathbf{E}}_1\hat{\mathbf{E}}_2 \cdots \hat{\mathbf{E}}_r)(\hat{\mathbf{E}}_{r+1}\hat{\mathbf{E}}_{r+2} \cdots \hat{\mathbf{E}}_{r+n}) \tag{3-22}$$

In Eq. (3-22) we have expressed the square matrix $\mathbf{A}$ as the product of elementary matrices. Recall that the only assumption we have made about $\mathbf{A}$ is that $\det(\mathbf{A}) \neq 0$.

We have therefore proved:

Theorem 3-13 Any square matrix $\mathbf{A}$ for which $\det(\mathbf{A}) \neq 0$ may be expressed as the product of elementary matrices.

Let us now consider the case $\det(\mathbf{A}) = 0$. We modify the argument used to prove Theorem 3-13 as follows: If $\det(\mathbf{A}) = 0$, then $\mathbf{A}$ is reducible by elementary operations to an upper triangular matrix with at least one zero along its diagonal. This upper triangular matrix, in turn, is then reducible by elementary operations to a matrix which has at least one row containing only zero elements. (The reader is asked to prove these last two statements in the Exercises.) Thus, if $\det(\mathbf{A}) = 0$, we can write

$$\mathbf{A} = \mathbf{E}_1\mathbf{E}_2 \cdots \mathbf{E}_k\mathbf{Z} \tag{3-23}$$

where as before the $\mathbf{E}_j$'s denote elementary matrices, and $\mathbf{Z}$ denotes a matrix with at least one row of zeros.

We are now ready to consider the calculation of the determinant of the product of several matrices. First, we shall prove:

Theorem 3-14 If $\mathbf{E}_1, \mathbf{E}_2, \ldots, \mathbf{E}_k$ are elementary matrices and $\mathbf{B}$ is any square matrix, then

$$\det(\mathbf{E}_1\mathbf{E}_2 \cdots \mathbf{E}_k\mathbf{B}) = \det(\mathbf{E}_1)\det(\mathbf{E}_2)\cdots\det(\mathbf{E}_k)\det(\mathbf{B}) \tag{3-24}$$

PROOF Let

$$\mathbf{A} = \mathbf{E}_1\mathbf{E}_2 \cdots \mathbf{E}_k\mathbf{B}$$

Then, $\mathbf{A}$ is a matrix obtained from $\mathbf{B}$ by performing a sequence of elementary row operations on $\mathbf{B}$ and on the matrices resulting from each such operation. Thus, we may easily compare det ($\mathbf{A}$) and det ($\mathbf{B}$), by examining the specific elementary operations used. For each elementary operation of type (*a*) employed, the determinant of the resulting matrix is unchanged; for each elementary operation of type (*b*) employed, the determinant of the resulting matrix changes sign; for each elementary operation of type (*c*) employed, the determinant of the resulting matrix is multiplied by a scalar. Now if we examine the right-hand side of Eq. (3-24), we see that this is exactly what happens: For each elementary operation of type (*a*), there is a corresponding elementary matrix whose determinant is unity (by Theorem 3-11), for each elementary operation of type (*b*) there is a corresponding elementary matrix whose determinant is -1, and for each elementary operation of type (*c*) there is a corresponding elementary matrix whose determinant is the appropriate multiple α. Thus, the right-hand side of Eq. (3-24) is precisely equal to det ($\mathbf{A}$). QED

Theorem 3-14 may seem to have rather limited application, but we shall immediately put it to good use to help us prove:

Theorem 3-15 For any square matrices $\mathbf{A}$, $\mathbf{B}$,

$$\det(\mathbf{AB}) = \det(\mathbf{A})\det(\mathbf{B})$$

PROOF We consider two cases separately:

Case 1: $\det(\mathbf{A}) = 0$

Thus, we wish to show that $\det(\mathbf{AB}) = 0$. But by Eq. (3-23) we can write **A** as the product of elementary matrices and **Z**, where **Z** is a matrix with at least one row of zeros. Hence,

$$\begin{aligned}\det(\mathbf{AB}) &= \det([\mathbf{E}_1\mathbf{E}_2 \cdots \mathbf{E}_k\mathbf{Z}]\mathbf{B}) \\ &= \det([\mathbf{E}_1\mathbf{E}_2 \cdots \mathbf{E}_k][\mathbf{ZB}])\end{aligned}$$

Now, letting $\mathbf{ZB} = \mathbf{C}$, we observe that **C** is a matrix with at least one row of zeros, and hence $\det(\mathbf{C}) = 0$. Moreover, by Theorem 3-14,

$$\begin{aligned}\det(\mathbf{E}_1\mathbf{E}_2 \cdots \mathbf{E}_k\mathbf{C}) &= \det(\mathbf{E}_1)\det(\mathbf{E}_2)\cdots\det(\mathbf{E}_k)\det(\mathbf{C}) \\ &= 0\end{aligned}$$

Thus, $\det(\mathbf{AB}) = 0$, as we wished to show.

Case 2: $\det(\mathbf{A}) \neq 0$

By Theorem 3-13, we may write

$$\mathbf{A} = \mathbf{E}_1\mathbf{E}_2 \cdots \mathbf{E}_k$$

where the $\mathbf{E}_j$'s are elementary matrices. Then,

$$\begin{aligned}\det(\mathbf{AB}) &= \det(\mathbf{E}_1\mathbf{E}_2 \cdots \mathbf{E}_k\mathbf{B}) \\ &= \det(\mathbf{E}_1)\det(\mathbf{E}_2)\cdots\det(\mathbf{E}_k)\det(\mathbf{B}) \qquad \text{by Theorem 3-14} \\ &= \det(\mathbf{E}_1\mathbf{E}_2 \cdots \mathbf{E}_k\mathbf{I})\det(\mathbf{B}) \qquad \text{by Theorem 3-14} \\ &= \det(\mathbf{E}_1\mathbf{E}_2 \cdots \mathbf{E}_k)\det(\mathbf{B}) \\ &= \det(\mathbf{A})\det(\mathbf{B})\end{aligned}$$

QED

EXAMPLE 3-16 Let

$$\mathbf{A} = \begin{bmatrix} 3 & 1 & 0 \\ -1 & 2 & 4 \\ 5 & 6 & -2 \end{bmatrix} \qquad \mathbf{B} = \begin{bmatrix} 1 & -3 & 2 \\ -8 & 0 & 9 \\ 4 & 5 & -7 \end{bmatrix}$$

Then $\det(\mathbf{A}) = -66$ see Example 3-3

$\det(\mathbf{B}) = -65$ see Example 3-6

$\det(\mathbf{A})\det(\mathbf{B}) = (-66)(-65) = 4290$

$$\begin{aligned}\mathbf{AB} &= \begin{bmatrix} 3 & 1 & 0 \\ -1 & 2 & 4 \\ 5 & 6 & -2 \end{bmatrix}\begin{bmatrix} 1 & -3 & 2 \\ -8 & 0 & 9 \\ 4 & 5 & -7 \end{bmatrix} \\ &= \begin{bmatrix} -5 & -9 & 15 \\ -1 & 23 & -12 \\ -51 & -25 & 78 \end{bmatrix}\end{aligned}$$

$$\det(\mathbf{AB}) = \begin{vmatrix} -5 & -9 & 15 \\ -1 & 23 & -12 \\ -51 & -25 & 78 \end{vmatrix}$$
$$= \begin{vmatrix} -5 & -9 & 15 \\ 0 & \frac{124}{5} & -15 \\ 0 & \frac{334}{5} & -75 \end{vmatrix}$$
$$= \begin{vmatrix} -5 & -9 & 15 \\ 0 & \frac{124}{5} & -15 \\ 0 & 0 & -\frac{4290}{124} \end{vmatrix}$$
$$= (-5)(\tfrac{124}{5})(-\tfrac{4290}{124})$$
$$= 4290 \qquad ////$$

REFERENCES

1. HOHN, FRANZ E.: "Elementary Matrix Algebra," 2d ed., Macmillan, New York, 1964.
2. NOBLE, BEN: "Applied Linear Algebra," Prentice-Hall, Englewood Cliffs, N.J., 1969.

EXERCISES

1 Compute the determinant of each of the following matrices, by cofactor expansion and by the method of Sec. 3-3:

$$(a)\ \begin{bmatrix} 1 & -1 & 2 & 4 \\ 1 & 0 & 2 & 5 \\ 2 & -1 & 6 & 8 \\ 3 & -1 & 8 & 5 \end{bmatrix} \qquad (b)\ \begin{bmatrix} 1 & -1 & 2 & 4 \\ -2 & 1 & 0 & 9 \\ 3 & -2 & 4 & -6 \\ -3 & 2 & 2 & 1 \end{bmatrix}$$

$$(c)\ \begin{bmatrix} 5 & 3 & 6 & 4 \\ 1 & 2 & 1 & 1 \\ 1 & 1 & -1 & 1 \\ 4 & 0 & 4 & 4 \end{bmatrix}$$

2 Compute the determinant of each of the following matrices by the method of Sec. 3-3:

$$(a)\ \begin{bmatrix} 1 & 2 & -1 & 3 & -2 \\ 1 & 3 & 0 & 2 & 0 \\ 1 & 5 & -2 & 3 & 5 \\ 2 & 3 & -3 & 0 & -4 \\ 3 & 1 & -8 & 8 & 1 \end{bmatrix} \qquad (b)\ \begin{bmatrix} 3 & 2 & 3 & 0 & 8 \\ 2 & 3 & 5 & 3 & 1 \\ 1 & 1 & 1 & 2 & 3 \\ 1 & 2 & 4 & 1 & 1 \\ 2 & 0 & -1 & -1 & 4 \end{bmatrix}$$

$$(c)\quad \begin{bmatrix} 1 & 1 & 1 & 1 & 1 \\ 0 & 2 & 3 & -1 & 0 \\ 2 & 3 & 6 & 1 & 2 \\ 1 & 2 & 5 & 2 & 2 \\ 3 & 5 & 2 & 0 & 1 \end{bmatrix} \qquad (d)\quad \begin{bmatrix} 0 & 1 & 4 & 7 & 1 \\ 0 & 1 & 4 & 6 & 3 \\ 0 & 3 & 1 & 2 & 8 \\ 1 & 2 & 4 & 5 & -3 \\ 4 & 3 & -2 & 0 & 2 \end{bmatrix}$$

3 Prepare a flow chart of the method of Sec. 3-3, including row interchanges when necessary and a test for a zero determinant (i.e., when row interchanges are necessary but not possible).

4 Prove that when computing det (**A**) by the method of Sec. 3-3, if a stage is reached such that $\hat{a}_{kk} = 0$ and $\hat{a}_{qk} = 0$ for $q = k+1, k+2, \ldots, n$ [see Eq. (3-16)], then det $(\mathbf{A}) = 0$.

5 Prove: (*a*) $\mathbf{E}_{kq}(\alpha) = \mathbf{F}_{qk}(\alpha)$

(*b*) $\mathbf{E}_{kq} = \mathbf{F}_{kq}$

(*c*) $\mathbf{E}_k(\alpha) = \mathbf{F}_k(\alpha)$

6 Prove Theorem 3-10: The inverse of an elementary matrix is also an elementary matrix of the same type.

7 Find the inverse of each of the following elementary matrices:

$$(a)\quad \begin{bmatrix} 1 & 0 & 0 & 0 \\ 0 & 1 & 0 & 0 \\ 0 & 0 & 1 & 0 \\ 0 & -3 & 0 & 1 \end{bmatrix} \qquad (b)\quad \begin{bmatrix} 1 & 0 & 0 & 0 \\ 0 & 5 & 0 & 0 \\ 0 & 0 & 1 & 0 \\ 0 & 0 & 0 & 1 \end{bmatrix}$$

$$(c)\quad \begin{bmatrix} 0 & 0 & 0 & 1 \\ 0 & 1 & 0 & 0 \\ 0 & 0 & 1 & 0 \\ 1 & 0 & 0 & 0 \end{bmatrix} \qquad (d)\quad \begin{bmatrix} 1 & 0 & 0 & 2 \\ 0 & 1 & 0 & 0 \\ 0 & 0 & 1 & 0 \\ 0 & 0 & 0 & 1 \end{bmatrix}$$

8 Prove Theorem 3-12: (*a*) The matrix obtained by performing an elementary row operation on a given matrix **A** is equal to the product of **A** premultiplied by the corresponding elementary matrix. (*b*) The matrix obtained by performing an elementary column operation on a given matrix **A** is equal to the product of **A** postmultiplied by the corresponding elementary matrix.

9 Prove that if det $(\mathbf{A}) = 0$:

(*a*) **A** is reducible by elementary operations to an upper triangular matrix with at least one zero along its diagonal.

(*b*) **A** is reducible by elementary operations to a matrix which has at least one row containing only zero elements.

10 Express each of the matrices of Exercise 1 as the product of elementary matrices (or else show that its determinant is zero).

11 Prove by induction: The determinant of the product of k square matrices is equal to the product of the determinants of the k matrices. That is, if $\mathbf{A}_1, \mathbf{A}_2, \ldots, \mathbf{A}_k$ are square matrices, det $(\mathbf{A}_1\mathbf{A}_2 \cdots \mathbf{A}_k) = \det(\mathbf{A}_1)\det(\mathbf{A}_2)\cdots\det(\mathbf{A}_k)$.

12 Consider two points (a_1, b_1) and (a_2, b_2) in a two-dimensional Euclidean plane. These two points, along with the origin (0, 0) and the point $(a_1 + a_2, b_1 + b_2)$, determine a parallelogram, as shown in the figure below. Show that the area of this parallelogram is equal to the determinant of

$$\mathbf{A} = \begin{bmatrix} a_1 & b_1 \\ a_2 & b_2 \end{bmatrix}$$

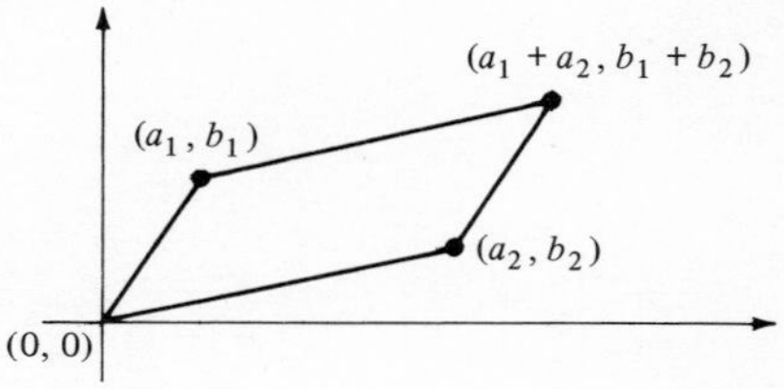

13 Consider the generalization of Exercise 12 to three-dimensional Euclidean space: The three points (a_1, b_1, c_1), (a_2, b_2, c_2), and (a_3, b_3, c_3) generate a parallelopiped. Show that the volume of this parallelopiped is equal to the determinant of

$$\mathbf{A} = \begin{bmatrix} a_1 & b_1 & c_1 \\ a_2 & b_2 & c_2 \\ a_3 & b_3 & c_3 \end{bmatrix}$$

14 If $\mathbf{A} = [a_{ij}]_{n \times n}$, show that $\sum_{j=1}^{n} a_{pj} \mathbf{A}_{kj} = 0$ and that $\sum_{i=1}^{n} a_{ip} \mathbf{A}_{ik} = 0$, if $p \neq k$. [*Hint:* See proof of Theorem 3-8.]

15 If $\mathbf{A} = [a_{ij}]$ and $\mathbf{B} = [b_{ij}]$ are $n \times n$ matrices, and if $b_{ij} = A_{ji}$ (where A_{ji} is the cofactor of a_{ji}), then $\mathbf{B}$ is called the *adjoint matrix* of $\mathbf{A}$. Prove that, if $\det(\mathbf{A}) \neq 0$, then $\mathbf{A}^{-1} = \dfrac{1}{\det(\mathbf{A})} \mathbf{B}$. [*Hint:* Let $\mathbf{C} = \dfrac{1}{\det(\mathbf{A})} \mathbf{B}$ and show that $\mathbf{AC} = \mathbf{CA} = \mathbf{I}_n$ by applying the results of Exercise 14.]

16 Exercise 15 establishes that if $\det(\mathbf{A}) \neq 0$, then $\mathbf{A}^{-1}$ exists. Use Theorem 3-15 to prove the converse:
If $\mathbf{A}^{-1}$ exists, then $\det(\mathbf{A}) \neq 0$.

17 Prove that if $\mathbf{A}^{-1}$ exists, then $\det(\mathbf{A}^{-1}) = 1/\det(\mathbf{A})$.

18 Suppose that $\mathbf{B}$ is any $n \times n$ matrix, $\mathbf{C}$ is any $m \times m$ matrix, and $\mathbf{D}$ is any $n \times m$ matrix, and let

$$\mathbf{A} = \left[\begin{array}{c|c} \mathbf{B}_{n \times n} & \mathbf{D}_{n \times m} \\ \hline \mathbf{O}_{m \times n} & \mathbf{C}_{m \times m} \end{array}\right]$$

Prove that $\det(\mathbf{A}) = \det(\mathbf{B}) \det(\mathbf{C})$.

19 Using the flow chart prepared for Exercise 3, write a computer program to compute the determinant of a given square matrix. Test your program on each of the matrices given in Exercise 2.

4

SOLUTIONS OF SIMULTANEOUS EQUATIONS: n EQUATIONS IN n VARIABLES

4-1 INTRODUCTION

In this chapter we shall be concerned exclusively with systems of linear algebraic equations, in which the number of variables equals the number of equations. Thus, we will be considering systems of equations of the following form:

$$\begin{aligned} a_{11}x_1 + a_{12}x_2 + \cdots + a_{1n}x_n &= b_1 \\ a_{21}x_1 + a_{22}x_2 + \cdots + a_{2n}x_n &= b_2 \\ \vdots \qquad\qquad\qquad & \quad \vdots \\ a_{n1}x_1 + a_{n2}x_2 + \cdots + a_{nn}x_n &= b_n \end{aligned} \tag{4-1}$$

Equation (4-1) can be written in matrix notation as simply

$$\mathbf{Ax} = \mathbf{b} \tag{4-2}$$

where $\mathbf{A} = [a_{ij}]_{n \times n}$, $\mathbf{x} = [x_1 \quad x_2 \quad \cdots \quad x_n]^T$, and $\mathbf{b} = [b_1 \quad b_2 \quad \cdots \quad b_n]^T$.

The matrix $\mathbf{A}$ is called the *coefficient matrix* of the system of equations (4-1). A system of equations is completely defined by its coefficient matrix $\mathbf{A}$ and right-hand-side vector $\mathbf{b}$. A convenient way of storing the two pieces of information $\mathbf{A}$ and $\mathbf{b}$ is to form a matrix with n rows and $(n + 1)$ columns, the first n columns containing $\mathbf{A}$ and the $(n+1)$st column containing $\mathbf{b}$. Such a matrix is called the *augmented matrix* of the system of equations (4-1) and in partitioned form is written $[\mathbf{A}, \mathbf{b}]$.

Throughout this chapter we shall assume that $\det(\mathbf{A}) \neq 0$; the case $\det(\mathbf{A}) = 0$ will be studied in Chap. 6. In considering systems of equations in which the coefficient matrix $\mathbf{A}$ is square, the situation which is of most practical interest is the case $\det(\mathbf{A}) \neq 0$, since this fact implies that the system of equations possesses a solution and that the solution is unique. The above statement will be proved in Theorem 4-3. We shall learn in Chap. 6 that the case $\det(\mathbf{A}) = 0$ implies that the system of equations either has no solution or possesses an infinity of solutions.

EXAMPLE 4-1 (*a*) Consider the system of equations

$$\begin{aligned} x + y &= 2 \\ x + 2y &= 3 \end{aligned}$$

The coefficient matrix is $\mathbf{A} = \begin{bmatrix} 1 & 1 \\ 1 & 2 \end{bmatrix}$ and $\det(\mathbf{A}) = 1$. Subtracting the first equation from the second yields $y = 1$; substitution of this result into the first equation yields $x = 1$. The fact that this solution is unique may be seen by graphing the two equations; the solution $x = y = 1$ occurs at the intersection of the two lines $x + y = 2$ and $x + 2y = 3$.

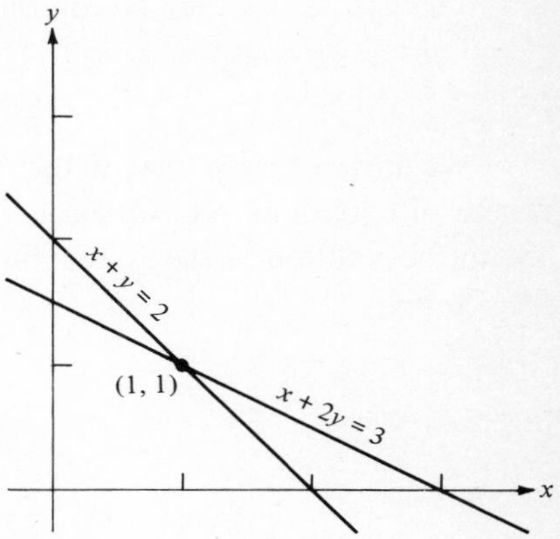

(*b*) Consider the system of equations

$$\begin{aligned} x + y &= 2 \\ 2x + 2y &= 3 \end{aligned}$$

The coefficient matrix is $\mathbf{A} = \begin{bmatrix} 1 & 1 \\ 2 & 2 \end{bmatrix}$ and $\det(\mathbf{A}) = 0$. Graphing these two equations reveals that the above system has no solution, since the two lines are parallel: no values of x and y satisfy both equations simultaneously.

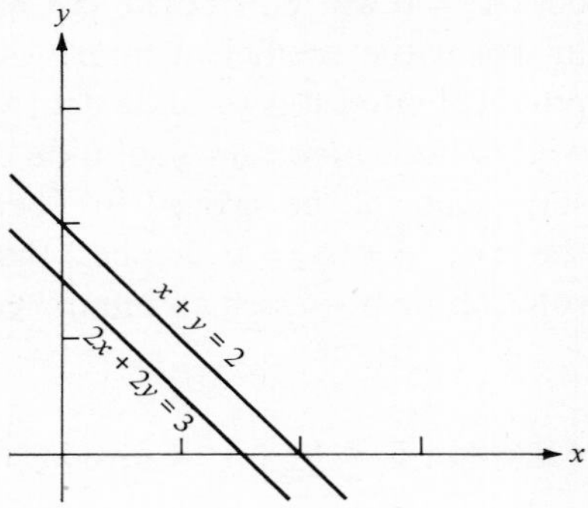

(*c*) Consider the system of equations

$$x + \ \ y = 2$$
$$2x + 2y = 4$$

The coefficient matrix $\mathbf{A}$ is the same as in (*b*); hence $\det(\mathbf{A}) = 0$. However, graphing these two equations (or simple observation) reveals that the two lines are coincident; that is, the system actually consists of only one straight line, $x + y = 2$, and so the system has an infinity of solutions: any pair of values (x, y) which lies on the line $x + y = 2$ is a solution to the system of equations. ////

We now observe that if the inverse of the coefficient matrix $\mathbf{A}$ for a system of equations $\mathbf{Ax} = \mathbf{b}$ exists, then we can—theoretically—immediately obtain the solution to the system by premultiplying both sides of the equation $\mathbf{Ax} = \mathbf{b}$ by $\mathbf{A}^{-1}$:

$$\mathbf{Ax} = \mathbf{b}$$
$$\mathbf{A}^{-1}(\mathbf{Ax}) = \mathbf{A}^{-1}\mathbf{b}$$
$$(\mathbf{A}^{-1}\mathbf{A})\mathbf{x} = \mathbf{A}^{-1}\mathbf{b}$$
$$\mathbf{Ix} = \mathbf{A}^{-1}\mathbf{b}$$
$$\mathbf{x} = \mathbf{A}^{-1}\mathbf{b}$$

Thus, if $\mathbf{A}^{-1}$ exists, the solution to $\mathbf{Ax} = \mathbf{b}$ is simply $\mathbf{A}^{-1}\mathbf{b}$. But we have not yet developed a method for computing $\mathbf{A}^{-1}$. Consider, then:

Theorem 4-1 If $\det(\mathbf{A}) \neq 0$, then $\mathbf{A}^{-1} = [\alpha_{ij}]$, where

$$\alpha_{ij} = \frac{A_{ji}}{\det(\mathbf{A})}$$

(A_{ji} is the cofactor of the element a_{ji} of $\mathbf{A}$, as defined in Sec. 3-1.)

PROOF See Exercise 15 of Chap. 3. QED

Theorem 4-1 provides us with a formula for computing $\mathbf{A}^{-1}$. However, note that we would have to compute n^2 determinants, each of order $(n - 1)$, to obtain $\mathbf{A}^{-1}$ by this formula. Since each such determinant requires approximately† $\frac{1}{3}n^3$ additions and $\frac{1}{3}n^3$ multiplications (using the method of Sec. 3-3), the total number of arithmetic operations required to compute $\mathbf{A}^{-1}$ by the formula of Theorem 4-1 is roughly $\frac{1}{3}n^5$ additions and $\frac{1}{3}n^5$ multiplications. Moreover, in order to obtain the desired solution to $\mathbf{Ax} = \mathbf{b}$, we would still have to perform the matrix multiplication $\mathbf{A}^{-1}\mathbf{b}$.

However, if we are not actually interested in $\mathbf{A}^{-1}$, but rather only want the solution to $\mathbf{Ax} = \mathbf{b}$, then we can apply Theorem 4-1 to obtain:

Theorem 4-2 **Cramer's rule** Given a system of equations $\mathbf{Ax} = \mathbf{b}$, if $\det(\mathbf{A}) \neq 0$, then the solution $x_1, x_2, \ldots, x_n$ may be obtained from

$$x_i = \frac{\det(\mathbf{A}^{(i)})}{\det(\mathbf{A})} \qquad i = 1, 2, \ldots, n$$

where $\mathbf{A}^{(i)}$ is the matrix obtained from $\mathbf{A}$ by replacing the ith column of $\mathbf{A}$ by the elements $b_1, b_2, \ldots, b_n$ (the right-hand side of $\mathbf{Ax} = \mathbf{b}$).

PROOF By Theorem 4-1, if $\det(\mathbf{A}) \neq 0$, then $\mathbf{A}^{-1}$ exists and the solution to $\mathbf{Ax} = \mathbf{b}$ is simply $\mathbf{x} = \mathbf{A}^{-1}\mathbf{b}$. If $\mathbf{A}^{-1} = [\alpha_{ij}]$, then in component form we can write $\mathbf{x} = \mathbf{A}^{-1}\mathbf{b}$ as

$$x_i = \sum_{j=1}^{n} \alpha_{ij} b_j \tag{4-3}$$

Again, by Theorem 4-1, $\alpha_{ij} = A_{ji}/\det(\mathbf{A})$. Thus,

$$\begin{aligned} x_i &= \sum_{j=1}^{n} \frac{A_{ji}}{\det(\mathbf{A})} b_j \\ &= \frac{1}{\det(\mathbf{A})} \sum_{j=1}^{n} b_j A_{ji} \end{aligned} \tag{4-4}$$

† For very large n, the terms involving powers of n below n^3 are quite small compared to the n^3 term, and we will often ignore them in comparing different methods.

Observe that $\sum_{j=1}^{n} b_j A_{ji}$ is the cofactor expansion by the ith column of the determinant of $\mathbf{A}^{(i)}$, where $\mathbf{A}^{(i)}$ is defined in the statement of the theorem; i.e.,

$$\det(\mathbf{A}^{(i)}) = \sum_{j=1}^{n} b_j A_{ji}$$

Hence,

$$x_i = \frac{\det(\mathbf{A}^{(i)})}{\det(\mathbf{A})} \qquad \text{QED}$$

Theorem 4-2 is often called *Cramer's rule* for the solution of $\mathbf{Ax} = \mathbf{b}$. It requires the calculation of $(n + 1)$ determinants, each of order n, or approximately $\frac{1}{3}n^4$ multiplications and $\frac{1}{3}n^4$ additions. For large n, Cramer's rule represents a significant saving of computational effort over the method of directly applying Theorem 4-1.

However, it will be the purpose of the rest of this chapter to develop methods for solving $\mathbf{Ax} = \mathbf{b}$ which are far superior to Cramer's rule. One might suspect that such methods exist by noting that the $(n + 1)$ determinants of Cramer's rule are very similar to one another, and so there is probably a large amount of repetition in the separate calculation of each.

We conclude this section with:

Theorem 4-3 The system of n equations in n variables, $\mathbf{Ax} = \mathbf{b}$, possesses a unique solution if $\det(\mathbf{A}) \neq 0$.

PROOF If $\det(\mathbf{A}) \neq 0$, then $\mathbf{A}^{-1}$ exists by Theorem 4-1. Thus, any solution $\mathbf{x}_0$ to $\mathbf{Ax} = \mathbf{b}$ must satisfy $\mathbf{x}_0 = \mathbf{A}^{-1}\mathbf{b}$. But, in Chap. 2 we proved that if $\mathbf{A}^{-1}$ exists, then it is unique for any square matrix $\mathbf{A}$. Therefore, $\mathbf{x}_0 = \mathbf{A}^{-1}\mathbf{b}$ is also unique. QED

Theorem 4-4 For any square matrix $\mathbf{A}$, $\mathbf{A}^{-1}$ exists if and only if $\det(\mathbf{A}) \neq 0$.

PROOF (*a*) If $\det(\mathbf{A}) \neq 0$, then we have already proved (Theorem 4-2) that $\mathbf{A}^{-1}$ exists.

(*b*) If $\mathbf{A}^{-1}$ exists, then we may write $\mathbf{AA}^{-1} = \mathbf{I}_n$. Thus,

$$\det(\mathbf{AA}^{-1}) = \det(\mathbf{I}_n)$$
$$= 1$$

Moreover, by Theorem 3-15,

$$\det(\mathbf{AA}^{-1}) = \det(\mathbf{A})\det(\mathbf{A}^{-1})$$

Hence,

$$\det(\mathbf{A}) \det(\mathbf{A}^{-1}) = 1 \qquad (4\text{-}5)$$

Since both $\det(\mathbf{A})$ and $\det(\mathbf{A}^{-1})$ are scalars, Eq. (4-5) states that the product of two scalars is nonzero; therefore, neither scalar can be zero:

$$\det(\mathbf{A}) \neq 0 \qquad \det(\mathbf{A}^{-1}) \neq 0$$

We have thus shown that if $\mathbf{A}^{-1}$ exists, $\det(\mathbf{A}) \neq 0$. QED

Theorem 4-4 states that the two conditions "$\mathbf{A}^{-1}$ exists" and "$\det(\mathbf{A}) \neq 0$" are equivalent. Similarly, the two conditions "$\mathbf{A}^{-1}$ does not exist" and "$\det(\mathbf{A}) = 0$" must also be equivalent. A square matrix whose determinant is zero is said to be a *singular* matrix; a square matrix whose determinant is nonzero is said to be *nonsingular*. We summarize this paragraph in the following table:

EQUIVALENT STATEMENTS ABOUT A SQUARE MATRIX **A**

1*a*. **A** is nonsingular	1*b*. **A** is singular
2*a*. $\mathbf{A}^{-1}$ exists	2*b*. $\mathbf{A}^{-1}$ does not exist
3*a*. $\det(\mathbf{A}) \neq 0$	3*b*. $\det(\mathbf{A}) = 0$

For any square matrix **A**, either the statements (*a*) are true or else the statements (*b*) are true. If any one of the statements (*a*) is true, then all the statements (*a*) are true; if any of the statements (*b*) is true, then all the statements (*b*) are true. These equivalent statements will hereinafter be used interchangeably, depending on the context of the discussion. The reader should, therefore, familiarize himself thoroughly with the meaning of the table.

4-2 ELIMINATION METHODS

In this section we shall investigate two methods for solving the system of equations $\mathbf{Ax} = \mathbf{b}$. Both methods are computationally superior to the methods discussed in the previous section, as we shall determine in Sec. 4-4. Moreover, both methods shall prove to be of great value in the theoretical discussions of Chaps. 6, 7, and 8.

The two methods—called Gaussian elimination and Gauss-Jordan elimination—are quite similar to each other, and both are based on the elementary operations discussed in Chap. 3. In particular, we observe that, given the system of equations (4-1), if we reorder the equations, we obtain a new system of equations which is essentially the same as the original system of equations, in the sense that both systems possess the same solutions. Any two systems which possess identical solutions are said to be *equivalent systems.* Thus, reordering the equations yields an equivalent system. Moreover, "reordering equations" corresponds to interchanging *rows* of the augmented matrix $[\mathbf{A}, \mathbf{b}]$. Similarly, we observe that performing the other two types of elementary *row* operations on $[\mathbf{A}, \mathbf{b}]$ yields an equivalent system of equations. Finally, we recall from Chap. 3 that the method of Sec. 3-3 employs only elementary row operations on $\mathbf{A}$ to obtain an upper triangular matrix $\mathbf{T}$; if we were to perform this same sequence of elementary row operations on $[\mathbf{A}, \mathbf{b}]$, we would then obtain a matrix $[\mathbf{T}, \hat{\mathbf{b}}]$. This matrix is the augmented matrix for a system of equations $\mathbf{Tx} = \hat{\mathbf{b}}$ which is equivalent to the original system of equations $\mathbf{Ax} = \mathbf{b}$.

Let us write the system $\mathbf{Tx} = \hat{\mathbf{b}}$ in component form:

$$\begin{aligned} t_{11}x_1 + t_{12}x_2 + t_{13}x_3 + \cdots + t_{1,n-1}x_{n-1} + t_{1n}x_n &= \hat{b}_1 \\ t_{22}x_2 + t_{23}x_3 + \cdots + t_{2,n-1}x_{n-1} + t_{2n}x_n &= \hat{b}_2 \\ t_{33}x_3 + \cdots + t_{3,n-1}x_{n-1} + t_{3n}x_n &= \hat{b}_3 \\ &\vdots \\ t_{n-1,n-1}x_{n-1} + t_{n-1,n}x_n &= \hat{b}_{n-1} \\ t_{nn}x_n &= \hat{b}_n \end{aligned} \tag{4-6}$$

Recall also that no $t_{ii} = 0$ since det ($\mathbf{A}$), and hence det ($\mathbf{T}$), is nonzero. The system of equations (4-6) is quite simple to solve, if we solve for the variables in the sequence $x_n, x_{n-1}, \ldots, x_2, x_1$:

From Eq. (4-6) we obtain† $x_n^* = \hat{b}_n/t_{nn}$. Substitution of this result into the $(n-1)$st equation of (4-6) yields

$$x_{n-1}^* = \frac{\hat{b}_{n-1} - t_{n-1,n}\,x_n^*}{t_{n-1,n-1}}$$

Thus, we have used the nth and $(n-1)$st equations to determine x_n^*, x_{n-1}^*, respectively. We can then substitute these two values into the $(n-2)$nd equation and solve directly for x_{n-2}^*. Continuing in this manner, we will eventually obtain values for all the variables.

The method we have just described is called *Gaussian elimination;* we summarize it below:

† We shall denote the *value* of the variable x_j by x_j^* to distinguish between those variables to which we have already calculated values and those for which we have not yet solved.

Gaussian Elimination

I. *Step k* ($k = 1, 2, \ldots, n-1$): Add multiples of row k (of the augmented matrix) to rows $k+1, k+2, \ldots, n$ so that the elements of column k in rows $k+1, k+2, \ldots, n$ become zero. (If the (k, k) element is zero, interchange row k with some row q, $q > k$.)

These first $(n-1)$ steps will yield a system of the form (4-6).

II. *Step n:* From the nth equation of (4-6), solve for x_n^*.

Step k ($k = n+1, n+2, \ldots, 2n-1$): Let $p = 2n - k$. Substitute the previously obtained values $x_n^*, x_{n-1}^*, \ldots, x_{p+1}^*$ into the pth equation of (4-6) and solve for x_p^*.

Part I (steps $1, 2, \ldots, n-1$) is often called the "forward pass" of Gaussian elimination; Part II (steps $n, n+1, \ldots, 2n-1$) is often called the "backward pass."

EXAMPLE 4-2 Consider the system of equations $\mathbf{Ax} = \mathbf{b}$, where

$$\mathbf{A} = \begin{bmatrix} 1 & 1 & 1 & 1 \\ 2 & -1 & 3 & 0 \\ 0 & 2 & 0 & 3 \\ -1 & 0 & 2 & 1 \end{bmatrix} \qquad \mathbf{b} = \begin{bmatrix} 3 \\ 3 \\ 1 \\ 0 \end{bmatrix}$$

Then,

$$[\mathbf{A}, \mathbf{b}] = \left[\begin{array}{cccc:c} 1 & 1 & 1 & 1 & 3 \\ 2 & -1 & 3 & 0 & 3 \\ 0 & 2 & 0 & 3 & 1 \\ -1 & 0 & 2 & 1 & 0 \end{array}\right]$$

Step 1 yields the augmented matrix

$$\left[\begin{array}{cccc:c} 1 & 1 & 1 & 1 & 3 \\ 0 & -3 & 1 & -2 & -3 \\ 0 & 2 & 0 & 3 & 1 \\ 0 & 1 & 3 & 2 & 3 \end{array}\right]$$

Similarly, steps 2 and 3 yield, respectively, the following augmented matrices:

$$\left[\begin{array}{cccc:c} 1 & 1 & 1 & 1 & 3 \\ 0 & -3 & 1 & -2 & -3 \\ 0 & 0 & \frac{2}{3} & \frac{5}{3} & -1 \\ 0 & 0 & \frac{10}{3} & \frac{4}{3} & 2 \end{array}\right] \qquad \left[\begin{array}{cccc:c} 1 & 1 & 1 & 1 & 3 \\ 0 & -3 & 1 & -2 & -3 \\ 0 & 0 & \frac{2}{3} & \frac{5}{3} & -1 \\ 0 & 0 & 0 & -7 & 7 \end{array}\right]$$

The latter matrix above is in the desired form $[\mathbf{T}, \hat{\mathbf{b}}]$. For this system, then, Eq. (4-6) becomes

$$\begin{aligned} x_1 + x_2 + x_3 + x_4 &= 3 \\ -3x_2 + x_3 + -2x_4 &= -3 \\ \tfrac{2}{3}x_3 + \tfrac{5}{3}x_4 &= -1 \\ -7x_4 &= 7 \end{aligned}$$

Thus, we begin the backward pass with step n:

$$x_4^* = -1$$

For step $(n + 1)$, $p = 2n - (n + 1) = n - 1 = 3$; we substitute x_4^* into the third equation above and solve for x_3^*:

$$\begin{aligned} x_3^* &= \frac{-1 - \frac{5}{3}x_4^*}{(\frac{2}{3})} \\ &= \tfrac{3}{2}[-1 - \tfrac{5}{3}(-1)] \\ &= 1 \end{aligned}$$

Next, we perform step $(n + 2)$: $p = 2n - (n + 2) = n - 2 = 2$;

$$\begin{aligned} x_2^* &= \frac{-3 - x_3^* + 2x_4^*}{-3} \\ &= \frac{-3 - 1 + 2(-1)}{-3} \\ &= 2 \end{aligned}$$

Finally, for $p = 2n - (n + 3) = 1$, we obtain

$$\begin{aligned} x_1^* &= 3 - x_2^* - x_3^* - x_4^* \\ &= 3 - 2 - 1 - (-1) \\ &= 1 \end{aligned}$$

Hence, the desired solution is $x_1^* = 1$, $x_2^* = 2$, $x_3^* = 1$, $x_4^* = -1$. ////

EXAMPLE 4-3 In the proof of Theorem 3-13 (which precedes the statement of the theorem) we noted that the upper triangular matrix $\mathbf{T}$ could be reduced to an identity matrix by a sequence of elementary row operations, provided $\det(\mathbf{T}) \neq 0$.

Let us apply this procedure to the **T** of Example 4-2; more precisely, we shall include the $\hat{\mathbf{b}}$ column and perform elementary row operations on $[\mathbf{T}, \hat{\mathbf{b}}]$ as follows:

$$[\mathbf{T}, \hat{\mathbf{b}}] = \left[\begin{array}{cccc:c} 1 & 1 & 1 & 1 & 3 \\ 0 & -3 & 1 & -2 & -3 \\ 0 & 0 & \frac{2}{3} & \frac{5}{3} & -1 \\ 0 & 0 & 0 & -7 & 7 \end{array}\right]$$

Step 4 (since it took us three steps to obtain $[\mathbf{T}, \hat{\mathbf{b}}]$ from $[\mathbf{A}, \mathbf{b}]$) is to multiply row 4 by $-\frac{1}{7}$, yielding

$$\left[\begin{array}{cccc:c} 1 & 1 & 1 & 1 & 3 \\ 0 & -3 & 1 & -2 & -3 \\ 0 & 0 & \frac{2}{3} & \frac{5}{3} & -1 \\ 0 & 0 & 0 & 1 & -1 \end{array}\right]$$

Step 5 is to add multiples of row 4 to rows 3, 2, and 1 so that the elements of column 4 in rows 3, 2, and 1 become zero:

$$\left[\begin{array}{cccc:c} 1 & 1 & 1 & 0 & 4 \\ 0 & -3 & 1 & 0 & -5 \\ 0 & 0 & \frac{2}{3} & 0 & \frac{2}{3} \\ 0 & 0 & 0 & 1 & -1 \end{array}\right]$$

Step 6 is to multiply row 3 by $\frac{3}{2}$ and then to add multiples of the resulting row 3 to rows 2 and 1 so that the elements of column 3 in rows 2 and 1 become zero:

$$\left[\begin{array}{cccc:c} 1 & 1 & 0 & 0 & 3 \\ 0 & -3 & 0 & 0 & -6 \\ 0 & 0 & 1 & 0 & 1 \\ 0 & 0 & 0 & 1 & -1 \end{array}\right]$$

Step 7 is to multiply row 2 by $-\frac{1}{3}$ and then to add multiples of the resulting row 2 to row 1 so that the element in column 2, row 1 becomes zero:

$$\left[\begin{array}{cccc:c} 1 & 0 & 0 & 0 & 1 \\ 0 & 1 & 0 & 0 & 2 \\ 0 & 0 & 1 & 0 & 1 \\ 0 & 0 & 0 & 1 & -1 \end{array}\right]$$

The result above is an augmented matrix for a system of equations which is equivalent to the original system $\mathbf{Ax} = \mathbf{b}$ of Example 4-2, since we employed only elementary row operations to obtain this matrix. This new system of equations is:

$$x_1 = 1$$
$$x_2 = 2$$
$$x_3 = 1$$
$$x_4 = -1$$

What we have actually done in this example is to perform the backward pass of Gaussian elimination by using elementary row operations. The reader should make certain he understands that the two approaches (of Example 4-2 and this example) are indeed virtually identical. ////

The above example illustrates that we can describe Gaussian elimination completely in terms of elementary row operations. In the forward pass, we begin with $[\mathbf{A}, \mathbf{b}]$ and perform elementary row operations until we obtain $[\mathbf{T}, \hat{\mathbf{b}}]$. In the backward pass we begin with $[\mathbf{T}, \hat{\mathbf{b}}]$ and perform elementary operations, finally obtaining $[\mathbf{I}_n, \mathbf{x}^*]$, where $\mathbf{x}^* = [x_1^*, x_2^*, \ldots, x_n^*]^T = \mathbf{A}^{-1}\mathbf{b}$.

How does Gaussian elimination compare with Cramer's rule? We can get a rough idea by noting that the forward pass is almost exactly the same as the method of Sec. 3-3 for computing det $(\mathbf{A})$; the only difference is the addition of the $(n + 1)$st column $\mathbf{b}$. This results in one more multiplication and one more addition at each step. Thus, there are still approximately $\frac{1}{3}n^3$ additions and $\frac{1}{3}n^3$ multiplications in the forward pass. Moreover, the backward pass can be considered as a "mirror reflection" of the forward pass, at least in terms of the sequence of operations. Thus, we can estimate that the backward pass also requires approximately $\frac{1}{3}n^3$ additions and $\frac{1}{3}n^3$ multiplications.

Hence, Gaussian elimination requires roughly $\frac{2}{3}n^3$ multiplications and $\frac{2}{3}n^3$ additions, while Cramer's rule requires about $\frac{1}{3}n^4$ multiplications and $\frac{1}{3}n^4$ additions. For large n, Gaussian elimination is clearly the better of the two. We shall study these comparisons more closely in Sec. 4-4.

Let us now consider the following variation of Gaussian elimination, called the *Gauss-Jordon elimination method*:

Gauss-Jordan Elimination

Step k ($k = 1, 2, \ldots\ n$): Add multiples of row k to every row (except row k) so that the elements in column k become zero.

These first n steps will produce an augmented matrix $[\mathbf{D}, \hat{\mathbf{b}}]$ in which $\mathbf{D}$ is a diagonal matrix.

Step $(n + 1)$: If the diagonal elements of **D** are denoted by $d_1, d_2, \ldots, d_n$, then calculate

$$x_j^* = \frac{\hat{b}_j}{d_j} \qquad j = 1, 2, \ldots, n$$

EXAMPLE 4-4 Consider again the system $\mathbf{Ax} = \mathbf{b}$ of Example 4-2:

$$[\mathbf{A}, \hat{\mathbf{b}}] = \left[\begin{array}{cccc:c} 1 & 1 & 1 & 1 & 3 \\ 2 & -1 & 3 & 0 & 3 \\ 0 & 2 & 0 & 3 & 1 \\ -1 & 0 & 2 & 1 & 0 \end{array}\right]$$

Adding multiples of row 1 to rows 2, 3, and 4 yields:

$$\left[\begin{array}{cccc:c} 1 & 1 & 1 & 1 & 3 \\ 0 & -3 & 1 & -2 & -3 \\ 0 & 2 & 0 & 3 & 1 \\ 0 & 1 & 3 & 2 & 3 \end{array}\right]$$

Adding multiples of row 2 to rows 1, 3, and 4 yields:

$$\left[\begin{array}{cccc:c} 1 & 0 & \frac{4}{3} & \frac{1}{3} & 2 \\ 0 & -3 & 1 & -2 & -3 \\ 0 & 0 & \frac{2}{3} & \frac{5}{3} & -1 \\ 0 & 0 & \frac{10}{3} & \frac{4}{3} & 2 \end{array}\right]$$

Adding multiples of row 3 to rows 1, 2, and 4 yields:

$$\left[\begin{array}{cccc:c} 1 & 0 & 0 & -3 & 4 \\ 0 & -3 & 0 & -\frac{9}{2} & -\frac{3}{2} \\ 0 & 0 & \frac{2}{3} & \frac{5}{3} & -1 \\ 0 & 0 & 0 & -7 & 7 \end{array}\right]$$

Adding multiples of row 4 to rows 1, 2, and 3 yields:

$$\left[\begin{array}{cccc:c} 1 & 0 & 0 & 0 & 1 \\ 0 & -3 & 0 & 0 & -6 \\ 0 & 0 & \frac{2}{3} & 0 & \frac{2}{3} \\ 0 & 0 & 0 & -7 & 7 \end{array}\right]$$

Finally, performing step $(n + 1)$ yields $x_1 = 1$, $x_2 = 2$, $x_3 = 1$, $x_4 = -1$.

////

At first glance, one might suspect that Gauss-Jordan elimination is even more efficient than Gaussian elimination. In fact, the opposite is true, as we shall see in Sec. 4-4.

Indeed, for solving a general system of n equations in n variables, there is no method which requires fewer arithmetic operations than Gaussian elimination. It is, therefore, the most important method we shall study for solving systems of equations. We shall devote all of Chap. 5 to a detailed investigation of the computational aspects of Gaussian elimination.

4-3 COMPUTATION OF $\mathbf{A}^{-1}$

In our discussions of Cramer's rule and the two elimination methods of the previous section we have observed that the way *not* to solve a system of equations $\mathbf{Ax} = \mathbf{b}$ is to first calculate $\mathbf{A}^{-1}$ and then compute $\mathbf{x} = \mathbf{A}^{-1}\mathbf{b}$. However, there are occasions when it is desirable to calculate $\mathbf{A}^{-1}$. For example, suppose the system $\mathbf{Ax} = \mathbf{b}$ represents a manufacturing system, in which:

x_j = quantity of product j to be produced $\quad j = 1, 2, \ldots, n$
b_i = quantity of raw material i available $\quad i = 1, 2, \ldots, n$
a_{ij} = quantity of raw material i required to produce one unit of product j $\quad i = 1, 2, \ldots, n \quad$ and $j = 1, 2, \ldots, n$

Then, the total quantity of raw material i used in the production of $x_1, x_2, \ldots, x_n$ is

$$a_{i1}x_1 + a_{i2}x_2 + \cdots + a_{in}x_n \qquad i = 1, 2, \ldots, n$$

Setting this quantity equal to the total quantity of resource i available, for $i = 1, 2, \ldots, n$, yields the system $\mathbf{Ax} = \mathbf{b}$. Suppose further that each day the company wishes to determine its production run for the day (i.e., values for $x_1, x_2, \ldots, x_n$) and moreover that the amounts of the resources available b_i change each day. Thus, the company must essentially solve a different system of equations $\mathbf{Ax} = \mathbf{b}$ each day, with the only difference occurring in the "resource vector" $\mathbf{b}$. If $\mathbf{A}^{-1}$ were known, the company would merely have to calculate $\mathbf{A}^{-1}\mathbf{b}$ each day (for each different $\mathbf{b}$) instead of re-solving $\mathbf{Ax} = \mathbf{b}$.

Clearly, in the above situation, it would be desirable to calculate $\mathbf{A}^{-1}$ rather than to re-solve the system of equations each day. In this section, we shall discover how to apply Gaussian elimination (or Gauss-Jordan elimination) to the problem of computing $\mathbf{A}^{-1}$ in a way which is substantially more efficient than direct application of the formula for $\mathbf{A}^{-1}$ given in Theorem 4-1.

Before doing so, however, let us consider a situation similar to that described in the previous paragraphs.

Suppose that the manufacturing company discussed above knows the amounts of each resource which will be available for each of the next p days. The company then wishes to solve p systems of equations $\mathbf{A}\mathbf{x}_k = \mathbf{b}_k$ (where $\mathbf{b}_k$ is the resource vector for the kth day and $\mathbf{x}_k$ is the corresponding solution vector†) $k = 1, 2, \ldots, p$. There are two approaches one could employ to solve these p systems of equations: The first approach is to first calculate $\mathbf{A}^{-1}$ and then compute $\mathbf{x}_k = \mathbf{A}^{-1}\mathbf{b}_k$, $k = 1, 2, \ldots, p$. The second approach is to simultaneously solve all p systems by Gaussian elimination.

Let us consider this latter method. Each system of equations $\mathbf{A}\mathbf{x}_k = \mathbf{b}_k$ has the same coefficient matrix, and since the elementary operations employed in Gaussian elimination depend only on the coefficient matrix, we would apply the same sequence of elementary operations in solving each of the p systems. In order to eliminate the repetition of these operations p times on the same coefficient matrix, we instead apply elementary operations to the $n \times (n + p)$ matrix $[\mathbf{A} \quad \mathbf{b}_1 \quad \mathbf{b}_2 \quad \cdots \quad \mathbf{b}_p]$, which may be thought of as the augmented matrix for the p systems of equations $\mathbf{A}\mathbf{x}_k = \mathbf{b}_k$, $k = 1, 2, \ldots, p$. The result of performing Gaussian elimination on this matrix will be a matrix of the form $[\mathbf{I}_n \quad \mathbf{x}_1 \quad \mathbf{x}_2 \quad \ldots \quad \mathbf{x}_p]$.

EXAMPLE 4-5 Consider the systems of equations $\mathbf{A}\mathbf{x}_1 = \mathbf{b}_1$, $\mathbf{A}\mathbf{x}_2 = \mathbf{b}_2$, $\mathbf{A}\mathbf{x}_3 = \mathbf{b}_3$, where

$$\mathbf{A} = \begin{bmatrix} 1 & 1 & 1 & 1 \\ 2 & -1 & 3 & 0 \\ 0 & 2 & 0 & 3 \\ -1 & 0 & 2 & 1 \end{bmatrix} \qquad \mathbf{b}_1 = \begin{bmatrix} 3 \\ 3 \\ 1 \\ 0 \end{bmatrix} \qquad \mathbf{b}_2 = \begin{bmatrix} 0 \\ -1 \\ 0 \\ -3 \end{bmatrix} \qquad \mathbf{b}_3 = \begin{bmatrix} 4 \\ 4 \\ 5 \\ 2 \end{bmatrix}$$

(Note that $\mathbf{A}$ and $\mathbf{b}_1$ yield the system of equations solved in Example 4-2.) We form

$$[\mathbf{A}, \mathbf{b}_1, \mathbf{b}_2, \mathbf{b}_3] = \left[\begin{array}{rrrr|rrr} 1 & 1 & 1 & 1 & 3 & 0 & 4 \\ 2 & -1 & 3 & 0 & 3 & -1 & 4 \\ 0 & 2 & 0 & 3 & 1 & 0 & 5 \\ -1 & 0 & 2 & 1 & 0 & -3 & 2 \end{array}\right]$$

Performing the steps of Gaussian elimination yields the following sequence of matrices:

† The reader should not confuse the notations $\mathbf{b}_1, \mathbf{b}_2, \ldots$, or $\mathbf{x}_1, \mathbf{x}_2, \ldots$, representing *vectors*, with the notation b_1, b_2, etc., denoting *components* of a vector. The distinction should be clear from the context of the discussion and from the use of boldface type to denote vectors.

$$\left[\begin{array}{cccc:ccc} 1 & 1 & 1 & 1 & 3 & 0 & 4 \\ 0 & -3 & 1 & -2 & -3 & -1 & -4 \\ 0 & 2 & 0 & 3 & 1 & 0 & 5 \\ 0 & 1 & 3 & 2 & 3 & -3 & 6 \end{array}\right] \to \left[\begin{array}{cccc:ccc} 1 & 1 & 1 & 1 & 3 & 0 & 4 \\ 0 & -3 & 1 & -2 & -3 & -1 & -4 \\ 0 & 0 & \frac{2}{3} & \frac{5}{3} & -1 & -\frac{2}{3} & \frac{7}{3} \\ 0 & 0 & \frac{10}{3} & \frac{4}{3} & 2 & -\frac{10}{3} & \frac{14}{3} \end{array}\right]$$

$$\to \left[\begin{array}{cccc:ccc} 1 & 1 & 1 & 1 & 3 & 0 & 4 \\ 0 & -3 & 1 & -2 & -3 & -1 & -4 \\ 0 & 0 & \frac{2}{3} & \frac{5}{3} & -1 & -\frac{2}{3} & \frac{7}{3} \\ 0 & 0 & 0 & -7 & 7 & 0 & -7 \end{array}\right] \to \left[\begin{array}{cccc:ccc} 1 & 1 & 1 & 0 & 4 & 0 & 3 \\ 0 & -3 & 1 & 0 & -5 & -1 & -2 \\ 0 & 0 & \frac{2}{3} & 0 & \frac{2}{3} & -\frac{2}{3} & \frac{2}{3} \\ 0 & 0 & 0 & 1 & -1 & 0 & 1 \end{array}\right]$$

$$\to \left[\begin{array}{cccc:ccc} 1 & 1 & 0 & 0 & 3 & 1 & 2 \\ 0 & -3 & 0 & 0 & -6 & 0 & -3 \\ 0 & 0 & 1 & 0 & 1 & -1 & 1 \\ 0 & 0 & 0 & 1 & -1 & 0 & 1 \end{array}\right] \to \left[\begin{array}{cccc:ccc} 1 & 0 & 0 & 0 & 1 & 1 & 1 \\ 0 & 1 & 0 & 0 & 2 & 0 & 1 \\ 0 & 0 & 1 & 0 & 1 & -1 & 1 \\ 0 & 0 & 0 & 1 & -1 & 0 & 1 \end{array}\right]$$

Hence, the three systems of equations have solutions, respectively:

$$\mathbf{x}_1 = \begin{bmatrix} 1 \\ 2 \\ 1 \\ -1 \end{bmatrix} \qquad \mathbf{x}_2 = \begin{bmatrix} 1 \\ 0 \\ -1 \\ 0 \end{bmatrix} \qquad \mathbf{x}_3 = \begin{bmatrix} 1 \\ 1 \\ 1 \\ 1 \end{bmatrix}$$ ////

We shall discuss the computational efficiency of the above method in the next section. Now, however, let us consider how we may apply this method to the problem of calculating $\mathbf{A}^{-1}$. To do so, we need only to observe that the problem of finding $\mathbf{A}^{-1}$ may be considered as a problem of solving n systems of equations, each involving n equations in n variables, each with the same coefficient matrix $\mathbf{A}$. We write

$$\mathbf{A}\mathbf{A}^{-1} = \mathbf{I}_n \tag{4-7}$$

Let $\boldsymbol{\alpha}_j$ denote the jth column of $\mathbf{A}^{-1}$ and $\mathbf{e}_j$ denote the jth column of $\mathbf{I}_n$. Thus, Eq. (4-7) becomes

$$\begin{aligned} \mathbf{A}[\boldsymbol{\alpha}_1 \quad \boldsymbol{\alpha}_2 \quad \cdots \quad \boldsymbol{\alpha}_n] &= [\mathbf{e}_1 \quad \mathbf{e}_2 \quad \cdots \quad \mathbf{e}_n] \\ [\mathbf{A}\boldsymbol{\alpha}_1 \quad \mathbf{A}\boldsymbol{\alpha}_2 \quad \cdots \quad \mathbf{A}\boldsymbol{\alpha}_n] &= [\mathbf{e}_1 \quad \mathbf{e}_2 \quad \cdots \quad \mathbf{e}_n] \end{aligned} \tag{4-8}$$

The jth column of each side of Eq. (4-8) yields the equations

$$\mathbf{A}\boldsymbol{\alpha}_j = \mathbf{e}_j \qquad j = 1, 2, \ldots, n \tag{4-9}$$

Thus, solving the n systems of equations $\mathbf{A}\boldsymbol{\alpha}_j = \mathbf{e}_j$, $j = 1, 2, \ldots, n$ yields the n columns of $\mathbf{A}^{-1} = [\boldsymbol{\alpha}_1 \quad \boldsymbol{\alpha}_2 \quad \cdots \quad \boldsymbol{\alpha}_n]$. As discussed previously, these systems may be solved simultaneously, by applying Gaussian elimination to the matrix $[\mathbf{A}, \mathbf{I}_n]$, yielding the matrix $[\mathbf{I}_n, \mathbf{A}^{-1}]$.

EXAMPLE 4-6 Let us compute $\mathbf{A}^{-1}$ for the coefficient matrix of Example 4-5:

$$\mathbf{A} = \begin{bmatrix} 1 & 1 & 1 & 1 \\ 2 & -1 & 3 & 0 \\ 0 & 2 & 0 & 3 \\ -1 & 0 & 2 & 1 \end{bmatrix} \qquad [\mathbf{A}, \mathbf{I}_n] = \left[\begin{array}{cccc:cccc} 1 & 1 & 1 & 1 & 1 & 0 & 0 & 0 \\ 2 & -1 & 3 & 0 & 0 & 1 & 0 & 0 \\ 0 & 2 & 0 & 3 & 0 & 0 & 1 & 0 \\ -1 & 0 & 2 & 1 & 0 & 0 & 0 & 1 \end{array}\right]$$

$$\rightarrow \left[\begin{array}{cccc:cccc} 1 & 1 & 1 & 1 & 1 & 0 & 0 & 0 \\ 0 & -3 & 1 & -2 & -2 & 1 & 0 & 0 \\ 0 & 2 & 0 & 3 & 0 & 0 & 1 & 0 \\ 0 & 1 & 3 & 2 & 1 & 0 & 0 & 1 \end{array}\right] \rightarrow \left[\begin{array}{cccc:cccc} 1 & 1 & 1 & 1 & 1 & 0 & 0 & 0 \\ 0 & -3 & 1 & -2 & -2 & 1 & 0 & 0 \\ 0 & 0 & \frac{2}{3} & \frac{5}{3} & -\frac{4}{3} & \frac{2}{3} & 1 & 0 \\ 0 & 0 & \frac{10}{3} & \frac{4}{3} & \frac{1}{3} & \frac{1}{3} & 0 & 1 \end{array}\right]$$

$$\rightarrow \left[\begin{array}{cccc:cccc} 1 & 1 & 1 & 1 & 1 & 0 & 0 & 0 \\ 0 & -3 & 1 & -2 & -2 & 1 & 0 & 0 \\ 0 & 0 & \frac{2}{3} & \frac{5}{3} & -\frac{4}{3} & \frac{2}{3} & 1 & 0 \\ 0 & 0 & 0 & -7 & 7 & -3 & -5 & 1 \end{array}\right] \rightarrow \left[\begin{array}{cccc:cccc} 1 & 1 & 1 & 0 & 2 & -\frac{3}{7} & -\frac{5}{7} & \frac{1}{7} \\ 0 & -3 & 1 & 0 & -4 & \frac{13}{7} & \frac{10}{7} & -\frac{2}{7} \\ 0 & 0 & \frac{2}{3} & 0 & \frac{1}{3} & -\frac{1}{21} & -\frac{4}{21} & \frac{5}{21} \\ 0 & 0 & 0 & 1 & -1 & \frac{3}{7} & \frac{5}{7} & -\frac{1}{7} \end{array}\right]$$

$$\rightarrow \left[\begin{array}{cccc:cccc} 1 & 1 & 0 & 0 & \frac{3}{2} & -\frac{5}{14} & -\frac{3}{7} & -\frac{3}{14} \\ 0 & -3 & 0 & 0 & -\frac{9}{2} & \frac{27}{14} & \frac{12}{7} & -\frac{9}{14} \\ 0 & 0 & 1 & 0 & \frac{1}{2} & -\frac{1}{14} & -\frac{2}{7} & \frac{5}{14} \\ 0 & 0 & 0 & 1 & -1 & \frac{3}{7} & \frac{5}{7} & -\frac{1}{7} \end{array}\right] \rightarrow \left[\begin{array}{cccc:cccc} 1 & 0 & 0 & 0 & 0 & \frac{2}{7} & \frac{1}{7} & -\frac{3}{7} \\ 0 & 1 & 0 & 0 & \frac{3}{2} & -\frac{9}{14} & -\frac{4}{7} & \frac{3}{14} \\ 0 & 0 & 1 & 0 & \frac{1}{2} & -\frac{1}{14} & -\frac{2}{7} & \frac{5}{14} \\ 0 & 0 & 0 & 1 & -\frac{6}{7} & \frac{3}{7} & \frac{5}{7} & -\frac{1}{7} \end{array}\right]$$

Thus,

$$\mathbf{A}^{-1} = \begin{bmatrix} 0 & \frac{2}{7} & \frac{1}{7} & -\frac{3}{7} \\ \frac{3}{2} & -\frac{9}{14} & -\frac{4}{7} & \frac{3}{14} \\ \frac{1}{2} & -\frac{1}{14} & -\frac{2}{7} & \frac{5}{14} \\ -1 & \frac{3}{7} & \frac{5}{7} & -\frac{1}{7} \end{bmatrix} = \left(\tfrac{1}{14}\right) \begin{bmatrix} 0 & 4 & 2 & -6 \\ 21 & -9 & -8 & 3 \\ 7 & -1 & -4 & 5 \\ -14 & 6 & 10 & -2 \end{bmatrix}$$

The reader should verify this result by computing $\mathbf{AA}^{-1}$ or $\mathbf{A}^{-1}\mathbf{A}$.

////

Let us summarize the results of this section:

1 To solve the system of equations $\mathbf{Ax} = \mathbf{b}$ perform Gaussian elimination† on the $n \times (n + 1)$ augmented matrix $[\mathbf{A}, \mathbf{b}]$, thus obtaining $[\mathbf{I}_n, \mathbf{x}^*]$, $\mathbf{x}^* = \mathbf{A}^{-1}\mathbf{b}$.

2 To solve the p systems of equations $\mathbf{Ax}_k = \mathbf{b}_k$, $k = 1, 2, \ldots, p$, perform Gaussian elimination† on the $n \times (n + p)$ matrix $[\mathbf{A}, \mathbf{b}_1, \mathbf{b}_2, \ldots, \mathbf{b}_p]$, thus obtaining the matrix $[\mathbf{I}_n \quad \mathbf{x}_1^* \quad \mathbf{x}_2^* \quad \cdots \quad \mathbf{x}_p^*]$ where $\mathbf{x}_k^* = \mathbf{A}^{-1}\mathbf{b}_k$, $k = 1, 2, \ldots, p$.

3 To calculate $\mathbf{A}^{-1}$, perform Gaussian elimination† on the $n \times 2n$ matrix $[\mathbf{A}, \mathbf{I}_n]$, thus obtaining the matrix $[\mathbf{I}_n, \mathbf{A}^{-1}]$.

† One could also use Gauss-Jordan elimination here, since both methods are nothing more than systematic applications of elementary row operations.

4-4 COMPUTATIONAL CONSIDERATIONS

Thus far in this chapter we have discovered:

1 Two methods for computing the inverse of a nonsingular matrix **A**:
(*a*) The direct application of the formula of Theorem 4-1
(*b*) The application of Gaussian elimination (or Gauss-Jordan elimination) to the matrix $[\mathbf{A}, \mathbf{I}_n]$, yielding $[\mathbf{I}_n, \mathbf{A}^{-1}]$.

2 Four methods of solving a system of equations $\mathbf{Ax} = \mathbf{b}$:
(*a*) The calculation of $\mathbf{A}^{-1}$ followed by the matrix product $\mathbf{A}^{-1}\mathbf{b}$
(*b*) Cramer's rule (Theorem 4-2)
(*c*) Gaussian elimination
(*d*) Gauss-Jordan elimination

In this section we shall attempt to put the above collection of methods into some perspective by comparing the relative computational efficiency of each. Again, we shall restrict our notions about relative efficiency to a comparison of the number of arithmetic operations required by each method.

We begin by computing the number of arithmetic operations required to solve $\mathbf{Ax} = \mathbf{b}$ by Gaussian elimination. We recall that the forward pass of Gaussian elimination is identical with the method of Sec. 3-3 for computing det (**A**), except that in Gaussian elimination each row operation is performed on one extra column: **b**. In Sec. 3-3 we calculated the number of operations necessary to obtain det (**A**) to be

$$\left(\frac{n^3}{3} - \frac{n^2}{2} + \frac{n}{6}\right) \qquad \text{additions}$$

$$\left(\frac{n^3}{3} - \frac{n}{3}\right) \qquad \text{multiplications}$$

excluding step n (multiplying the n diagonal terms together) which is not part of the Gaussian elimination method. In addition, at the kth step we now have $(n - k)$ more multiplications and $(n - k)$ more additions: For each row p, $p = k + 1, k + 2, \ldots, n$, we multiply $\hat{b}_k$ by the constant $(-\hat{a}_{pk}/\hat{a}_{kk})$ and add the result to $\hat{b}_p$. (The computation of the above constant has already been included in our calculations.) Since there are $(n - 1)$ such steps, we thus have

$$\sum_{k=1}^{n-1} (n - k) = \frac{n(n - 1)}{2} \qquad \text{more additions}$$

$$\frac{n(n - 1)}{2} \qquad \text{more multiplications}$$

Therefore, the forward pass of Gaussian elimination requires

$$\left(\frac{n^3}{3} - \frac{n}{3}\right) \qquad \text{additions}$$

$$\left(\frac{n^3}{3} + \frac{n^2}{2} - \frac{5n}{6}\right) \qquad \text{multiplications}$$

In solving for x_p^* in the backward pass, we must perform $(n - p)$ multiplications, one division, and $(n - p)$ additions:

$(n - p)$ multiplications

$$x_p^* = [\hat{b}_p - t_{p,p+1}x_{p+1}^* - t_{p,p+2}x_{p+2}^* - \cdots - t_{pn}x_n^*]/t_{pp}$$

$(n - p)$ additions — one division

Combining the multiplications and divisions and summing p from 1 to n yields a total of

$$\sum_{p=1}^{n} (n - p) = \frac{n(n-1)}{2} \qquad \text{additions}$$

$$\sum_{p=1}^{n} (n - p + 1) = \frac{n(n+1)}{2} \qquad \text{multiplications}$$

for the backward pass.

Thus, Gaussian elimination requires a total of

$$\left(\frac{n^3}{3} + \frac{n^2}{2} - \frac{5n}{6}\right) \qquad \text{additions}$$

$$\left(\frac{n^3}{3} + n^2 - \frac{n}{3}\right) \qquad \text{multiplications}$$

Before turning to the calculation of the number of operations required for Gauss-Jordan elimination, we note that our earlier estimate for Gaussian elimination (given in Sec. 4-2) was off by a factor of 2, since at that time we estimated roughly $\frac{2}{3}n^3$ multiplications and $\frac{2}{3}n^3$ additions were necessary for Gaussian elimination!

Number of Operations in Gauss-Jordan Elimination

As described in Sec. 4-2, the kth step of Gauss-Jordan elimination, $k = 1, 2, \ldots, n$, is to add multiples of row k to every other row, so that the elements in the kth column of these rows become zero. The kth row of the augmented matrix will have the following form at the kth step:

$$0 \quad 0 \quad \cdots \quad 0 \quad \hat{a}_{kk} \quad \hat{a}_{k,\,k+1} \quad \cdots \quad \hat{a}_{kn} \quad \hat{b}_k$$

Thus, the kth step consists of (1) computing the $(n - 1)$ multipliers

$$m_p = \frac{-\hat{a}_{pk}}{\hat{a}_{kk}} \qquad \begin{array}{l} p = 1, 2, \ldots, n \\ p \neq k \end{array}$$

and (2) adding m_p times row k to row p, $p = 1, 2, \ldots, n$; $p \neq k$. This requires multiplying m_p by each of the $(n - k + 1)$ elements $\hat{a}_{k,k+1}$, $\hat{a}_{k,k+2}, \ldots, \hat{a}_{kn}, \hat{b}_k$ and adding each of the results to the corresponding element in row p.

Since there are $(n - 1)$ rows p, (2) requires $(n - 1)(n - k + 1)$ additions and the same number of multiplications. In addition, (1) requires $(n - 1)$ divisions (counted as multiplications).

Moreover, the last step of Gauss-Jordan requires another n divisions. Thus, altogether, Gauss-Jordan elimination has

$$\sum_{k=1}^{n} (n - 1)(n - k + 1) = \left(\frac{n^3}{2} - \frac{n}{2}\right) \qquad \text{additions}$$

$$\sum_{k=1}^{n} [(n - 1)(n - k + 1) + (n - 1)] + n = \left(\frac{n^3}{2} + n^2 - \frac{n}{2}\right) \qquad \text{multiplications}$$

Observe that, for large n, Gauss-Jordan elimination requires roughly 50 percent more arithmetic operations than Gaussian elimination does, to solve one system of equations $\mathbf{Ax} = \mathbf{b}$. Therefore, if we merely wish to solve one such system, Gaussian elimination is clearly the preferable of the two methods. However, there are other situations in which Gaussian elimination loses its advantage, such as in the calculation of $\mathbf{A}^{-1}$. In fact, in Chap. 8 we shall investigate an application of Gauss-Jordan—linear programming—for which Gaussian elimination is entirely unsuited.

Let us now determine the number of operations required to calculate $\mathbf{A}^{-1}$ by Gaussian elimination.

$\mathbf{A}^{-1}$ by Gaussian Elimination

Although we are now performing elementary row operations on the $n \times 2n$ matrix $[\mathbf{A}, \mathbf{I}_n]$, the following calculations are not computationally equivalent to solving *any* n systems of equations (whose $n \times 2n$ augmented matrix would be $[\mathbf{A} \;\; \mathbf{b}_1 \;\; \mathbf{b}_2 \;\; \cdots \;\; \mathbf{b}_n]$). In the calculation of $\mathbf{A}^{-1}$, we can take advantage of the large amount of zeros in $\mathbf{I}_n$ and of the 1's along its diagonal, by observing that "addition to zero or of zero" and "multiplication by 1" are not really arithmetic operations (at least not in a cleverly written computer program!). Thus, at the first step, in adding multiples of row 1 of $[\mathbf{A}, \mathbf{I}_n]$ to the other $(n - 1)$ rows, the last $(n - 1)$ elements of row 1 are zero and do not involve arithmetic operations. In the second step—adding multiples of row 2 to rows 3, 4, ..., n—the last $(n - 2)$ elements of row 2 are zero, etc.

At the kth step of the forward pass, the augmented matrix has the form:

$$\left[\begin{array}{cccccccc|cccccccc}
\hat{a}_{11} & \hat{a}_{12} & \cdots & \hat{a}_{1,k-1} & \hat{a}_{1k} & \hat{a}_{1,k+1} & \cdots & \hat{a}_{1n} & 1 & 0 & \cdots & 0 & 0 & 0 & \cdots & 0 \\
0 & \hat{a}_{22} & \cdots & \hat{a}_{2,k-1} & \hat{a}_{2k} & \hat{a}_{2,k+1} & \cdots & \hat{a}_{2n} & \hat{e}_{21} & 1 & \cdots & 0 & 0 & 0 & \cdots & 0 \\
\vdots & & & & & & & & & & & & & & & \\
0 & 0 & \cdots & 0 & \hat{a}_{kk} & \hat{a}_{k,k+1} & \cdots & \hat{a}_{kn} & \hat{e}_{k1} & \hat{e}_{k2} & \cdots & \hat{e}_{k,k-1} & 1 & 0 & \cdots & 0 \\
0 & 0 & \cdots & 0 & \hat{a}_{k+1,k} & \hat{a}_{k+1,k+1} & \cdots & \hat{a}_{k+1,n} & \hat{e}_{k+1,1} & \hat{e}_{k+1,2} & \cdots & \hat{e}_{k+1,k-1} & 0 & 1 & \cdots & 0 \\
\vdots & \vdots & & \vdots & & & & & \vdots & & & & & & & \\
0 & 0 & \cdots & 0 & \hat{a}_{nk} & \hat{a}_{n,k+1} & \cdots & \hat{a}_{nn} & \hat{e}_{n1} & \hat{e}_{n2} & \cdots & \hat{e}_{n,k-1} & 0 & 0 & \cdots & 1
\end{array}\right]$$

To perform the kth step of the forward pass, we must:

1 Compute $(n - k)$ multiples $m_p = -\hat{a}_{pk}/\hat{a}_{kk}$, $p = k + 1, k + 2, \ldots, n$, requiring $(n - k)$ divisions.

2 Add m_p times row k to row p, $p = k + 1, k + 2, \ldots, n$. Specifically, we multiply the $\{(n - k) + (k - 1)\}$ elements $\hat{a}_{k,k+1}, \hat{a}_{k,k+2}, \ldots, \hat{a}_{kn}, \hat{e}_{k1}, \hat{e}_{k2}, \ldots, \hat{e}_{k,k-1}$ by m_p, thus performing $(n - 1)$ multiplications for each row p. We then add the results to the corresponding elements of row p. (We must also set $\hat{e}_{pk} = m_p$, $p = k + 1, k + 2, \ldots, n$, but these are not arithmetic operations.) Hence, (2) requires $(n - 1)$ multiplications and a like number of additions for each of the $(n - k)$ rows $k + 1, k + 2, \ldots, n$, or a total of $(n - k)(n - 1)$ additions and $(n - k)(n - 1)$ additions.

Combining the number of operations required in (1) and (2) and summing the k steps, $k = 1, 2, \ldots, n - 1$, yields

$$\sum_{k=1}^{n-1} [(n - k)(n - 1)] = \frac{n(n - 1)^2}{2} \qquad \text{additions}$$

$$\sum_{k=1}^{n-1} [(n - k)(n - 1) + (n - k)] = \frac{n^2(n - 1)}{2} \qquad \text{multiplications}$$

to perform the forward pass.

At the completion of the forward pass, the augmented matrix will be of the form $[\mathbf{T}, \mathbf{E}]$, where $\mathbf{T}$ is the usual upper triangular matrix and $\mathbf{E}$ is a lower triangular matrix with 1's along the diagonal (and zeros above the diagonal). For the first step of the backward pass, we divide row n by t_{nn} and then add multiples of row n to each of the other $(n - 1)$ rows. This step involves (1) n divisions (plus setting $\hat{t}_{nn}$ to 1); (2) for each row p, $p = 1, 2, \ldots, n - 1$, the multiplication of each of the last n elements of row n

by $m_p = -\hat{t}_{pn}$ and the addition of the results to the corresponding elements in row p. Note however, that the last $(n - p)$ elements of row p are zero, thus involving no actual addition. Therefore, the first step of the backward pass requires

$$\sum_{p=1}^{n-1} \{n - (n - p)\} = \frac{n(n-1)}{2} \qquad \text{additions}$$

$$n(n - 1) + n = n^2 \qquad \text{multiplications}$$

In the remaining steps of the backward pass, there are no more zeros left in the last n columns of the augmented matrix which we can take advantage of. In each of the next $(n - 2)$ steps, we (1) divide a row k by $\hat{t}_{kk}$, involving n divisions (and we also set $\hat{t}_{kk} = 1$), since the kth row will be of the form

$$0 \quad 0 \quad \cdots \quad \hat{t}_{kk} \quad 0 \quad \cdots \quad 0 \quad \hat{e}_{k1} \quad \hat{e}_{k2} \quad \cdots \quad \hat{e}_{kn}$$

and (2) add the multiple $m_p = -\hat{t}_{pk}$ times row k to row p, $p = k - 1$, $k - 2, \ldots, 1$. This involves $n(k - 1)$ multiplications and $n(k - 1)$ additions. In total, the kth step involves

$$n(k - 1) = nk - n \qquad \text{additions}$$

$$n(k - 1) + n = nk \qquad \text{multiplications}$$

This kth step is performed for each of the rows $k = n - 1, n - 2, \ldots, 2$; finally, the last step is to divide row 1 by $\hat{t}_{11}$, involving n extra multiplications.

Adding up all the operations described above for the backward pass, we arrive at a total number of operations for the backward pass of

$$\frac{n(n-1)}{2} + \sum_{k=2}^{n-1} n(k - 1) = \frac{n(n-1)^2}{2} \qquad \text{additions}$$

$$n^2 + \sum_{k=2}^{n-1} nk + n = \frac{n^2(n+1)}{2} \qquad \text{multiplications}$$

Finally, combining the forward and backward pass, we see that computing $\mathbf{A}^{-1}$ by means of Gaussian elimination requires

$$\frac{n(n-1)^2}{2} + \frac{n(n-1)^2}{2} = (n^3 - 2n^2 + n) \qquad \text{additions}$$

$$\frac{n^2(n-1)}{2} + \frac{n^2(n+1)}{2} = n^3 \qquad \text{multiplications}$$

We conclude this section with several comments and a table summarizing the results.

Procedure	Number of additions	Number of multiplications
1. Computing det (**A**)	$\frac{n^3}{3} - \frac{n^2}{2} + \frac{n}{6}$	$\frac{n^3}{3} + \frac{2n}{3} - 1$
2. Solving **Ax** = **b** by Gaussian elimination	$\frac{n^3}{3} + \frac{n^2}{2} - \frac{5n}{6}$	$\frac{n^3}{3} + n^2 - \frac{n}{3}$
3. Solving **Ax** = **b** by Gauss-Jordan elimination	$\frac{n^3}{2} - \frac{n}{2}$	$\frac{n^3}{2} + n^2 - \frac{n}{2}$
4. Computing $\mathbf{A}^{-1}$ by Gaussian elimination	$n^3 - 2n^2 + n$	n^3

Comment 1. Solving the system of equations $\mathbf{Ax} = \mathbf{b}$ by Gauss-Jordan elimination requires roughly 50 percent more arithmetic operations than Gaussian elimination.

Comment 2. Solving the system of equations $\mathbf{Ax} = \mathbf{b}$ by first computing $\mathbf{A}^{-1}$ requires roughly three times the number of arithmetic operations as using Gaussian elimination.

Comment 3. As the reader is asked to show in the Exercises, computing $\mathbf{A}^{-1}$ by Gauss-Jordan elimination requires roughly the same number of operations as computing $\mathbf{A}^{-1}$ by Gaussian elimination.

4-5 ITERATIVE METHODS FOR SOLVING $\mathbf{Ax} = \mathbf{b}$

The methods we have discussed so far in this chapter for solving the system of equations $\mathbf{Ax} = \mathbf{b}$ all fall in the general category of what are called "direct methods." That is, in each method (i.e., Cramer's rule, Gaussian elimination, Gauss-Jordan elimination) the desired solution is obtained after a finite—and predeterminable—number of steps. In this section we will consider an entirely different type of approach to the problem of solving $\mathbf{Ax} = \mathbf{b}$. This approach leads to the development of a class of methods known as "iterative methods" (or "iteration methods"). In these methods, one makes successive estimates of the solution to a given set of equations $\mathbf{Ax} = \mathbf{b}$ in such a way that eventually one of these estimates is determined to be the desired solution.

We begin by considering the following example:

$$\begin{aligned} 5x_1 + 2x_2 \qquad &= 7 \\ x_1 - 4x_2 + x_3 &= -2 \\ x_2 + 2x_3 &= 3 \end{aligned} \tag{4-10}$$

Suppose we rewrite the above system as follows:

$$\begin{aligned} x_1 &= \tfrac{7}{5} - \tfrac{2}{5}x_2 \\ x_2 &= \tfrac{1}{2} + \tfrac{1}{4}x_1 + \tfrac{1}{4}x_3 \\ x_3 &= \tfrac{3}{2} - \tfrac{1}{2}x_2 \end{aligned} \qquad \text{or} \qquad \begin{aligned} x_1 &= 1.4 - 0.4x_2 \\ x_2 &= 0.5 + 0.25x_1 + 0.25x_3 \\ x_3 &= 1.5 - 0.5x_2 \end{aligned} \tag{4-11}$$

Thus, in the first equation we have expressed x_1 in terms of the other variables, in the second equation we have expressed x_2 in terms of the other variables, and in the third equation we have expressed x_3 in terms of the other variables.

Now, suppose we have some reason to believe the solution to (4-10), and hence (4-11), is approximately $x_1 = 1.2$, $x_2 = 0.8$, $x_3 = 1.2$. If these values are indeed correct, then substitution of them into the right-hand side of (4-11) ought to yield the same values on the left-hand side of (4-11). On the other hand, if these values are not correct, then their substitution into the right-hand side of (4-11) will yield different values on the left-hand side. Let us proceed to perform this substitution:

$$x_1 = 1.4 - 0.4(0.8) = 1.08$$

$$x_2 = 0.5 + 0.25(1.2) + 0.25(1.2) = 1.10$$

$$x_3 = 1.5 - 0.5(0.8) = 1.10$$

Suppose, now, we decide to use these new values for x_1, x_2, and x_3 as a new guess, and substitute them into the right-hand side of (4-11). The result is

$$x_1 = 1.4 - 0.4(1.10) = 0.956$$

$$x_2 = 0.5 + 0.25(1.08) + 0.25(1.10) = 1.0545$$

$$x_3 = 1.5 - 0.5(1.10) = 0.95$$

Let us continue in this manner, substituting the newly obtained values of x_1, x_2, and x_3 into the right-hand side of (4-11), until the new set of values equals the old set. Below is a table summarizing these calculations (rounded to four decimal places):

Guess number	Values			Guess number	Values		
	x_1	x_2	x_3		x_1	x_2	x_3
1	1.2	0.8	1.2	8	0.9989	0.9988	0.9986
2	1.08	1.10	1.10	9	1.0004	0.9994	1.0006
3	0.956	1.0545	0.95	10	1.0002	1.0003	1.0003
4	0.9782	0.9765	0.9728	11	0.9999	1.0001	0.9999
5	1.0094	0.9878	1.0118	12	1.0000	1.0000	1.0000
6	1.0049	1.0053	1.0061	13	1.0000	1.0000	1.0000
7	0.9979	1.0028	0.9974				

Thus, 13 guesses later, we have arrived at the desired solution: $x_1 = x_2 = x_3 = 1.0000$. The reader may verify that this is indeed the correct solution.

The method we have used to solve the above example is called the *Jacobi iteration method.* In general, the method may be described as follows:

1 For each equation, $i = 1, 2, \ldots, n$, rewrite the ith equation, expressing the ith variable x_i in terms of the other variables. Thus, the ith equation

$$a_{i1}x_1 + a_{i2}x_2 + \cdots + a_{ii}x_i + \cdots + a_{in}x_n = b_i$$

becomes

$$x_i = \frac{b_i - a_{i1}x_1 - a_{i2}x_2 - \cdots - a_{i,\,i-1}x_{i-1} - a_{i,\,i+1}x_{i+1} - \cdots - a_{in}x_n}{a_{ii}}$$

or

$$x_i = \frac{b_i - \sum\limits_{\substack{j=1 \\ j \neq i}}^{n} a_{ij}x_j}{a_{ii}} \qquad i = 1, 2, \ldots, n \tag{4-12}$$

2 Choose an initial guess for the values of the variables, denoted by $x_1^{(0)}, x_2^{(0)}, \ldots, x_n^{(0)}$.

3 Substitute this guess into the right-hand side of Eqs. (4-12), yielding a new guess: $x_1^{(1)}, x_2^{(1)}, \ldots, x_n^{(1)}$. Let $k = 1$.

4 Substitute the kth guess $x_1^{(k)}, x_2^{(k)}, \ldots, x_n^{(k)}$ into the right-hand side of Eqs. (4-12), yielding the $(k+1)$st guess $x_1^{(k+1)}, x_2^{(k+1)}, \ldots, x_n^{(k+1)}$.

5 Compare the kth and $(k+1)$st guesses. If $x_i^{(k)} = x_i^{(k+1)}$, $i = 1, 2, \ldots, n$, then the method is completed, and the solution is $x_i^{(k)} = x_i^{(k+1)}$, $i = 1, 2, \ldots, n$. If any $x_i^{(k)} \neq x_i^{(k+1)}$, increment k by 1 and repeat step 4.

Each execution of step 4 is called an *iteration.*

Our use of Eqs. (4-12) in steps 3 and 4 suggests that we should write these equations as follows:

$$x_i^{(k+1)} = \frac{b_i - \sum\limits_{\substack{j=1 \\ j \neq i}}^{n} a_{ij}x_j^{(k)}}{a_{ii}} \tag{4-13}$$

We should also note that it might be necessary to reorder the original equations so that the diagonal coefficients a_{ii} are all nonzero. As long as $\det(\mathbf{A}) \neq 0$, it will always be possible to do so.

There are still several questions we should ask ourselves about the Jacobi iteration method:

1 Does the method always find the solution, or do situations exist in which step 4 will be repeated indefinitely?

2 Why would one choose to use this method in preference to Gaussian elimination?

The answer to the first question is that the method does not always obtain the solution. Therefore, we will have to investigate more carefully when we can successfully employ Jacobi iteration. We shall do so in the next section.

In order to answer the second question above, we must again consider the relative efficiency of Jacobi iteration and Gaussian elimination. Since we already know that the latter requires approximately $\frac{1}{3}n^3$ additions and $\frac{1}{3}n^3$ multiplications, we should choose to use Jacobi iteration only if it will require fewer such operations. However, the number of operations required by Jacobi iteration depends upon the number of iterations the method will need, to solve a given set of equations.

Let us count the number of operations required per iteration. First, we observe that in order to avoid having to repeat the division by a_{ii} at each iteration, Eqs. (4-13) should be rewritten as

$$x_i^{(k+1)} = \hat{b}_i - \sum_{\substack{j=1 \\ j \neq i}}^{n} \hat{a}_{ij} x_j^{(k)} \qquad i = 1, 2, \ldots, n \tag{4-14}$$

where

$$\hat{b}_i = \frac{b_i}{a_{ii}} \qquad i = 1, 2, \ldots, n$$

$$\hat{a}_{ij} = \frac{a_{ij}}{a_{ii}} \qquad \begin{array}{l} j = 1, 2, \ldots, n \qquad j \neq i \\ i = 1, 2, \ldots, n \end{array}$$

Thus, the computation of the coefficients $\hat{b}_i$, $\hat{a}_{ij}$ requires an initial effort of n^2 divisions.

Thereafter, we see from Eqs. (4-14) that each iteration requires $(n - 1)$ multiplications and $(n - 1)$ additions per equation, or a total of $n(n - 1)$ additions and $n(n - 1)$ multiplications per iteration.

If we employ a total of m iterations, then, the Jacobi iteration method requires

$$mn(n-1) = mn^2 - mn \qquad \text{additions}$$
$$mn(n-1) + n^2 = (m+1)n^2 - mn \qquad \text{multiplications}$$

Ignoring the $(-mn)$ term, we see that Jacobi iteration will be more efficient than Gaussian elimination if

$$(m+1) \leq \frac{n}{3} \tag{4-15}$$

since if the inequality (4-15) is satisfied, Jacobi iteration will require *fewer* than

$$\left(\frac{n}{3} - 1\right)n^2 = \frac{n^3}{3} - n^2 \qquad \text{additions}$$
$$\frac{n}{3}(n^2) = \frac{n^3}{3} \qquad \text{multiplications}$$

while Gaussian elimination requires

$$\left(\frac{n^3}{3} + \frac{n^2}{2} - \frac{5n}{6}\right) \qquad \text{additions}$$
$$\left(\frac{n^3}{3} + n^2 - \frac{n}{3}\right) \qquad \text{multiplications}$$

In the above calculations, we have assumed that all the elements $\hat{a}_{ij}$, $\hat{b}_i$ are nonzero. However, it is frequently true in practical applications that the coefficient matrix $\mathbf{A}$ has a large number of zero elements. Such a matrix is often called a *sparse matrix*. If we were to solve by Gaussian elimination a system of equations $\mathbf{Ax} = \mathbf{b}$ in which $\mathbf{A}$ was sparse, we would not be able to take advantage of all the zero elements, since as soon as we add multiples of row 1 to the other rows, we will have eliminated (possibly) many of these zeros. But, the Jacobi iteration method uses the original set of equations [expressed in the form (4-14)] at each iteration. Thus, all the original zeros are retained throughout, and we can therefore take advantage of these zeros to reduce the amount of computational effort per iteration.

If p is the actual number of nonzero elements in $\mathbf{A}$, then Jacobi iteration requires p divisions initially, and $(p-n)$ additions and $(p-n)$ multiplications per iteration, or a total of

$$\begin{aligned} m(p-n) &= mp - mn \qquad \text{additions} \\ m(p-n) + p &= mp + p - mn \qquad \text{multiplications} \end{aligned} \tag{4-16}$$

Clearly, $p \leq n^2$. The number $\alpha = p/n^2$ represents the fraction of nonzero elements in **A**. That is, 100α percent of the elements of **A** are nonzero. 100α percent is often called the *density* of **A**. Ignoring the initial p divisions and the $-mn$ terms in (4-16), we see that Jacobi iteration requires roughly

$$mp = m\alpha n^2 \qquad \text{additions}$$

$$mp = m\alpha n^2 \qquad \text{multiplications}$$

Hence, Jacobi iteration will be more efficient than Gaussian elimination if

$$m\alpha \leq \tfrac{1}{3}n$$

or

$$m \leq \frac{n}{3\alpha} \qquad 0 < \alpha \leq 1 \qquad \text{(4-17)}$$

Frequently, α will be in the range of 0.05 to 0.30 or so. In such cases a great many iterations can be performed and Jacobi iteration will still be more efficient than Gaussian elimination, as indicated in the following table:

MAXIMUM DESIRABLE NUMBER OF ITERATIONS (JACOBI)

n	$\alpha = 0.05$	$\alpha = 0.10$	$\alpha = 0.20$	$\alpha = 0.30$
5	33	16	8	5
10	66	33	16	11
20	133	66	33	22
40	266	133	66	43
60	340	170	85	56
75	500	250	125	83
100	666	333	166	111

Thus, the more sparse the coefficient matrix, the more likely it will be that Jacobi iteration will be more efficient than Gaussian elimination. Moreover, it turns out that Jacobi iteration generally requires fewer iterations when the coefficient matrix is sparse.

We summarize the above discussion by concluding that one ought to consider using Jacobi iteration instead of Gaussian elimination to solve a given system of equations $\mathbf{Ax} = \mathbf{b}$ when **A** is sparse.

Many other iteration methods similar to Jacobi iteration can be obtained by varying the way we arrange the original equations $\mathbf{Ax} = \mathbf{b}$ into an iteration format. Consider the following:

Suppose we separate **A** into two matrices thus:

$$\mathbf{A} = \mathbf{G} + \mathbf{H} \qquad \text{(4-18)}$$

Suppose also that $\mathbf{H}^{-1}$ exists. Then the system $\mathbf{Ax} = \mathbf{b}$ can be expressed

$$\begin{aligned}(\mathbf{G} + \mathbf{H})\mathbf{x} &= \mathbf{b} \\ \mathbf{Gx} + \mathbf{Hx} &= \mathbf{b} \\ \mathbf{Hx} &= \mathbf{b} - \mathbf{Gx} \\ \mathbf{x} &= \mathbf{H}^{-1}\mathbf{b} - \mathbf{H}^{-1}\mathbf{Gx} \qquad (4\text{-}19)\end{aligned}$$

If we now let $\mathbf{x}^{(k)}$ denote the vector of guesses at the kth iteration, then we could use (4-19) instead of (4-14) to obtain the $(k+1)$st guess, $\mathbf{x}^{(k+1)}$:

$$\mathbf{x}^{(k+1)} = \mathbf{H}^{-1}\mathbf{b} - \mathbf{H}^{-1}\mathbf{Gx}^{(k)} \qquad (4\text{-}20)$$

For each different way we choose $\mathbf{H}$ and $\mathbf{G}$, we obtain essentially a different iteration method.

If, for example, we let

$$\begin{aligned}\mathbf{G} &= [g_{ij}] \qquad g_{ij} = \begin{cases} a_{ij} & i \neq j \\ 0 & i = j \end{cases} \\ \mathbf{H} &= [h_{ij}] \qquad h_{ij} = \begin{cases} 0 & i \neq j \\ a_{ii} & i = j \end{cases}\end{aligned} \qquad (4\text{-}21)$$

then clearly $\mathbf{G} + \mathbf{H} = \mathbf{A}$ and $\mathbf{H}^{-1}$ exists if all $a_{ii} \neq 0$. Moreover, the ith row of the matrix equation (4-20) is

$$\begin{aligned}x_i^{(k+1)} &= \frac{1}{a_{ii}} b_i - \frac{1}{a_{ii}} \sum_{j=1}^{n} g_{ij} x_j^{(k)} \\ &= \frac{1}{a_{ii}} b_i - \frac{1}{a_{ii}} \sum_{\substack{j=1 \\ j \neq i}}^{n} a_{ij} x_j^{(k)} \\ &= [b_i - \sum_{\substack{j=1 \\ j \neq i}}^{n} a_{ij} x_j^{(k)}]/a_{ii}\end{aligned}$$

In other words, defining $\mathbf{G}$ and $\mathbf{H}$ as in (4-21) leads to the Jacobi iteration method!

Another frequently used iteration method, called the *Gauss-Seidel iteration method*, is derived by defining

$$\begin{aligned}\mathbf{G} &= [g_{ij}] \qquad g_{ij} = \begin{cases} a_{ij} & i < j \\ 0 & i \geq j \end{cases} \\ \mathbf{H} &= [h_{ij}] \qquad h_{ij} = \begin{cases} 0 & i < j \\ a_{ij} & i \geq j \end{cases}\end{aligned}$$

Again, clearly $\mathbf{G} + \mathbf{H} = \mathbf{A}$ and $\mathbf{H}^{-1}$ exists if all $a_{ii} \neq 0$ (since $\mathbf{H}$ is lower triangular with no zeros along its diagonal and hence det $(\mathbf{H}) \neq 0$). We thus obtain:

$$\mathbf{x}^{(k+1)} = \mathbf{H}^{-1}\mathbf{b} - \mathbf{H}^{-1}\mathbf{G}\mathbf{x}^{(k)}$$

However, the calculation of $\mathbf{H}^{-1}$ is somewhat messy; to avoid this difficulty, we employ the following trick:

Let $\mathbf{H} = \hat{\mathbf{H}} + \mathbf{D}$, where

$$\hat{\mathbf{H}} = [\hat{h}_{ij}] \qquad \hat{h}_{ij} = \begin{cases} 0 & i \leq j \\ a_{ij} & i > j \end{cases}$$

$$\mathbf{D} = [d_{ij}] \qquad d_{ij} = \begin{cases} 0 & i \neq j \\ a_{ii} & i = j \end{cases}$$

Now, we substitute $\mathbf{H} = \hat{\mathbf{H}} + \mathbf{D}$ into

$$\mathbf{H}\mathbf{x}^{(k+1)} = \mathbf{b} - \mathbf{G}\mathbf{x}^{(k)}$$

and obtain

$$(\hat{\mathbf{H}} + \mathbf{D})\mathbf{x}^{(k+1)} = \mathbf{b} - \mathbf{G}\mathbf{x}^{(k)}$$

$$\mathbf{D}\mathbf{x}^{(k+1)} = \mathbf{b} - \mathbf{G}\mathbf{x}^{(k)} - \hat{\mathbf{H}}\mathbf{x}^{(k+1)}$$

$$\mathbf{x}^{(k+1)} = \mathbf{D}^{-1}\mathbf{b} - \mathbf{D}^{-1}\mathbf{G}\mathbf{x}^{(k)} - \mathbf{D}^{-1}\hat{\mathbf{H}}\mathbf{x}^{(k+1)} \tag{4-22}$$

Observe that:

1 Since $\mathbf{D}$ is diagonal, $\mathbf{D}^{-1}$ is also diagonal and trivial to compute.
2 $\mathbf{D}^{-1}\mathbf{G}$ is upper triangular.
3 $\mathbf{D}^{-1}\hat{\mathbf{H}}$ is lower triangular.
4 The ith row of the matrix equation (4-22) is

$$x_i^{(k+1)} = \begin{cases} \dfrac{1}{a_{11}} b_1 - \dfrac{1}{a_{11}} \sum\limits_{j=2}^{n} a_{1j} x_j^{(k)} & i = 1 \\ \dfrac{1}{a_{ii}} b_i - \dfrac{1}{a_{ii}} \sum\limits_{j=1}^{i-1} a_{ij} x_j^{(k+1)} - \dfrac{1}{a_{ii}} \sum\limits_{j=i+1}^{n} a_{ij} x_j^{(k)} & i = 2, \ldots, n-1 \\ \dfrac{1}{a_{nn}} b_n - \dfrac{1}{a_{nn}} \sum\limits_{j=1}^{n-1} a_{nj} x_j^{(k+1)} & i = n \end{cases}$$

Thus, for $i = 2, \ldots, n$ the first summation contains only the variables $x_1, x_2, \ldots, x_{i-1}$. Hence, we can use the $(k + 1)$st estimate of these variables, along with the kth values of $x_{i+1}, x_{i+2}, \ldots, x_n$, to obtain $x_i^{(k+1)}$. In other words, the Gauss-Seidel iteration method always uses the most recently computed values of each variable in every calculation. Jacobi iteration, on the other hand, uses the kth values for *all* the variables to obtain the $(k+1)$st values for *all* the variables.

EXAMPLE 4-7 Let us solve the system of equations (4-10) by Gauss-Seidel iteration. The iteration equations are:

$$\begin{aligned}
x_1^{(k+1)} &= 1.4 \qquad\qquad\qquad -0.4x_2^{(k)} \\
x_2^{(k+1)} &= 0.5 + 0.25x_1^{(k+1)} \qquad\qquad +0.25x_3^{(k)} \\
x_3^{(k+1)} &= 1.5 \qquad\qquad\qquad -0.5x_2^{(k+1)}
\end{aligned}$$

If we again start with an initial guess of $x_1^{(0)} = 1.2$, $x_2^{(0)} = 0.8$, $x_3^{(0)} = 1.2$, we obtain the results given in the table below (again rounded to four decimal places).

Iteration number (k)	$x_1^{(k)}$	$x_2^{(k)}$	$x_3^{(k)}$
0	1.2	0.8	1.2
1	1.08	1.07	0.965
2	0.9720	0.9843	1.0079
3	1.0063	1.0036	0.9982
4	0.9986	0.9992	1.0004
5	1.0003	1.0002	0.9999
6	0.9999	1.0000	1.0000
7	1.0000	1.0000	1.0000
8	1.0000	1.0000	1.0000

////

Although Gauss-Seidel converged to the solution much faster than Jacobi iteration in this example, there is no way, in general, of telling in advance which method will be faster on any given problem. This fact may run counter to one's intuition, which might suggest that Gauss-Seidel should be better, since it always uses the most recent estimates for each variable. We repeat, however, that such is not necessarily the case.

We close this section with one additional comment. In practical computing situations, one is not usually interested in (or able to achieve) an *exact* solution to a large system of equations. Typically, it is desired to determine the solution accurate to some specified number of decimal points (e.g., four in our example) or accurate to some specified number of significant digits. Thus, the "stopping rule" mentioned in step 5 of the description of the

Jacobi iteration method (which is also applicable to other iteration methods) should be modified to one of the following:

A. If $|x_i^{(k+1)} - x_i^{(k)}| < \varepsilon \qquad i = 1, 2, \ldots, n$

terminate the method. Otherwise, continue.

B. If $\left|\dfrac{x_i^{(k+1)} - x_i^{(k)}}{x_i^{(k)}}\right| < \varepsilon \qquad i = 1, 2, \ldots, n$

terminate the method. Otherwise, continue.

If, for example, one desired accuracy to within three decimal places, the ε in A above would be chosen as $\varepsilon = 5 \times 10^{-4}$. If one desired accuracy to four significant digits, then B would be employed as a stopping rule, with $\varepsilon = 5 \times 10^{-5}$

4-6 CONVERGENCE CONSIDERATIONS

As we have indicated earlier, one cannot always use an iteration method with the assurance that a solution will be obtained. When such a method does reach a solution, it is said to have *converged;* when the difference between successive estimates appears to be increasing, the method will not converge and is said to be *diverging*.

In this section we shall investigate the convergence properties of iteration methods. We begin by examining the iteration formula (4-20):

$$\mathbf{x}^{(k+1)} = \mathbf{H}^{-1}\mathbf{b} - \mathbf{H}^{-1}\mathbf{G}\mathbf{x}^{(k)} \qquad (4\text{-}20)$$

To simplify the notation, let us define

$$\begin{aligned} \mathbf{r} &= \mathbf{H}^{-1}\mathbf{b} \\ \mathbf{Q} &= -\mathbf{H}^{-1}\mathbf{G} \end{aligned} \qquad (4\text{-}23)$$

Thus, Eq. (4-20) becomes

$$\mathbf{x}^{(k+1)} = \mathbf{r} + \mathbf{Q}\mathbf{x}^{(k)}$$

Now, if we arbitrarily select an initial guess $\mathbf{x}^{(0)}$, we obtain successively

$$\begin{aligned} \mathbf{x}^{(1)} &= \mathbf{r} + \mathbf{Q}\mathbf{x}^{(0)} \\ \mathbf{x}^{(2)} &= \mathbf{r} + \mathbf{Q}\mathbf{x}^{(1)} \\ &= \mathbf{r} + \mathbf{Q}(\mathbf{r} + \mathbf{Q}\mathbf{x}^{(0)}) \\ &= \mathbf{r} + \mathbf{Q}\mathbf{r} + \mathbf{Q}^2\mathbf{x}^{(0)} \end{aligned}$$

$$\begin{aligned}
\mathbf{x}^{(3)} &= \mathbf{r} + \mathbf{Q}\mathbf{x}^{(2)} \\
&= \mathbf{r} + \mathbf{Q}(\mathbf{r} + \mathbf{Q}\mathbf{r} + \mathbf{Q}^2\mathbf{x}^{(0)}) \\
&= \mathbf{r} + \mathbf{Q}\mathbf{r} + \mathbf{Q}^2\mathbf{r} + \mathbf{Q}^3\mathbf{x}^{(0)} \\
&\vdots \\
\mathbf{x}^{(k)} &= \mathbf{r} + \mathbf{Q}\mathbf{r} + \mathbf{Q}^2\mathbf{r} + \cdots + \mathbf{Q}^{k-1}\mathbf{r} + \mathbf{Q}^k\mathbf{x}^{(0)} \\
&= (\mathbf{I}_n + \mathbf{Q} + \mathbf{Q}^2 + \cdots + \mathbf{Q}^{k-1})\mathbf{r} + \mathbf{Q}^k\mathbf{x}^{(0)} \qquad (4\text{-}24)
\end{aligned}$$

Thus, Eq. (4-24) expresses the kth estimate of the solution, $\mathbf{x}^{(k)}$, in terms of the initial estimate, $\mathbf{x}^{(0)}$.

Now, we observe the identity

$$(\mathbf{I}_n - \mathbf{Q})(\mathbf{I}_n + \mathbf{Q} + \mathbf{Q}^2 + \cdots + \mathbf{Q}^{k-1}) = \mathbf{I}_n - \mathbf{Q}^k \qquad (4\text{-}25)$$

Thus, if $(\mathbf{I}_n - \mathbf{Q})^{-1}$ exists, then

$$\mathbf{I}_n + \mathbf{Q} + \mathbf{Q}^2 + \cdots + \mathbf{Q}^{k-1} = (\mathbf{I}_n - \mathbf{Q})^{-1}(\mathbf{I}_n - \mathbf{Q}^k) \qquad (4\text{-}26)$$

[We shall prove later in this section that $(\mathbf{I}_n - \mathbf{Q})^{-1}$ does in fact exist.]

Substitution of (4-26) into (4-24) yields

$$x^{(k)} = (\mathbf{I}_n = \mathbf{Q})^{-1}(\mathbf{I}_n - \mathbf{Q}^k)\mathbf{r} + \mathbf{Q}^k\mathbf{x}^{(0)} \qquad (4\text{-}27)$$

Now, ideally, we would like the convergence of the iteration method to be independent of our initial guess $\mathbf{x}^{(0)}$, since usually we won't have any idea of how to choose a "good" $\mathbf{x}^{(0)}$. From (4-27), we see that if

$$\lim_{k\to\infty} \mathbf{Q}^k = \mathbf{O}_{n\times n} \qquad (4\text{-}28)$$

then

$$\begin{aligned}
\lim_{k\to\infty} \mathbf{x}^{(k)} &= (\mathbf{I}_n - \mathbf{Q})^{-1}(\mathbf{I}_n - \mathbf{O}_{n\times n})\mathbf{r} + \mathbf{O}_{n\times n}\mathbf{x}^{(0)} \\
&= (\mathbf{I}_n - \mathbf{Q})^{-1}\mathbf{r} \\
&= (\mathbf{I}_n + \mathbf{H}^{-1}\mathbf{G})^{-1}(\mathbf{H}^{-1}\mathbf{b}) \qquad \text{by Eq. (4-23)} \\
&= [(\mathbf{I}_n + \mathbf{H}^{-1}\mathbf{G})^{-1}\mathbf{H}^{-1}]\mathbf{b} \\
&= [\mathbf{H}(\mathbf{I}_n + \mathbf{H}^{-1}\mathbf{G})]^{-1}\mathbf{b} \qquad \text{by Eq. (2-33)} \\
&= (\mathbf{H} + \mathbf{G})^{-1}\mathbf{b} \\
&= \mathbf{A}^{-1}\mathbf{b} \qquad \text{since } \mathbf{A} = \mathbf{H} + \mathbf{G}, \text{ by definition}
\end{aligned}$$

Hence, if Eq. (4-28) holds, then in the limit, as $k \to \infty$, the successive estimates $\mathbf{x}^{(k)}$ do indeed approach the exact solution $\mathbf{x}^* = \mathbf{A}^{-1}\mathbf{b}$.

Before turning to the question of under what circumstances Eq. (4-28) does hold, let us verify the validity of the assumption we have used in this discussion that $(\mathbf{I}_n - \mathbf{Q})^{-1}$ exists:

$$\begin{aligned}\mathbf{I}_n - \mathbf{Q} &= \mathbf{I}_n + \mathbf{H}^{-1}\mathbf{G}\\ &= \mathbf{H}^{-1}\mathbf{H} + \mathbf{H}^{-1}\mathbf{G}\\ &= \mathbf{H}^{-1}(\mathbf{H} + \mathbf{G})\\ &= \mathbf{H}^{-1}\mathbf{A}\end{aligned}$$

Therefore,

$$\begin{aligned}\det(\mathbf{I}_n - \mathbf{Q}) &= \det(\mathbf{H}^{-1}\mathbf{A})\\ &= \det(\mathbf{H}^{-1})\det(\mathbf{A})\end{aligned}$$

Now, since we have assumed that $\det(\mathbf{A}) \neq 0$, and since $\det(\mathbf{H}^{-1}) \neq 0$ (because we have also assumed that $\mathbf{H}^{-1}$ exists), it is obvious that $\det(\mathbf{I}_n - \mathbf{Q}) \neq 0$. Thus, by Theorem 4-4, $(\mathbf{I}_n - \mathbf{Q})^{-1}$ exists.

At this point, in order to help motivate the material covered in Chaps. 9 and 10, we shall extract two theorems from these chapters. First, we define:

For any $n \times n$ matrix $\mathbf{Q}$, a scalar λ such that $\det(\mathbf{Q} - \lambda\mathbf{I}_n) = 0$ is called an *eigenvalue* of $\mathbf{Q}$. As we shall see in Chap. 9, every square matrix $\mathbf{Q}$ has at least one eigenvalue and at most n distinct eigenvalues.

We now quote:

Theorem 9-20 If $\mathbf{Q}$ is any nth-order matrix with distinct eigenvalues $\lambda_1, \lambda_2, \ldots, \lambda_p$, then

$$\lim_{k \to \infty} \mathbf{Q}^k = \mathbf{O}_{n \times n}$$

if and only if

$$\max_{j = 1, 2, \ldots, p} |\lambda_j| < 1 \tag{4-29}$$

Thus, Theorem 9-20, combined with our prior discussion in this section, tells us that an iteration method (4-20) will converge if the eigenvalue of $\mathbf{Q}$ largest in magnitude is less than 1. In Chap. 10 we shall learn a method for finding this eigenvalue with relatively little computational effort.

However, we have an even easier way to test the convergence of the iteration method, and this is to employ:

Theorem 10-2 If $\mathbf{Q} = [q_{ij}]_{n \times n}$ and if λ_1 is the eigenvalue of $\mathbf{Q}$ largest in magnitude, then

$$|\lambda_1| \le \max_{k=1,2,\ldots,n} \left\{ \sum_{j=1}^{n} |q_{kj}| \right\} \tag{4-30}$$

Combining Theorems 9-20 and 10-2 yields:

Theorem 4-5 The iteration method $\mathbf{x}^{(k+1)} = \mathbf{r} + \mathbf{Q}\mathbf{x}^{(k)}$ will converge† if

$$\max_{i=1,2,\ldots,n} \left\{ \sum_{j=1}^{n} |q_{ij}| \right\} < 1 \tag{4-31}$$

Equation (4-31) states that the sum of the absolute values of the elements in row i must be less than 1, for each i, $i = 1, 2, \ldots, n$.

Finally, we recall that in the Jacobi iteration method,

$$\mathbf{Q} = -\mathbf{H}^{-1}\mathbf{G}$$

$$q_{ij} = \begin{cases} 0 & i = j \\ -\dfrac{a_{ij}}{a_{ii}} & i \ne j \end{cases} \tag{4-32}$$

Thus,

$$\sum_{j=1}^{n} |q_{ij}| = \sum_{\substack{j=1 \\ j \ne i}}^{n} \left| -\frac{a_{ij}}{a_{ii}} \right| \tag{4-33}$$

and if (4-31) holds, then

$$\sum_{\substack{j=1 \\ j \ne i}}^{n} \left| -\frac{a_{ij}}{a_{ii}} \right| < 1 \qquad i = 1, 2, \ldots, n \tag{4-34}$$

Upon multiplying both sides of (4-34) by $|a_{ii}|$, we obtain

$$\sum_{\substack{j=1 \\ j \ne i}}^{n} |a_{ij}| < |a_{ii}| \qquad i = 1, 2, \ldots, n \tag{4-35}$$

The inequalities (4-35) provide us with a simple test as to whether the Jacobi iteration method will converge or not. Moreover, note that these inequalities involve only the elements of the original $\mathbf{A}$ matrix.

† Note that the iteration method may converge even if (4-31) is not satisfied. Consider, for example, the matrix $\mathbf{Q} = \begin{bmatrix} 1 & 1 \\ -1 & -1 \end{bmatrix}$, whose eigenvalues are $\lambda_1 = \lambda_2 = 0$. The resulting iteration method converges, by Theorem 9-20. Thus Theorem 4-5 provides a sufficient but not necessary condition for convergence.

Each inequality (4-35) states that the absolute value of the diagonal element of a row is greater than the sum of the absolute values of the other elements of that row. A matrix whose rows satisfy (4-35) is said to be *diagonally dominant*. Jacobi iteration will converge if the coefficient matrix **A** is diagonally dominant.

It is often possible to achieve diagonal dominance for a given matrix by judicious interchanges of rows. In the example below, we illustrate this point.

EXAMPLE 4-8 Consider the following system:

$$\begin{aligned} 3x_1 - 5x_2 + 47x_3 + 20x_4 &= 18 \\ 56x_1 + 23x_2 + 11x_3 - 19x_4 &= 36 \\ 2x_1 - 4x_2 + 8x_3 + 18x_4 &= 25 \\ 17x_1 + 65x_2 - 13x_3 + 7x_4 &= 84 \end{aligned}$$

1. Determine the largest (in magnitude) coefficient. In this case, it is $a_{42} = 65$. Since a_{42} is in the second column, interchange row 2 and row 4, so that 65 becomes a diagonal element. The new augmented matrix is:

$$\left[\begin{array}{rrrr:r} 3 & -5 & 47 & 20 & 18 \\ 17 & 65 & 13 & 7 & 84 \\ 2 & -4 & 8 & 18 & 25 \\ 56 & 23 & 11 & -19 & 36 \end{array}\right]$$

2. Cross out the row and column containing 65, and select the largest remaining coefficient (excluding the last column). Here, it is 56. Interchange row 1 and row 4:

$$\left[\begin{array}{rrrr:r} 56 & 23 & 11 & -19 & 36 \\ 17 & 65 & -13 & 7 & 84 \\ 2 & -4 & 8 & 18 & 25 \\ 3 & -5 & 47 & 20 & 18 \end{array}\right]$$

3. Now, cross out the row and column containing 56, and select the next largest coefficient. Here, it is 47. Interchange row 3 and row 4:
4. The augmented matrix is now

$$\left[\begin{array}{rrrr:r} 56 & 23 & 11 & -19 & 36 \\ 17 & 65 & -13 & 7 & 84 \\ 3 & -5 & 47 & 20 & 18 \\ 2 & -4 & 8 & 18 & 25 \end{array}\right]$$

Checking for diagonal dominance:

$$\begin{aligned}
&\text{Row 1: } |23| + |11| + |-19| = 53 < |56| \\
&\text{Row 2: } |17| + |-13| + |7| = 37 < |65| \\
&\text{Row 3: } |-3| + |-5| + |20| = 28 < |47| \\
&\text{Row 4: } |2| + |-4| + |8| = 14 < |18|
\end{aligned}$$

Hence, Jacobi iteration *will* converge for the system represented by the above augmented matrix. ////

REFERENCES

1. FADDEEV, D. K., and V. N. FADDEEVA: "Computational Methods of Linear Algebra," Freeman, San Francisco, 1963.
2. FOX, L.: "An Introduction to Numerical Linear Algebra," Oxford University Press, New York, 1965.
3. FRANKLIN, J. N.: "Matrix Theory," Prentice-Hall, Englewood Cliffs, N.J., 1968.
4. ISAACSON, E., and H. B. KELLER: "Analysis of Numerical Methods, Wiley, New York, 1966.
5. NOBLE, BEN: "Applied Linear Algebra," Prentice-Hall, Englewood Cliffs, N.J., 1969.

EXERCISES

1 Solve by Gaussian elimination:

$$\begin{bmatrix} 1 & 2 & 4 & -1 \\ 1 & 3 & 5 & 0 \\ 1 & 2 & 5 & 1 \\ 0 & 1 & 1 & 0 \end{bmatrix} \begin{bmatrix} x_1 \\ x_2 \\ x_3 \\ x_4 \end{bmatrix} = \begin{bmatrix} 6 \\ 9 \\ 9 \\ 2 \end{bmatrix}$$

2 Solve by Gaussian elimination:

$$\begin{bmatrix} 1 & 3 & -1 & 0 \\ 2 & 6 & 5 & 3 \\ 2 & 5 & 3 & 4 \\ 3 & 0 & 0 & 7 \end{bmatrix} \begin{bmatrix} x_1 \\ x_2 \\ x_3 \\ x_4 \end{bmatrix} = \begin{bmatrix} -5 \\ 4 \\ 1 \\ 0 \end{bmatrix}$$

3 Solve by Gaussian elimination:

$$\begin{bmatrix} 5 & 1 & 1 & 0 \\ 0 & 4 & 0 & 1 \\ 0 & 0 & 2 & 1 \\ -2 & 1 & 0 & 4 \end{bmatrix} \begin{bmatrix} x_1 \\ x_2 \\ x_3 \\ x_4 \end{bmatrix} = \begin{bmatrix} 3 \\ 4 \\ -4 \\ -1 \end{bmatrix}$$

4 Solve by Gaussian elimination:

$$\begin{bmatrix} 0 & 0 & 3 & 1 \\ 5 & 1 & 1 & 0 \\ 0 & 4 & 2 & 1 \\ 1 & 1 & 0 & -3 \end{bmatrix} \begin{bmatrix} x_1 \\ x_2 \\ x_3 \\ x_4 \end{bmatrix} = \begin{bmatrix} 2 \\ 7 \\ 5 \\ -5 \end{bmatrix}$$

5 Solve the system of Exercise 1 by Gauss-Jordan elimination.

6 Solve the system of Exercise 2 by Gauss-Jordan elimination.

7 Solve the system of Exercise 3 by Gauss-Jordan elimination.

8 Solve the system of Exercise 4 by Gauss-Jordan elimination.

9 Prepare a detailed flow chart of Gaussian elimination. Include tests for row interchanges if necessary and for the possibility of a singular coefficient matrix.

10 Prepare a detailed flow chart of Gauss-Jordan elimination, including the tests described in Exercise 9.

11 Determine the number of arithmetic operations required to compute $\mathbf{A}^{-1}$ using Gauss-Jordan elimination.

12 Determine the number of arithmetic operations required to solve the p systems of simultaneous equations $\mathbf{Ax}_k = \mathbf{b}_k$, $k = 1, 2, \ldots, p$, by Gaussian elimination.

13 Determine the number of arithmetic operations required to solve the p systems of simultaneous equations $\mathbf{Ax}_k = \mathbf{b}_k$, $k = 1, 2, \ldots, p$, by Gauss-Jordan elimination.

14 Find the inverse of the coefficient matrix of Exercise 1 by (*a*) Gaussian elimination; (*b*) Gauss-Jordan elimination.

15 Find the inverse of the coefficient matrix of Exercise 2 by (*a*) Gaussian elimination; (*b*) Gauss-Jordan elimination.

16 Test the coefficient matrix of Exercise 2 to see if it is diagonally dominant. If not, try modifying it to obtain a diagonally dominant coefficient matrix.

17 Test the coefficient matrix of Exercise 3 to see if it is diagonally dominant. If not, try modifying it to obtain a diagonally dominant coefficient matrix.

18 Test the coefficient matrix of Exercise 4 to see if it is diagonally dominant. If not, try modifying it to obtain a diagonally dominant coefficient matrix.

19 Solve the system of Exercise 3 by the Jacobi iteration method.

20 Solve the system of Exercise 4 by the Jacobi iteration method.

21 Prepare a flow chart of the Jacobi iteration method.

22 Solve the system of Exercise 3 by the Gauss-Seidel iteration method.

23 Solve the system of Exercise 4 by the Gauss-Seidel iteration method.

24 Prepare a flow chart of the Gauss-Seidel iteration method.

25 Write a computer program to solve a system of equations by Gaussian elimination (see Exercise 9). Test your program by solving the systems of Exercises 1 and 2.

5

GAUSSIAN ELIMINATION AND COMPUTER ARITHMETIC

5-1 COMPUTER ARITHMETIC

In Chap. 4 we stated that Gaussian elimination is generally the most efficient and widely used method for solving systems of linear equations. Inevitably, when one wishes to solve all but the smallest such system, a digital computer is employed to perform the calculations. In using such a machine, we must be concerned not only with the efficiency of our method, but also with the accuracy of the resulting computations. All too frequently the latter is ignored, with disastrous consequences. The problem of computer accuracy is unfortunately quite complex, and a comprehensive treatment of this subject is quite beyond the scope of this text. However, it is very important for the user of digital computers to be at least aware of the computational difficulties he may encounter and to have some familiarity with the possible "remedies" which exist. Such is the intent of this chapter. The reader interested in pursuing the subject in greater depth is invited to consult Forsythe and Moler [1], Stark [3], and Wilkinson [4]. As in Chap. 4, we shall restrict our attention to the case in which the coefficient matrix **A** is nonsingular.

In this section we introduce the concept of "computer arithmetic," and in subsequent sections we shall investigate its consequences for Gaussian elimination. Much of these discussions is also applicable to Gauss-Jordan elimination, as well.

The process by which a digital computer performs arithmetic operations is not quite the same as that employed by the ordinary *Homo sapiens* using scratch paper. The key difference is based upon the fact that a computer has only a finite number of storage locations in which to place—or store—a given number. For example, the computer may be constructed in such a way that it can store only numbers with up to seven digits, or ten digits, etc. If such a computer were given a number with more than this number of digits, it would not be able to store the complete number and would somehow have to "shorten" the number before storing it. It is this process of "shortening" numbers which introduces errors in computer calculations. We shall not go into the details of how a computer stores numbers, but rather shall consider the concept of *floating-point arithmetic*, which is essentially how a computer performs arithmetic.

For the purposes of the ensuing discussion, we shall assume a hypothetical computer which can store numbers with up to four *significant digits;* by this we mean any number N that can be expressed in the form

$$N = \pm 0.a_1 a_2 a_3 a_4 \cdot 10^p \tag{5-1}$$

where a_1, a_2, a_3, a_4 are integers between 0 and 9 and p is any one- or two-digit integer† (positive or negative):

$$0 \le a_i \le 9 \qquad i = 1, 2, 3, 4$$
$$-99 \le p \le +99$$

Such a number N is called a *floating-point number*. Thus, the numbers 0.1234×10^2, 0.31×10^{-10}, 0.5×10^3 are examples of allowable floating-point numbers.

Arithmetic performed by such a computer is called 4-*digit arithmetic*. If a computer using 4-digit arithmetic is given a number with five (or more) significant digits, such as $(\pm 0.a_1 a_2 a_3 a_4 a_5 \ldots) \cdot 10^p$, it will either *truncate* the number, resulting in $(\pm 0.a_1 a_2 a_3 a_4) \cdot 10^p$ or *round* the number, resulting in

$$(\pm 0.a_1 a_2 a_3 a_4) \cdot 10^p \qquad \text{if } a_5 < 5$$
$$(\pm 0.a_1 a_2 a_3 \hat{a}_4) \cdot 10^p \qquad \text{if } a_5 \ge 5$$

and where‡ $\quad \hat{a}_4 = a_4 + 1$

† The restriction on the range of p is a result of the way computers store floating-point numbers. This range varies somewhat with different computers.

‡ If $a_4 = 9$, then $\hat{a}_4 = 0$ and $\hat{a}_3 = a_3 + 1$, unless a_3 also equals 9, in which case the obvious changes must be made.

Thus, if $N = 0.12346 \cdot 10^3$, the truncated number is $0.1234 \cdot 10^3$. Let us assume that our hypothetical computer *rounds* all numbers to four significant digits. Numerical errors resulting from this process (whether by rounding or truncation) are called *roundoff errors*.

EXAMPLE 5-1 Let $a = 0.1234$, $b = 0.3011$, $c = 0.2287$. Then, $ab = (0.1234)(0.3011) = 0.03715574$.

However, our hypothetical computer would calculate $ab = 0.3716 \cdot 10^{-1}$; hence the roundoff error is $0.3716 \cdot 10^{-1} - 0.3715574 \cdot 10^{-1} = 0.4260 \cdot 10^{-5}$. Also, suppose we calculate $(ab)c$ and $a(bc)$, in 4-digit arithmetic:

$$(ab)c = (0.3716 \cdot 10^{-1})(0.2287) = 0.8498 \cdot 10^{-2}$$

$$a(bc) = (0.1234)[(0.3011) \cdot (0.2287)] = (0.1234)(0.6886 \cdot 10^{-1}) = 0.8497 \cdot 10^{-2}$$

Hence, the associative law of scalar multiplication is no longer valid: The product of three (or more) numbers depends upon the order in which the intermediate multiplications are performed. ////

Roundoff errors resulting from multiplication are not nearly as serious as those resulting from addition (and subtraction). Before considering the latter, we define:

Given any number α, we shall denote by $\hat{\alpha}$ the resulting number which is stored in a computer. Thus, in our hypothetical computer, $\hat{\alpha} = \alpha$ if α is of the form (5-1); otherwise $\hat{\alpha} \neq \alpha$. The following quantities will facilitate our discussion:

$$\text{Absolute error} \equiv \text{error}_{\text{abs}} = \hat{\alpha} - \alpha$$

$$\text{Relative error} \equiv \text{error}_{\text{rel}} = \frac{\hat{\alpha} - \alpha}{\alpha}$$

In most cases the error of interest is the relative error, since an error of 10^{-3}, for example, is significant if the quantity α is also of magnitude 10^{-3}, but such an error may be negligible if α is of magnitude 10^7 or so. However, if α is zero, then obviously we cannot discuss relative error.

We now describe three of the greatest sources of relatively large roundoff errors.

1 Addition of Large Numbers to Small Numbers

Suppose we wish to add $a = 0.1234 \cdot 10^{-1}$ and $b = 0.6789 \cdot 10^3$. If $c = a + b$, the result in 4-digit arithmetic would be $c = 0.6789 \cdot 10^3$; hence, the term $a = 0.1234 \cdot 10^{-1}$ has no effect on the sum. Of course, the relative error in

$\hat{c}$ is fairly small. However, if we are adding many numbers together, such errors may accumulate. Consider the following:

Let

$$S = \sum_{k=1}^{\infty} \frac{1}{3^k} = \frac{1}{3} + \frac{1}{9} + \frac{1}{27} + \cdots$$

Since $3S = 1 + \frac{1}{3} + \frac{1}{9} + \cdots = 1 + S$, it is obvious that $S = 0.5000$. If we calculate the four-digit versions of $\frac{1}{3}$, $\frac{1}{9}$, etc., we obtain:

k	$\frac{1}{3^k}$ (rounded to four digits)
1	0.3333
2	0.1111
3	0.03703
4	0.01234
5	0.004113
6	0.001371
7	0.0004570
8	0.0001523
9	0.00005077
10	0.00001692

If we add the first 10 terms in S from $k = 1$ to $k = 10$ in 4-digit arithmetic (rounding each intermediate sum to four significant digits), we find that $\hat{S} = 0.4996$, so that $\text{error}_{\text{rel}} = (\hat{S} - S)/S = -8 \cdot 10^{-4}$. Moreover, if we were to continue adding terms to $\hat{S}$, the latter would remain unchanged, as each additional term is too small to affect the four-digit sum.

However, if we add the first 10 terms in S from $k = 10$ to $k = 1$ (i.e., adding the smaller numbers first), we find that $\hat{S} = 0.4999$, with $\text{error}_{\text{rel}} = -2 \cdot 10^{-4}$. Thus, the relative error is four times as large when S is calculated by the first method. Furthermore, we can improve this result by adding, say, the first 15 terms from $k = 15$ to $k = 1$ (which yields $\hat{S} = 0.5000$).

The type of error described above is frequently called *accumulation error.*

2 Subtraction of Two Nearly Equal Numbers

Suppose we wish to subtract $b_1 = 0.12345$ from $a = 0.12334$. In our hypothetical computer, we would actually subtract $\hat{b}_1 = 0.1235$ from $\hat{a} = 0.1233$, obtaining

$$\hat{c}_1 = \hat{b}_1 - \hat{a} = 0.2000 \cdot 10^{-3}$$

Thus, $\hat{c}_1$ really has only one significant digit. Moreover, suppose that there was a small error in b_1, and the true value of b_1 is $b_2 = 0{,}12344$.

Then, $\hat{b}_2 = 0.1234$ and $\hat{c}_2 = 0.1000 \cdot 10^{-3}$. The relative error in $\hat{c}_1$ (compared to $\hat{c}_2$) is then

$$\text{Error}_{\text{rel}} = \frac{\hat{c}_1 - \hat{c}_2}{\hat{c}_2} = \frac{0.2000 \cdot 10^{-3} - 0.1000 \cdot 10^{-3}}{0.1000 \cdot 10^{-3}} = 1.0$$

This is a rather remarkable error, being of substantially greater magnitude than the original numbers themselves.

3 Division by Very Small Numbers

Frequently, as the result of previous roundoff errors, a number which is supposed to be zero is actually computed as a very small but nonzero number; for example, 10^{-15}. Suppose, now, that we wish to divide $a = 0.1234 \cdot 10^{90}$ by 10^{-15}. The result is

$$b = \frac{a}{10^{-15}} = 0.1234 \cdot 10^{105}$$

However, since the exponent of 10 above is greater than 99, our hypothetical computer will not obtain this result. Such a calculation is called an *overflow;* a calculation resulting in an exponent too small (i.e., less than -99) is called an *underflow*. An overflow is caused by attempting to divide by a very small number. As we shall see, such calculations are extremely common in Gaussian elimination, unless special precautions are made to avoid them.

5-2 PIVOTING IN GAUSSIAN ELIMINATION

We shall now investigate some of the implications of computer arithmetic as far as Gaussian elimination is concerned. These are perhaps best illustrated by an example.

EXAMPLE 5-2 Consider the system

$$\begin{bmatrix} 0.1 \cdot 10^{-4} & 1 \\ 1 & 1 \end{bmatrix} \begin{bmatrix} x_1 \\ x_2 \end{bmatrix} = \begin{bmatrix} 1 \\ 2 \end{bmatrix}$$

We note that the correct solution, to four significant digits, is $x_1 = x_2 = 1$. Solving by Gaussian elimination, with 4-digit arithmetic:

Add $-0.1 \cdot 10^6$ times row 1 of the augmented matrix to row 2, yielding

$$\left[\begin{array}{cc|c} 0.1 \cdot 10^{-4} & 1 & 1 \\ 0 & -0.1 \cdot 10^6 & -0.1 \cdot 10^6 \end{array}\right]$$

The backward pass then yields

$$\left[\begin{array}{cc:c} 0.1\cdot 10^{-4} & 0 & 0 \\ 0 & 1 & 1 \end{array}\right] \rightarrow \left[\begin{array}{cc:c} 1 & 0 & 0 \\ 0 & 1 & 1 \end{array}\right]$$

or $x_1 = 0$, $x_2 = 1$. Hence, something drastically wrong has occurred.

Essentially, the difficulty is the roundoff error caused by:

1 Dividing by a very small number to obtain the multiplier $m_1 = -0.1 \cdot 10^6$:

$$m_1 = \frac{-1}{0.1 \cdot 10^{-4}} = -0.1 \cdot 10^6$$

2 Adding the resulting very large number to relatively small numbers; i.e., adding $-0.1 \cdot 10^6$ to $+1$ and also to $+2$. In 4-digit arithmetic, the 1 and 2 are lost in these additions.

Consider now the same system of equations with the two equations interchanged:

$$\begin{bmatrix} 1 & 1 \\ 0.1 \cdot 10^{-4} & 1 \end{bmatrix} \begin{bmatrix} x_1 \\ x_2 \end{bmatrix} = \begin{bmatrix} 2 \\ 1 \end{bmatrix}$$

We thus obtain the following sequence of augmented matrices:

$$\left[\begin{array}{cc:c} 1 & 1 & 2 \\ 0 & 1 & 1 \end{array}\right] \rightarrow \left[\begin{array}{cc:c} 1 & 0 & 1 \\ 0 & 1 & 1 \end{array}\right]$$

Hence, we have calculated the desired result, $x_1 = x_2 = 1$. ////

In the above example, the key to all the computational problems was caused by division by a very small number (small in comparison with the other coefficients). To avoid this difficulty, most computer programs for Gaussian elimination employ what is known as a *pivoting strategy*. Recall that at the kth step of the forward pass, we calculate the multipliers:

$$m_p = -\frac{\hat{a}_{pk}}{\hat{a}_{kk}} \qquad p = k+1, k+2, \ldots, n \tag{5-2}$$

The coefficient $\hat{a}_{kk}$ is called the *pivot element*. There are two basic types of pivoting strategies which are commonly used:

A. *Partial Pivoting:* At the kth step, we find

$$|\hat{a}_{rk}| = \max_{p=k, \ldots, n} \{|\hat{a}_{pk}|\} \tag{5-3}$$

and choose $\hat{a}_{rk}$ as the pivot element by interchanging rows r and k. This is actually what we did in Example 5-2 above.

B. *Complete Pivoting:* At the kth step, we find

$$|\hat{a}_{qr}| = \max_{\substack{t=k,\ldots,n \\ p=k,\ldots,n}} \{|\hat{a}_{tp}|\} \tag{5-4}$$

and choose $\hat{a}_{qr}$ as the pivot element by interchanging both row q with row k and column r with column† k.

The complete pivoting strategy requires substantially more comparisons at each step to determine the pivot element than partial pivoting does; on the other hand, the latter in general will yield slightly less accurate results. However, practical computing experience seems to indicate that for most purposes partial pivoting provides sufficient accuracy to warrant its use.

In summary, pivoting strategies are designed to reduce as much as possible the likelihood of a division by a small number. We cannot overemphasize the importance of including a pivoting strategy when using Gaussian elimination.

5-3 SCALING A SYSTEM OF EQUATIONS

As we have seen throughout this chapter, roundoff errors are often more severe if there are large differences in the magnitudes of the numbers involved. Thus, given a system of equations $\mathbf{Ax} = \mathbf{b}$, the first thing we should do before applying Gaussian elimination is to make some attempt to "scale" the equations (and/or the variables). However, we must be extremely careful when trying such scaling, as the following examples illustrate.

EXAMPLE 5-3 Consider the system

$$\begin{bmatrix} 10 & 10^6 \\ 1 & 1 \end{bmatrix} \begin{bmatrix} x_1 \\ x_2 \end{bmatrix} = \begin{bmatrix} 10^6 \\ 2 \end{bmatrix}$$

This is the system of Example 5-2 with the first equation multiplied (scaled) by 10^6. Employing the partial pivoting strategy (in 4-digit arithmetic) results in the following sequence of augmented matrices:

$$\left[\begin{array}{cc:c} 10 & 10^6 & 10^6 \\ 0 & -10^5 & -10^5 \end{array}\right] \rightarrow \left[\begin{array}{cc:c} 10 & 0 & 0 \\ 0 & 1 & 1 \end{array}\right] \rightarrow \left[\begin{array}{cc:c} 1 & 0 & 0 \\ 0 & 1 & 1 \end{array}\right]$$

† Since each column corresponds to a variable, we must remember to interchange the corresponding variables each time we interchange two columns.

Hence, even with a pivoting strategy, we have obtained the same miserable result as we did in Example 5-2 before we interchanged rows. Moreover, suppose we also scale the second column of the above system, by defining $x_2' = 10^6 x_2$, so that the system becomes

$$\begin{bmatrix} 10 & 1 \\ 1 & 10^{-6} \end{bmatrix} \begin{bmatrix} x_1 \\ x_2' \end{bmatrix} = \begin{bmatrix} 10^6 \\ 2 \end{bmatrix}$$

The reader may easily verify that even the use of the complete pivoting strategy on this latter system will yield

$$\left[\begin{array}{cc:c} 10 & 1 & 10^6 \\ 0 & -10^{-1} & -10^5 \end{array}\right] \to \left[\begin{array}{cc:c} 10 & 0 & 0 \\ 0 & 1 & 10^6 \end{array}\right] \to \left[\begin{array}{cc:c} 1 & 0 & 0 \\ 0 & 1 & 10^6 \end{array}\right]$$

or $x_1 = 0$, $x_2' = 10^6$ (and $x_2 = 1$), instead of the desired result $x_1 = 1$, $x_2' = 10^6$. ////

EXAMPLE 5-4 Consider the system

$$\begin{bmatrix} 1 & 1 & 1 \\ 0.2 & -1 \cdot 10^{-7} & 1 \cdot 10^{-7} \\ 0.1 & 3 \cdot 10^{-7} & 2 \cdot 10^{-7} \end{bmatrix} \begin{bmatrix} x_1 \\ x_2 \\ x_3 \end{bmatrix} = \begin{bmatrix} 0 \\ 0.2 \\ 0.3 \end{bmatrix} \tag{5-5}$$

Since a_{11} is the largest element, either partial or complete pivoting will yield the following augmented matrix after one step of Gaussian elimination:

$$\left[\begin{array}{ccc:c} 1 & 1 & 1 & 0 \\ 0 & -0.2 & -0.2 & 0.2 \\ 0 & -0.1 & -0.1 & 0.3 \end{array}\right] \tag{5-6}$$

It is clear that the system represented by (5-6) is inconsistent (i.e., possesses no solution)!

However, if we scale the second and third equations by multiplying both by 10^7, we obtain

$$\left[\begin{array}{ccc:c} 1 & 1 & 1 & 0 \\ 2 \cdot 10^6 & -1 & 1 & 2 \cdot 10^6 \\ 10^6 & 3 & 2 & 3 \cdot 10^6 \end{array}\right]$$

Using either pivoting strategy, we would interchange the first and second equations and obtain the following sequence:

$$\left[\begin{array}{ccc|c} 2\cdot 10^6 & -1 & 1 & 2\cdot 10^6 \\ 0 & 1 & 1 & -1 \\ 0 & 3.5 & 1.5 & 2\cdot 10^6 \end{array}\right] \to \left[\begin{array}{ccc|c} 2\cdot 10^6 & -1 & 1 & 2\cdot 10^6 \\ 0 & 3.5 & 1.5 & 10^6 \\ 0 & 1 & 1 & -1 \end{array}\right]$$

$$\to \left[\begin{array}{ccc|c} 2\cdot 10^6 & -1 & 1 & 2\cdot 10^6 \\ 0 & 3.5 & 1.5 & 10^6 \\ 0 & 0 & 0.5714 & -0.5714\cdot 10^6 \end{array}\right] \to \left[\begin{array}{ccc|c} 2\cdot 10^6 & -1 & 0 & 3\cdot 10^6 \\ 0 & 3.5 & 0 & 3.5\cdot 10^6 \\ 0 & 0 & 1 & -1\cdot 10^6 \end{array}\right]$$

$$\to \left[\begin{array}{ccc|c} 2\cdot 10^6 & 0 & 0 & 4\cdot 10^6 \\ 0 & 1 & 0 & 1\cdot 10^6 \\ 0 & 0 & 1 & -1\cdot 10^6 \end{array}\right] \to \left[\begin{array}{ccc|c} 1 & 0 & 0 & 2 \\ 0 & 1 & 0 & 10^6 \\ 0 & 0 & 1 & -10^6 \end{array}\right]$$

Hence, $x_1 = 2$, $x_2 = 10^6$, $x_3 = -10^6$, which satisfies the original set of equations (5-5), in 4-digit arithmetic. ////

The above examples show that scaling a system of equations affects the choice of pivot element at each step. Since pivoting strategies are employed in the first place to reduce roundoff errors, it should now be clear that great care should be taken when preparing a system of equations to be solved by Gaussian elimination.

In general, it is desirable for the coefficients of the system to be of the same magnitude. This is of course not always possible to accomplish. Perhaps the most frequently used—and generally successful—method of scaling is to *equilibrate* the augmented matrix:

A matrix is said to be equilibrated if the element of largest magnitude in each row and in each column is between† 10^{-1} and 1.0. However, as Example 5-4 demonstrates, equilibration is not always successful. Moreover, there is more than one way to equilibrate any given matrix.

EXAMPLE 5-5 (*a*) If

$$[\mathbf{A}, \mathbf{b}] = \left[\begin{array}{ccc|c} 10^{-6} & 1 & 1 & 0 \\ 2 & -1 & 1 & 2 \\ 1 & 3 & 2 & 3 \end{array}\right]$$

then the following are both equilibrated scaled augmented matrices corresponding to $[\mathbf{A}, \mathbf{b}]$:

† On computers whose number base is other than decimal (base 10), this range on the magnitude of the largest element is from β^{-1} to 1.0, where β is the number base. In all our discussions, we have assumed for simplicity that $\beta = 10$.

$$\begin{bmatrix} 10^{-6} & 1 & 1 & \vdots & 0 \\ 1 & -0.5 & 0.5 & \vdots & 1 \\ 0.1 & 0.3 & 0.2 & \vdots & 0.3 \end{bmatrix} \qquad \begin{bmatrix} 1 & 1 & 1 & \vdots & 0 \\ 0.2 & -10^{-7} & 10^{-7} & \vdots & 0.2 \\ 0.1 & 3 \cdot 10^{-7} & 2 \cdot 10^{-7} & \vdots & 0.3 \end{bmatrix}$$

(*b*) If

$$[\mathbf{A}, \mathbf{b}] = \begin{bmatrix} 1 & 1 & 0 & \vdots & 10^8 \\ 2 & 3 & -10^8 & \vdots & 2 \cdot 10^8 \\ 0 & 4 & 4 \cdot 10^8 & \vdots & -3 \cdot 10^8 \end{bmatrix}$$

then the following are corresponding equilibrated augmented matrices:

$$\begin{bmatrix} 10^{-8} & 10^{-8} & 0 & \vdots & 1 \\ 2 \cdot 10^{-9} & 3 \cdot 10^{-9} & -1 & \vdots & 0.2 \\ 0 & 4 \cdot 10^{-9} & 0.4 & \vdots & -0.3 \end{bmatrix} \qquad \begin{bmatrix} 0.1 & 0.1 & 0 & \vdots & 0.1 \\ 0.2 & 0.3 & -0.1 & \vdots & 0.2 \\ 0 & 0.4 & 0.4 & \vdots & -0.3 \end{bmatrix} \qquad ////$$

Unfortunately, there is no known method of optimally equilibrating a matrix. Hence, in actual practice, one may have to resort to a trial-and-error approach until satisfactory results (in terms of the solution obtained by Gaussian elimination) are achieved.

5-4 ILL-CONDITIONED SYSTEMS

In the previous few sections we have investigated the consequences of roundoff errors in Gaussian elimination on the calculated solution of a system of equations. Another not unrelated problem is that of uncertainty in the elements of $[\mathbf{A}, \mathbf{b}]$. These elements may have been determined by imprecise experimental measuring techniques or by previous arithmetic computation on a computer or both (e.g., the least-squares example, 2-13) and hence perhaps there is some small but unknown error in the values of some of these elements.

Let us again begin our discussion with an example.

EXAMPLE 5-6 Consider the system $\mathbf{Ax} = \mathbf{b}$, where

$$\mathbf{A} = \begin{bmatrix} 41 & 40 \\ 40 & 39 \end{bmatrix} \qquad \mathbf{b} = \begin{bmatrix} 81 \\ 79 \end{bmatrix}$$

Clearly, the exact solution is $x_1 = x_2 = 1$ (and this solution will be found using 4-digit arithmetic).

However, suppose that there is a small error in $\mathbf{b}$, and that the true right-hand side should be $\mathbf{b}' = \begin{bmatrix} 80.99 \\ 79.01 \end{bmatrix}$. Then, the *exact* solution to $\mathbf{Ax}' = \mathbf{b}'$

is $x_1' = 1.790$, $x_2' = 0.1900$. Hence relative errors of approximately $\frac{1}{8000} = 1.25 \cdot 10^{-4}$ in the right-hand side have produced relative errors of $-\frac{0.79}{1.79} \approx -0.44$ and $\frac{0.81}{0.19} \approx 4.3$ in x_1 and x_2, respectively! This is truly a disaster, for if we only know the elements of **b** rounded to three decimal places, then our solution is absolutely meaningless, even if the elements of **A** are exact and if all computations are exact. ////

A system of equations $\mathbf{Ax} = \mathbf{b}$ in which relatively small changes in the elements of $[\mathbf{A}, \mathbf{b}]$ are capable of producing relatively large changes in the solution $\mathbf{x}$ is called an *ill-conditioned system.* As we shall see, whether or not a system is ill-conditioned depends only on its coefficient matrix **A**. Therefore, we shall also refer to such a matrix as an ill-conditioned matrix.

A system of equations in which relatively small changes in the elements of $[\mathbf{A}, \mathbf{b}]$ are capable of producing only relatively small changes in $\mathbf{x}$ is called a *well-conditioned system.* Fortunately, most systems of equations which arise from practical applications are well-conditioned.†

In the above definitions, we have used the phrase "is capable of" rather than "will" because in an ill-conditioned system it is possible that certain small changes in the elements of $[\mathbf{A}, \mathbf{b}]$ produce relatively small changes in $\mathbf{x}$. For instance, in the system of Example 5-6, if $\mathbf{b}' = \begin{bmatrix} 81.01 \\ 79.01 \end{bmatrix}$, then $x_1' = 1.010$, $x_2' = 0.990$. However, the fact that these changes are small is merely accidental, and the system is still ill-conditioned.

Let us examine this phenomenon in greater detail. Suppose $\mathbf{x}$ is the solution to $\mathbf{Ax} = \mathbf{b}$ and $\mathbf{x}'$ is the solution to $\mathbf{Ax}' = \mathbf{b}'$, where

$$\mathbf{A} = \begin{bmatrix} a_{11} & a_{12} \\ a_{21} & a_{22} \end{bmatrix} \qquad \mathbf{b} = \begin{bmatrix} b_1 \\ b_2 \end{bmatrix} \qquad \mathbf{x} = \begin{bmatrix} x_1 \\ x_2 \end{bmatrix}$$

and

$$\mathbf{b}' = \mathbf{b} + \boldsymbol{\varepsilon} = \begin{bmatrix} b_1 \\ b_2 \end{bmatrix} + \begin{bmatrix} \varepsilon_1 \\ \varepsilon_2 \end{bmatrix}$$

Then,

$$\mathbf{A}^{-1} = \frac{1}{\det(\mathbf{A})} \begin{bmatrix} a_{22} & -a_{12} \\ -a_{21} & a_{11} \end{bmatrix}$$

† One exception is the coefficient matrix arising from a least-squares problem, which is typically ill-conditioned.

and since $\mathbf{x} = \mathbf{A}^{-1}\mathbf{b}$,

$$\begin{aligned}\mathbf{x}' &= \mathbf{A}^{-1}\mathbf{b}' \\ &= \mathbf{A}^{-1}(\mathbf{b} + \boldsymbol{\varepsilon}) \\ &= \mathbf{A}^{-1}\mathbf{b} + \mathbf{A}^{-1}\boldsymbol{\varepsilon} \\ &= \mathbf{x} + \mathbf{A}^{-1}\boldsymbol{\varepsilon} \qquad (5\text{-}7)\end{aligned}$$

Hence,

$$\begin{aligned}\mathbf{x}' - \mathbf{x} &= \mathbf{A}^{-1}\boldsymbol{\varepsilon} \\ &= \frac{1}{\det(\mathbf{A})}\begin{bmatrix} a_{22} & -a_{12} \\ -a_{21} & a_{11}\end{bmatrix}\begin{bmatrix}\varepsilon_1 \\ \varepsilon_2\end{bmatrix} \\ &= \begin{bmatrix} \dfrac{a_{22}}{\det(\mathbf{A})}\varepsilon_1 - \dfrac{a_{12}}{\det(\mathbf{A})}\varepsilon_2 \\[2ex] \dfrac{-a_{21}}{\det(\mathbf{A})}\varepsilon_1 + \dfrac{a_{11}}{\det(\mathbf{A})}\varepsilon_2 \end{bmatrix} \qquad (5\text{-}8)\end{aligned}$$

Equation (5-8) indicates that small changes in $\mathbf{b}$ (i.e., ε_1 and ε_2) can be greatly magnified (in terms of errors in $\mathbf{x}$) if any of the factors $a_{ij}/\det(\mathbf{A})$ is very large. These factors are precisely the elements of $\mathbf{A}^{-1}$. It should also be clear from Eq. (5-8) that scaling will be of no help whatsoever.

A more precise measure of the conditioning of a matrix $\mathbf{A}$ is called the *condition number* of $\mathbf{A}$, denoted by cond $(\mathbf{A})$. In order to define cond $(\mathbf{A})$, we first define:

1 The *norm* of any real n-component vector $\mathbf{x} = [x_1 \quad x_2 \quad \cdots \quad x_n]^T$, denoted by $\|x\|$, is

$$\|x\| = (\mathbf{x}^T\mathbf{x})^{1/2}$$

2 The norm of any real square matrix $\mathbf{A}$, $\|\mathbf{A}\|$, is

$$\|\mathbf{A}\| = \max_{x \neq 0} \frac{\|\mathbf{Ax}\|}{\|\mathbf{x}\|} \qquad (5\text{-}9)$$

We now define

$$\text{cond}(\mathbf{A}) = \|\mathbf{A}\| \, \|\mathbf{A}^{-1}\| \qquad (5\text{-}10)$$

The norm of a vector is merely its length, and the norm of a matrix $\|\mathbf{A}\|$ is the maximum of the ratios of the norms of all vectors $\mathbf{y} = \mathbf{Ax}$ with the norm of $\mathbf{x}$. Thus, if $\mathbf{A}$ has any very large elements, $\|\mathbf{A}\|$ will tend to be large also. If we think of $\mathbf{A}$ as being a transformation (or "mapping"), then

A transforms (or "maps") the vector **x** into the vector **y**; hence,

$$\frac{\|\mathbf{Ax}\|}{\|\mathbf{x}\|} = \frac{\|\mathbf{y}\|}{\|\mathbf{x}\|}$$

is a measure of how much this transformation "stretched"—or "shrunk"—the vector **x** in transforming it to **y**, and the norm of **A** is the maximum such stretch over all nonzero **x**.

It can be shown (see, e.g., Reference [1]) that cond $(\mathbf{A}) \geq 1$ for any matrix **A**; in general, the larger cond (**A**), the more ill-conditioned **A** is.

As an appetite whetter for the material in Chaps. 9 and 10, we note also that cond (**A**) is directly related to its eigenvalues. More specifically, it can be shown (see, e.g., Reference [1]) that cond (**A**) is equal to the square root of the ratio of the largest eigenvalue† of $\mathbf{AA}^T$ to its smallest eigenvalue.

We turn now to a geometric interpretation of ill-conditioned systems. If we graph the system of Example 5-6, we find that the two equations are nearly parallel lines. This illustrates why a small change in **b** can lead to a large change in **x**, as the intersection of the two lines can move quite a bit even for small changes.

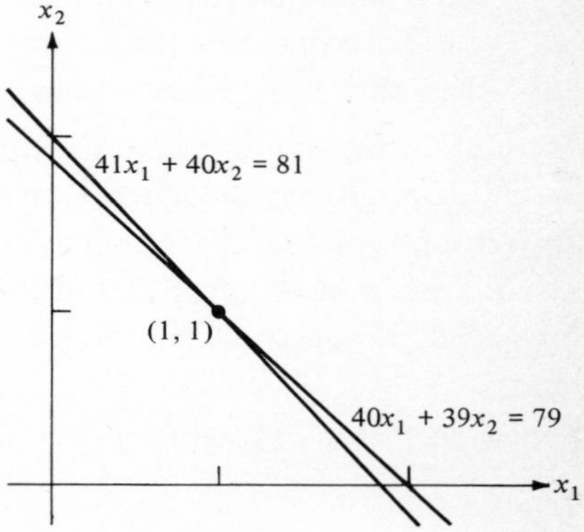

Unfortunately, for a large system of equations, we cannot visually determine whether or not it is ill-conditioned. Moreover, unless we calculate $\mathbf{A}^{-1}$ or the largest and smallest eigenvalues of **A**, we will also not be able to compute cond (**A**). Hence, if one suspects the existence of an ill-conditioned system, some additional investigation of this type should be performed in addition to merely computing a "solution"; as we have seen, such a "solution" may have no relation to the actual solution we desire.

† The eigenvalues of $\mathbf{AA}^T$ are always nonnegative real numbers for any matrix **A**.

5-5 ITERATIVE IMPROVEMENT OF A SOLUTION

In Secs. 5-2 and 5-3 we investigated the effects of roundoff error on Gaussian elimination and in Sec. 5-4 we discovered that solving an ill-conditioned system even with perfect arithmetic may produce disastrous results if there is uncertainty in any of the elements of the augmented matrix.

The question naturally arises: What happens if we must solve an ill-conditioned system on a computer and hence introduce roundoff errors into the calculations?

One commonly used method of reducing the effects of roundoff error on a digital computer is to instruct the computer to perform all calculations in *double-precision* arithmetic. Essentially, this means that the computer will double the allowable number of digits that it can store for any given number. In our hypothetical computer with 4-digit arithmetic, double precision would mean that all numbers could have as many as eight digits and all arithmetic calculations would thus be with eight-digit numbers.† Thus, we would normally expect to achieve greater accuracy through the use of double-precision calculations. The disadvantages of double-precision arithmetic are:

1 Each double-precision number requires twice as much storage space as its single-precision counterpart.

2 On most computers double-precision arithmetic is substantially slower than single-precision arithmetic.

Thus, we should not automatically use double-precision arithmetic, but should carefully consider whether our particular problem can be acceptably solved without the use of double-precision arithmetic. However, when in doubt, the author strongly suggests that double precision be employed (provided, of course, that sufficient storage space is available).

EXAMPLE 5-7 Consider the system whose augmented matrix is

$$[\mathbf{A}, \mathbf{b}] = \left[\begin{array}{cc:c} 1.019 & 1.556 & 2.520 \\ 0.2035 & 0.3112 & 0.5028 \end{array}\right]$$

The correct solution is $x_1 = 4$, $x_2 - 1$. Using 4-digit arithmetic and solving by Gaussian elimination yields:

$$\left[\begin{array}{cc:c} 1.019 & 1.556 & 2.520 \\ 0 & 0.0005 & -0.0004 \end{array}\right] \to \left[\begin{array}{cc:c} 1.019 & 1.556 & 2.520 \\ 0 & 1 & -1.200 \end{array}\right]$$

$$\to \left[\begin{array}{cc:c} 1.019 & 0 & 4.387 \\ 0 & 1 & -1.2 \end{array}\right] \to \left[\begin{array}{cc:c} 1 & 0 & 4.305 \\ 0 & 1 & -1.200 \end{array}\right]$$

† In most computers, double precision actually provides slightly more than twice the number of significant digits, as a result of the way computers store double-precision numbers.

Hence, 4-digit arithmetic did not yield a very good solution. However, if we solve the same system in 8-digit arithmetic, we obtain:

$$\left[\begin{array}{cc|c} 1.019 & 1.556 & 2.520 \\ 0.235 & 0.3112 & 0.5028 \end{array}\right] \rightarrow \left[\begin{array}{cc|c} 1.0190000 & 1.5560000 & 2.5200000 \\ 0 & 0.00045801 & -0.00045809 \end{array}\right]$$

$$\rightarrow \left[\begin{array}{cc|c} 1.0190000 & 0 & 4.0762718 \\ 0 & 1 & -1.0001747 \end{array}\right] \rightarrow \left[\begin{array}{cc|c} 1 & 0 & 4.0002667 \\ 0 & 1 & -1.0001747 \end{array}\right]$$

We see that the 8-digit arithmetic solution is accurate to four significant digits, a substantial improvement. ////

As we have noted above, it is not desirable to use double precision all the time. Below we shall discuss a method for improving a solution obtained by Gaussian elimination with single-precision arithmetic.

Suppose we have solved the system of equations $\mathbf{Ax} = \mathbf{b}$ by Gaussian elimination and obtained $\mathbf{x}^{(0)}$ as the solution. Suppose further that the true solution is $\mathbf{x}^*$, and let

$$\boldsymbol{\varepsilon}^{(0)} = \mathbf{x}^* - \mathbf{x}^{(0)} \tag{5-11}$$

Hence, $\mathbf{x}^* = \boldsymbol{\varepsilon}^{(0)} + \mathbf{x}^{(0)}$

and

$$\begin{aligned} \mathbf{b} &= \mathbf{Ax}^* \\ &= \mathbf{A}(\boldsymbol{\varepsilon}^{(0)} + \mathbf{x}^{(0)}) \\ &= \mathbf{A}\boldsymbol{\varepsilon}^{(0)} + \mathbf{Ax}^{(0)} \end{aligned}$$

or

$$\mathbf{A}\boldsymbol{\varepsilon}^{(0)} = \mathbf{b} - \mathbf{Ax}^{(0)} \tag{5-12}$$

The vector $\boldsymbol{\varepsilon}^{(0)}$, representing the errors in our solution $\mathbf{x}^{(0)}$, is of course unknown, since we do not know the true solution $\mathbf{x}^*$. However, the right-hand side of Eq. (5-12) is known. We define

$$\mathbf{r}^{(0)} = \mathbf{b} - \mathbf{Ax}^{(0)}$$

The vector $\mathbf{r}^{(0)}$ is called the *residual*. Thus we see that Eq. (5-12) is actually a system of equations whose variables are the components of $\boldsymbol{\varepsilon}^{(0)}$:

$$\mathbf{A}\boldsymbol{\varepsilon}^{(0)} = \mathbf{r}^{(0)}$$

If we solve this system by Gaussian elimination to obtain $\boldsymbol{\varepsilon}^{(0)}$, we will then have some estimate of the magnitude of the errors in $\mathbf{x}^{(0)}$. (We say "estimate" since there will of course be some errors introduced in the computation of $\boldsymbol{\varepsilon}^{(0)}$.)

We can then use our computed $\boldsymbol{\varepsilon}^{(0)}$ to obtain an improved estimate of $\mathbf{x}^*$, denoted by $\mathbf{x}^{(1)}$ and defined by

$$\mathbf{x}^{(1)} = \mathbf{x}^{(0)} + \boldsymbol{\varepsilon}^{(0)}$$

Moreover, we can repeat this process by solving the system of equations $\mathbf{A}\boldsymbol{\varepsilon}^{(1)} = \mathbf{r}^{(1)}$, where $\boldsymbol{\varepsilon}^{(1)} = \mathbf{x}^* - \mathbf{x}^{(1)}$ and $\mathbf{r}^{(1)} = \mathbf{b} - \mathbf{A}\mathbf{x}^{(1)}$.

We can continue in this manner, obtaining continually improved solutions $\mathbf{x}^{(k)}$ by solving sequentially the systems

$$\left.\begin{aligned} &\mathbf{A}\boldsymbol{\varepsilon}^{(k)} = \mathbf{r}^{(k)} \\ \text{where} \qquad &\mathbf{x}^{(k)} = \boldsymbol{\varepsilon}^{(k-1)} + \mathbf{x}^{(k-1)} \\ \text{and} \qquad &\mathbf{r}^{(k)} = \mathbf{b} - \mathbf{A}\mathbf{x}^{(k)} \end{aligned}\right\} k = 1, 2, \ldots \tag{5-13}$$

Our goal is to make the errors as small as possible, so that the above iterative procedure terminates at step p if the components of $\boldsymbol{\varepsilon}^{(p)}$ are all acceptably small. Thus, for example, if we desire a solution correct to four decimal places, then we continue calculating $\boldsymbol{\varepsilon}^{(k)}$ until all its components are less than $5 \cdot 10^{-5}$ in magnitude. The precise stopping rule one uses will depend on how one defines an acceptable solution, and this of course depends on the particular application.

The procedure we have just described is called *iterative improvement*.

We conclude the discussion with several comments:

Comment 1. Each system of equations (5-13) which we solve has the same coefficient matrix $\mathbf{A}$. Hence, it is not necessary to perform a complete Gaussian elimination each time. We need only to save the multipliers we have used at each step of the Gaussian elimination and repeat this same sequence of operations on the new right-hand side $\mathbf{r}^{(k)}$.

Comment 2. It is necessary that we use double precision arithmetic on all operations involving $\mathbf{r}^{(k)}$, since these numbers will be small compared to the elements of $\mathbf{A}$. However, it is usually not necessary to use double precision in performing the original Gaussian elimination on $\mathbf{A}\mathbf{x} = \mathbf{b}$.

Comment 3. The method of iterative improvement will almost always converge (i.e., the $\boldsymbol{\varepsilon}^{(k)}$ will become increasingly smaller) unless the system is extremely ill-conditioned. In the latter case, we should use double precision in solving $\mathbf{A}\mathbf{x} = \mathbf{b}$.

EXAMPLE 5-8 In solving the system of Example 5-7 in 4-digit arithmetic we obtained $\mathbf{x}^{(0)} = \begin{bmatrix} 4.305 \\ -1.200 \end{bmatrix}$. Hence, $\mathbf{r}^{(0)} = \mathbf{b} - \mathbf{A}\mathbf{x}^{(0)} = \begin{bmatrix} 4.050 \cdot 10^{-4} \\ 1.725 \cdot 10^{-4} \end{bmatrix}$

In solving that system we

1 Added $m_1 = -0.1997$ times row 1 to row 2
2 Divided row 2 by $d_1 = 5.000 \cdot 10^{-4}$
3 Added $m_2 = -1.556$ times row 2 to row 1
4 Divided row 1 by $d_2 = 1.019$

Let us perform this sequence of steps on $\mathbf{r}^{(0)}$, in 8-digit arithmetic:

1 $m_1(4.050 \cdot 10^{-4}) + 1.725 \cdot 10^{-4} = 0.916215 \cdot 10^{-4}$

2 $\dfrac{0.916215 \cdot 10^{-4}}{d_1} = 0.183243 = \varepsilon_2^{(0)}$

3 $m_2(0.183243) + 4.050 \cdot 10^{-4} = -0.28472111$

4 $\dfrac{-0.28472111}{d_2} = -0.27941228 = \varepsilon_1^{(0)}$

Hence,

$$\boldsymbol{\varepsilon}^{(0)} = \begin{bmatrix} -0.27941228 \\ 0.18324300 \end{bmatrix}$$

and

$$\mathbf{x}^{(1)} = \mathbf{x}^{(0)} + \boldsymbol{\varepsilon}^{(0)} = \begin{bmatrix} 4.0255877 \\ -1.0167570 \end{bmatrix}$$

$$\mathbf{r}^{(1)} = \mathbf{b} - \mathbf{A}\mathbf{x}^{(1)} = \begin{bmatrix} 0 \\ 0.768 \cdot 10^{-5} \end{bmatrix}$$

Repeating the sequence of steps 1 to 4 on $\mathbf{r}^{(1)}$ yields:

$$\boldsymbol{\varepsilon}^{(1)} = \begin{bmatrix} -0.023454524 \\ 0.015360000 \end{bmatrix}$$

Therefore,

$$\mathbf{x}^{(2)} = \mathbf{x}^{(1)} + \boldsymbol{\varepsilon}^{(1)} = \begin{bmatrix} 4.0021332 \\ -1.0013970 \end{bmatrix}$$

Upon performing one more iteration we obtain

$$\mathbf{r}^{(2)} = \mathbf{b} - \mathbf{A}\mathbf{x}^{(2)} = \begin{bmatrix} 0 \\ 0.64 \cdot 10^{-7} \end{bmatrix}$$

and

$$\boldsymbol{\varepsilon}^{(2)} = \begin{bmatrix} -1.9545437 \cdot 10^{-3} \\ 1.2800000 \cdot 10^{-3} \end{bmatrix}$$

$$\mathbf{x}^{(3)} = \mathbf{x}^{(2)} + \boldsymbol{\varepsilon}^{(2)} = \begin{bmatrix} 4.0001787 \\ -1.0001170 \end{bmatrix}$$

Thus, in three iterations we have found a solution accurate to four significant digits, beginning with a solution that had only one significant digit of accuracy. Since the original problem contains data with only four digits, we will terminate the procedure at this point and call $\mathbf{x}^{(3)}$ the desired solution. ////

REFERENCES

1. FORSYTHE, GEORGE E., and CLEVE B. MOLER: "Computer Solution of Linear Algebraic Systems," Prentice-Hall, Englewood Cliffs, N.J., 1967.
2. NOBLE, BEN: "Applied Linear Algebra," Prentice-Hall, Englewood Cliffs, N.J., 1969.
3. STARK, PETER A., "Introduction to Numerical Methods," Collier-Macmillan, Canada, Ltd., Toronto, 1970.
4. WILKINSON, J. H., "Rounding Errors in Algebraic Processes," Prentice-Hall, Englewood Cliffs, N.J., 1963.

EXERCISES

Perform all calculations in 3-digit arithmetic unless otherwise directed.

1 Consider the series

$$S = \sum_{k=0}^{\infty} 9^{-k}$$

(*a*) Determine the exact value of S.
(*b*) Add the first seven terms of S from largest to smallest (i.e., from $k = 0$ to $k = 6$).
(*c*) Add the first seven terms of S from smallest to largest.
(*d*) Compare the absolute and relative errors obtained in (*b*) and (*c*).

2 Consider the system

$$\begin{bmatrix} 10^{-4} & 1 & 1 \\ 0.1 & 2 & 1 \\ 1 & 10^{-4} & 0 \end{bmatrix} \begin{bmatrix} x_1 \\ x_2 \\ x_3 \end{bmatrix} = \begin{bmatrix} 1.0 \\ 1.1 \\ 1.0 \end{bmatrix}$$

(*a*) Solve the system without a pivoting strategy.
(*b*) Solve the system using partial pivoting.
(*c*) Solve the system using complete pivoting.

3 Explain how the method of iterative improvement can be used to estimate the accuracy of a given solution. Illustrate on the solution obtained in Exercise 2, part (*b*) or (*c*).

4 Consider the system $\mathbf{Ax} = \mathbf{b}$:

$$\begin{bmatrix} 0.51 & 0.50 \\ 0.50 & 0.49 \end{bmatrix} \begin{bmatrix} x_1 \\ x_2 \end{bmatrix} = \begin{bmatrix} 1.01 \\ 0.99 \end{bmatrix}$$

The correct solution is $x_1 = x_2 = 1$.
(*a*) Show graphically that the system is ill-conditioned.
(*b*) Demonstrate algebraically that the system is ill-conditioned by choosing small changes in **b** which produce a large charge in the solution.

5 Consider the system $\mathbf{Ax} = \mathbf{b}$:

$$\begin{bmatrix} 0.51 & 0.50 \\ 0.50 & -0.49 \end{bmatrix} \begin{bmatrix} x_1 \\ x_2 \end{bmatrix} = \begin{bmatrix} 1.03 \\ 1.99 \end{bmatrix}$$

Demonstrate that this system is well-conditioned by showing that small changes in **b** can produce only changes of the same magnitude in **x**.

6 Show that the system

$$\begin{bmatrix} 0.51+\varepsilon & 0.50 \\ 0.50 & 0.49 \end{bmatrix} \begin{bmatrix} x_1 \\ x_2 \end{bmatrix} = \begin{bmatrix} 1.01 \\ 0.99 \end{bmatrix}$$

is ill-conditioned by studying the possible effects on **x** of small changes in $a_{11} = 0.51 + \varepsilon$.

7 Solve the following system first in 4-digit arithmetic and then in double precision (8-digit arithmetic):

$$\begin{bmatrix} 1.28 & 0.98 \\ 0.13 & 0.10 \end{bmatrix} \begin{bmatrix} x_1 \\ x_2 \end{bmatrix} = \begin{bmatrix} 2.26 \\ 0.23 \end{bmatrix}$$

8 Beginning with the solution obtained by 4-digit arithmetic in Exercise 7, use the method of iterative improvement to obtain a solution exact to three significant digits.

9 Using the computer program for Gaussian elimination written in Chap. 4 (Exercise 25), solve the following system of equations:

$$[\mathbf{A}, \mathbf{b}] = \left[\begin{array}{cccc|c} 0.23145 & 3.27634 & 1.32173 & 4.12112 & 8.95064 \\ 0.34717 & 4.91451 & 1.98260 & 2.23141 & 9.47569 \\ 5.14788 & 3.27554 & 29.3977 & 1.75356 & 39.57468 \\ 3.57650 & 0.98522 & 10.9317 & 5.43274 & 20.92616 \end{array}\right]$$

Discuss your results (the correct answer is $x_1 = x_2 = x_3 = x_4 = 1$).

10 Rewrite your computer program for Gaussian elimination, incorporating a partial pivoting strategy. Then, re-solve the problem of Exercise 9 and compare the results.

11 Rewrite your computer program for Gaussian elimination, incorporating a complete pivoting strategy. Then, re-solve the problem of Exercise 9 and compare the results.

12 Let m_{ik} denote the multiple of row k added to row i in the kth step of the forward pass of Gaussian elimination, $k = 1, 2, \ldots, n$ and $i = k+1, k+2, \ldots, n$. Further, for $k = 1, 2, \ldots, n$, and $i = 1, 2, \ldots, k-1$, let m_{ik} denote the multiple of row $n-k+1$ added to row i in the kth step of the backward pass. Also, let m_{ii} be the number row i is divided by at the conclusion of Gaussian elimination. Explain how the matrix $\mathbf{M} = [m_{ik}]$ can be used efficiently with the method of iterative improvement.

13 Write a computer program for the method of iterative improvement and use it to solve the problem of Exercise 9, correct to six significant digits.

6

SOLUTIONS OF SIMULTANEOUS EQUATIONS: m EQUATIONS IN n VARIABLES

6-1 RANK

In Chap. 4 we investigated systems of simultaneous equations in which the coefficient matrix **A** was square and nonsingular. In this chapter we shall extend the results obtained in Chap. 4 to systems of equations in which either **A** is square and singular or **A** is not square (which implies the number of equations in the system is not equal to the number of variables). Actually, if we consider a general system of m equations in n variables, then the case we studied in Chap. 4 is just the special case $m = n$ and $\det(\mathbf{A}) \neq 0$.

If **A** is a general $m \times n$ matrix, then $\det(\mathbf{A})$ is undefined unless $m = n$; hence the first item on our agenda for generalizing the results of Chap. 4 might be to somehow generalize the concept of determinants for nonsquare matrices, in a meaningful way. Let us proceed to do so.

We begin by observing that, as far as a system of n equations in n variables, $\mathbf{Ax} = \mathbf{b}$, is concerned, the key property of $\det(\mathbf{A})$ is whether $\det(\mathbf{A})$ is zero or nonzero; that is, whether **A** is singular or nonsingular.

This fact suggests the following definition:

If **A** is an $m \times n$ matrix, then the *rank of* **A**, denoted rank (**A**), is defined to be the order of the largest nonsingular square submatrix of **A**.

A submatrix of **A** is a matrix obtained from **A** by deleting some of its rows and/or some of its columns. For example, if

$$\mathbf{A} = \begin{bmatrix} 1 & 0 & 1 & 2 \\ -1 & 2 & -2 & 3 \\ 2 & -2 & 3 & -1 \end{bmatrix}$$

then

$$\begin{bmatrix} 1 & 0 & 1 \\ -1 & 2 & -2 \\ 2 & -2 & 3 \end{bmatrix} \quad \begin{bmatrix} 1 & 0 & 2 \\ -1 & 2 & 3 \\ 2 & -2 & -1 \end{bmatrix}$$

$$\begin{bmatrix} 1 & 1 & 2 \\ -1 & -2 & 3 \\ 2 & 3 & -1 \end{bmatrix} \quad \begin{bmatrix} 0 & 1 & 2 \\ 2 & -2 & 3 \\ -2 & 3 & -1 \end{bmatrix}$$

are all the submatrices of **A** of order 3. The reader may verify that each of the above matrices is singular; hence, rank (**A**) is at most 2. Moreover, since the submatrix $\begin{bmatrix} 1 & 0 \\ -1 & 2 \end{bmatrix}$ is nonsingular, we see that rank (**A**) = 2.

If **A** is square of order n, then (a) rank (**A**) = n, if **A** is nonsingular, or (b) rank (**A**) = $(n - 1)$ or less, if **A** is singular.

Thus, our definition of the rank of a matrix is tied directly to the concept of determinants. Moreover, recall from Chap. 3 that if we perform any sequence of elementary operations (see Sec. 3-4) on a nonsingular matrix, the resulting matrix is also nonsingular. And if we perform any sequence of elementary operations on a singular matrix, the resulting matrix is also singular. Therefore, if we perform a sequence of elementary operations on any matrix **A**, not necessarily square, to obtain a new matrix **B**, then the largest nonsingular square submatrix of **B** will be of the same order as the largest nonsingular square submatrix of **A**. Hence, the rank of **B** will be equal to the rank of **A**. This is a very important result, and so we shall state it formally as:

Theorem 6-1 Performing elementary operations on a matrix **A** yields a matrix **B** such that rank (**A**) = rank (**B**).

Theorem 6-1 provides us with a tool for computing the rank of a given matrix, as illustrated in the following example:

EXAMPLE 6-1 Let

$$\mathbf{A} = \begin{bmatrix} 1 & 0 & -1 & 2 \\ 1 & 1 & 1 & -1 \\ 0 & -1 & -2 & 3 \\ 5 & 2 & -1 & 4 \\ -1 & 2 & 5 & -8 \end{bmatrix}$$

Adding appropriate multiples of row 1 to rows 2, 3, 4, and 5 yields

$$\begin{bmatrix} 1 & 0 & -1 & 2 \\ 0 & 1 & 2 & -3 \\ 0 & -1 & -2 & 3 \\ 0 & 2 & 4 & -6 \\ 0 & 2 & 4 & -6 \end{bmatrix}$$

Adding appropriate multiples of row 2 to rows 3, 4, and 5 yields

$$\begin{bmatrix} 1 & 0 & -1 & 2 \\ 0 & 1 & 2 & -3 \\ 0 & 0 & 0 & 0 \\ 0 & 0 & 0 & 0 \\ 0 & 0 & 0 & 0 \end{bmatrix}$$

It is clear that the rank of the above matrix is 2, since any submatrix of order 3 or more will contain at least one row of zeros. Hence, rank $(\mathbf{A}) = 2$, also ////

EXAMPLE 6-2 Let

$$\mathbf{A} = \begin{bmatrix} 1 & 1 & 0 & -1 & 1 \\ 1 & 1 & 1 & 3 & 2 \\ 2 & 2 & 1 & 2 & 3 \\ 1 & 1 & -2 & -9 & -1 \end{bmatrix}$$

Adding appropriate multiples of row 1 to rows 2, 3, and 4 yields

$$\begin{bmatrix} 1 & 1 & 0 & -1 & 1 \\ 0 & 0 & 1 & 4 & 1 \\ 0 & 0 & 1 & 4 & 1 \\ 0 & 0 & -2 & -8 & -2 \end{bmatrix}$$

$$\begin{bmatrix} 1 & 1 & 0 & -1 & 1 \\ 0 & 0 & 1 & 4 & 2 \\ 0 & 0 & 0 & 0 & 0 \\ 0 & 0 & 0 & 0 & 0 \end{bmatrix}$$

Again, it is now clear that rank $(\mathbf{A}) = 2$. ////

Theorem 6-2 If $\mathbf{A}$ is an $m \times n$ matrix of rank† r, then there exists a sequence of elementary row operations which will yield a matrix whose last $(m - r)$ rows contain only zeros.

† From the definition of rank, it is obvious that $r \le m$ and $r \le n$; i.e., $r \le$ minimum $\{m, n\}$.

PROOF Since $\mathbf{A}$ is of rank r, it must have at least one nonsingular submatrix of order r. Suppose for simplicity of notation that this submatrix consists of the first r rows and the first r columns of $\mathbf{A}$. Let us partition $\mathbf{A}$ as follows:

$$\mathbf{A} = \left[\begin{array}{c|c} \mathbf{B}_{r \times r} & \mathbf{C}_{r \times (n-r)} \\ \hline \mathbf{F}_{(m-r) \times r} & \mathbf{G}_{(m-r) \times (n-r)} \end{array}\right] \tag{6-1}$$

Thus, $\mathbf{B}$ is the nonsingular submatrix referred to above. Let us now perform the following variation of the forward pass of Gaussian elimination to $\mathbf{A}$:

For $k = 1, 2, \ldots, r$, add appropriate multiples of row k to rows $k+1, k+2, \ldots, r, \ldots, m$, so that the elements of column k in the above rows are all zero. (As in Gaussian elimination, it might be necessary to interchange some of the rows $1, 2, \ldots, r$ at various stages.) At the conclusion of step r, the resulting matrix $\hat{\mathbf{A}}$ will be of the form

$$\hat{\mathbf{A}} = \left[\begin{array}{c|c} \mathbf{T}_{r \times r} & \hat{\mathbf{C}}_{r \times (n-r)} \\ \hline \mathbf{O}_{(m-r) \times r} & \hat{\mathbf{G}}_{(m-r) \times (n-r)} \end{array}\right] \tag{6-2}$$

where $\mathbf{T}$ is upper triangular with all nonzero diagonal elements.

Now, to prove the theorem, we must show that $\hat{\mathbf{G}} = \mathbf{O}$. Let $\hat{\mathbf{G}} = [\hat{g}_{r+i,\,r+j}]$, so that the (i, j) element of $\hat{\mathbf{G}}$ is the $(r+i, r+j)$ element of $\hat{\mathbf{A}}$. Let $\hat{g}_{r+p,\,r+q}$ be any element of $\mathbf{G}$. Then, the $(r+1)$st-order submatrix $\mathbf{S}$ of $\hat{\mathbf{A}}$ formed by rows $1, 2, \ldots, r$ and $r+p$, and by columns $1, 2, \ldots, r$ and $r+q$ is

$$\mathbf{S} = \begin{bmatrix} \mathbf{T} & \hat{\mathbf{c}}_{r+q} \\ \mathbf{0} & \hat{g}_{r+p,\,r+q} \end{bmatrix} \tag{6-3}$$

where $\hat{\mathbf{c}}_{r+q}$ denotes the qth column of $\hat{\mathbf{C}}$. Since $\mathbf{S}$ is upper triangular, the determinant of $\mathbf{S}$ is equal to the product of its diagonal elements. Moreover, by Theorem 6-1, rank $(\mathbf{A}) =$ rank $(\hat{\mathbf{A}}) = r$ and therefore det $(\mathbf{S})$ must equal zero (since $\mathbf{S}$ is of order $r+1$). But, det $(\mathbf{S}) = 0$ if and only if $\hat{g}_{r+p,\,r+q} = 0$. Since we have assumed nothing about p and q, this argument is valid for all the elements of $\hat{\mathbf{G}}$. Therefore $\hat{\mathbf{G}} = \mathbf{O}$ and the theorem is proved.

QED

One additional comment about the above proof. Initially we assumed that the first r rows and the first r columns of $\mathbf{A}$ formed a nonsingular matrix. If such is not the case, then the proof of this theorem must be modified slightly. We can always interchange some of the rows of $\mathbf{A}$ so that the nonsingular submatrix of order r consists of the first r rows of the resulting matrix, since row interchanges are elementary row operations. However, it may also be necessary to interchange *columns* of $\mathbf{A}$ so that the first r rows and first r columns of the resulting matrix $\tilde{\mathbf{A}}$ form a nonsingular matrix. We can then apply the argument used in the proof of Theorem 6-2 to $\tilde{\mathbf{A}}$ to show that $\tilde{\mathbf{A}}$ can be reduced by elementary row operations to a matrix whose last $(m - r)$ rows contain only zeros. Then, if we reinterchange the columns (so that they correspond to the original columns of $\mathbf{A}$), we will have rigorously completed the proof of Theorem 6-2.

In fact, the discussion above leads us directly to:

Theorem 6-3 If $\mathbf{A}$ is an $m \times n$ matrix of rank r, then there exists a sequence of elementary row operations and (possibly) column interchanges which will yield a matrix of the following form:

$$\left[\begin{array}{c|c} \mathbf{I}_r & \mathbf{C}_{r \times (n-r)} \\ \hline \mathbf{O}_{(m-r) \times r} & \mathbf{O}_{(m-r) \times (n-r)} \end{array}\right]$$

The details of the proof of Theorem 6-3 are left as an exercise for the reader.

It is now time to turn our attention toward applying the results of this section to systems of equations.

6-2 SOLUTIONS TO SYSTEMS OF EQUATIONS

Consider a system of m equations in n variables, $\mathbf{Ax} = \mathbf{b}$. If the rank of $\mathbf{A}$ is r, then by Theorem 6-3, we can by a sequence of elementary row operations and (possibly) column interchanges, reduce the coefficient matrix $\mathbf{A}$ to

$$\hat{\mathbf{A}} = \begin{bmatrix} \mathbf{I}_r & \mathbf{C}_{r \times (n-r)} \\ \mathbf{O}_{(m-r) \times r} & \mathbf{O}_{(m-r) \times (n-r)} \end{bmatrix}$$

Moreover, if we were to perform the same sequence of elementary operations on the augmented matrix $[\mathbf{A}, \mathbf{b}]$, it is clear that we would obtain

$$[\hat{\mathbf{A}}, \hat{\mathbf{b}}] = \left[\begin{array}{cc|c} \mathbf{I}_r & \mathbf{C}_{r \times (n-r)} & \mathbf{f}_r \\ \mathbf{O}_{(m-r) \times r} & \mathbf{O}_{(m-r) \times (n-r)} & \mathbf{g}_{(m-r)} \end{array}\right] \tag{6-4}$$

where $\mathbf{f}$ and $\mathbf{g}$ are r-component and $(m - r)$-component vectors, respectively.

Recall from Chap. 4 that if we perform elementary *row* operations on an augmented matrix $[\mathbf{A}, \mathbf{b}]$, we obtain a new matrix which represents the augmented matrix of an equivalent system of equations; that is, a system of equations whose solutions are the same as those of the system $\mathbf{Ax} = \mathbf{b}$.

However, in obtaining the augmented matrix $[\hat{\mathbf{A}}, \hat{\mathbf{b}}]$ of Eq. (6-4), we noted that it might have been necessary to interchange some of the columns of $\mathbf{A}$. Since each of these columns corresponds to a variable, we can still consider $[\hat{\mathbf{A}}, \hat{\mathbf{b}}]$ to represent a system of equations equivalent to $[\mathbf{A}, \mathbf{b}]$ provided that we interchange the corresponding two *variables* whenever we interchange two columns. For example, if $\mathbf{x} = [x_1 \quad x_2 \quad \cdots \quad x_n]^T$ and in going from $[\mathbf{A}, \mathbf{b}]$ to $[\hat{\mathbf{A}}, \hat{\mathbf{b}}]$ it was necessary to interchange columns p and q, then the solution vector for $[\hat{\mathbf{A}}, \hat{\mathbf{b}}]$ would be

$$\hat{\mathbf{x}} = [x_1 \quad x_2 \quad \cdots \quad x_{p-1} \quad x_q \quad x_{p+1} \quad \cdots \quad x_{q-1} \quad x_p \quad x_{q+1} \quad \cdots \quad x_n]^T.$$

With the above in mind, it should be clear from Eq. (6-4) that the system $\hat{\mathbf{A}}\hat{\mathbf{x}} = \hat{\mathbf{b}}$ (and hence $\mathbf{Ax} = \mathbf{b}$) possesses solutions if and only if $\mathbf{g} = \mathbf{0}$. For, if any $g_i \neq 0$, the corresponding equation would be $\mathbf{0}^T\hat{\mathbf{x}} = g_i \neq 0$, which obviously has no solution. A system of equations which possesses no solutions is said to be *inconsistent;* a system of equations with at least one solution is said to be *consistent.*

Theorem 6-4 A system of equations $\mathbf{Ax} = \mathbf{b}$ is consistent if and only if rank $(\mathbf{A})$ = rank $[\mathbf{A}, \mathbf{b}]$.

PROOF Suppose rank $(\mathbf{A}) = r$. Then, by the discussion above, there exists an equivalent system of equations $\hat{\mathbf{A}}\hat{\mathbf{x}} = \hat{\mathbf{b}}$ such that $[\hat{\mathbf{A}}, \hat{\mathbf{b}}]$ is as given in Eq. (6-4). Moreover, we see from Eq. (6-4) that rank $[\hat{\mathbf{A}}, \hat{\mathbf{b}}] = r$ if and only if $\mathbf{g} = \mathbf{0}$; for suppose some $g_p \neq 0$, where g_p is in the $(r + p)$th row of $\mathbf{A}$, then the $(r + 1)$st-order submatrix of $\mathbf{A}$ formed by rows $1, 2, \ldots, r$ and $r + p$ and by columns $1, 2, \ldots, r$ and $n + 1$, is clearly nonsingular, which implies that rank $[\hat{\mathbf{A}}, \hat{\mathbf{b}}] = r + 1$. On the other hand, if $\mathbf{g} = \mathbf{0}$, then it is clear that rank $[\hat{\mathbf{A}}, \hat{\mathbf{b}}] = r$. Thus, since we have already discovered that $\mathbf{A}\hat{\mathbf{x}} = \hat{\mathbf{b}}$ is consistent if and only if $\hat{\mathbf{g}} = \mathbf{0}$, we have completed the required proof. QED

Note that Theorem 6-4 does not impose any restrictions on the order of the coefficient matrix $\mathbf{A}$ (and hence on the number of equations and/or variables); the theorem is valid for *any* system of equations. We will make great use of Theorems 6-3 and 6-4 in the remainder of this chapter.

By combining Theorems 6-3 and 6-4, we can easily prove:

Theorem 6-5 If $\mathbf{Ax} = \mathbf{b}$ is a system of m equations in n variables, and if rank $(\mathbf{A})$ = rank $[\mathbf{A}, \mathbf{b}] = r$, then:

(i) If $r = n$, then $\mathbf{Ax} = \mathbf{b}$ has a unique solution.

(ii) If $r < n$, then $\mathbf{Ax} = \mathbf{b}$ has an infinity of solutions.

PROOF Since $\mathbf{Ax} = \mathbf{b}$ is consistent, we can investigate its solutions by considering the equivalent system $\hat{\mathbf{A}}\hat{\mathbf{x}} = \hat{\mathbf{b}}$, where

$$[\hat{\mathbf{A}}, \hat{\mathbf{b}}] = \left[\begin{array}{cc:c} \mathbf{I}_r & \mathbf{C}_{r \times (n-r)} & \mathbf{f} \\ \mathbf{O}_{(m-r) \times r} & \mathbf{O}_{(m-r) \times (n-r)} & \mathbf{0} \end{array}\right] \tag{6-5}$$

Now, we let $\hat{\mathbf{x}} = [\hat{x}_1 \quad \hat{x}_2 \quad \cdots \quad \hat{x}_n]^T$ and partition $\hat{\mathbf{x}}$ as follows:

$$\hat{\mathbf{x}} = \begin{bmatrix} \hat{\mathbf{x}}_r \\ \hdashline \hat{\mathbf{x}}_{n-r} \end{bmatrix} \qquad \hat{\mathbf{x}}_r = \begin{bmatrix} \hat{x}_1 \\ \hat{x}_2 \\ \vdots \\ \hat{x}_r \end{bmatrix} \qquad \hat{\mathbf{x}}_{n-r} = \begin{bmatrix} \hat{x}_{r+1} \\ \hat{x}_{r+2} \\ \vdots \\ \hat{x}_n \end{bmatrix}$$

Hence,

$$\hat{\mathbf{A}}\hat{\mathbf{x}} = \begin{bmatrix} \mathbf{I}_r & \mathbf{C}_{r \times (n-r)} \\ \mathbf{O}_{(m-r) \times r} & \mathbf{O}_{(m-r) \times (n-r)} \end{bmatrix} \begin{bmatrix} \hat{\mathbf{x}}_r \\ \hat{\mathbf{x}}_{n-r} \end{bmatrix}$$

$$= \begin{bmatrix} (\mathbf{I}_r \hat{\mathbf{x}}_r & + \mathbf{C}_{r \times (n-r)} \hat{\mathbf{x}}_{n-r}) \\ (\mathbf{O}_{(m-r) \times r} \hat{\mathbf{x}}_r & + \mathbf{O}_{(m-r) \times (n-r)} \hat{\mathbf{x}}_{n-r}) \end{bmatrix} = \begin{bmatrix} \mathbf{f} \\ \mathbf{0} \end{bmatrix} \tag{6-6}$$

The first r components of each side of Eq. (6-6) yields

$$\hat{\mathbf{x}}_r + \mathbf{C}_{r \times (n-r)} \hat{\mathbf{x}}_{n-r} = \mathbf{f}$$

or

$$\hat{\mathbf{x}}_r = \mathbf{f} - \mathbf{C}_{r \times (n-r)} \hat{\mathbf{x}}_{n-r} \tag{6-7}$$

Since the last $(n - r)$ equations of (6-6) are satisfied by *any* $\hat{\mathbf{x}}$, the solutions to $\hat{\mathbf{A}}\hat{\mathbf{x}} = \hat{\mathbf{b}}$ are given by Eq. (6-7). That is, any $\hat{\mathbf{x}}$ which satisfies Eq. (6-7) is a solution to $\hat{\mathbf{A}}\hat{\mathbf{x}} = \hat{\mathbf{b}}$; and conversely, if $\hat{\mathbf{x}}$ is a solution to $\hat{\mathbf{A}}\hat{\mathbf{x}} = \hat{\mathbf{b}}$, then it satisfies Eq. (6-7).

Now, let us consider the two cases of the theorem:

Case (i). $r = n$; then, Eq. (6-7) reduces to $\hat{\mathbf{x}}_n = \mathbf{f}$, and hence $\hat{\mathbf{x}}_n = \mathbf{f}$ is the unique solution to $\hat{\mathbf{A}}\hat{\mathbf{x}} = \hat{\mathbf{b}}$. Since $\hat{\mathbf{A}}\hat{\mathbf{x}} = \hat{\mathbf{b}}$ and $\mathbf{Ax} = \mathbf{b}$ are equivalent systems, the latter must also possess exactly one solution. Again, we note that the components of $\hat{\mathbf{x}}$ are merely the components of $\mathbf{x}$, possibly reordered.

Case (ii). $r < n$; the right-hand side of Eq. (6-7) contains the last $(n - r)$ components of $\hat{\mathbf{x}}$. Each of these variables may be assigned a value arbitrarily. For example, if we let

$$\hat{x}_{r+j} = \alpha_j \qquad j = 1, 2, \ldots, n - r$$

the values of the first r variables $\hat{x}_1, \hat{x}_2, \ldots, \hat{x}_r$ are then determined; the ith component of each side of Eq. (6-7) becomes

$$\hat{x}_i = f_i - \sum_{k=1}^{n-r} c_{i,\,r+k}\alpha_k \qquad i = 1, 2, \ldots, r \tag{6-8}$$

where $\mathbf{C} = [c_{i,\,r+j}]$. Hence, since there are no restrictions on the ways the α_j's can be specified, there is in fact an infinity of solutions to $\hat{\mathbf{A}}\hat{\mathbf{x}} = \hat{\mathbf{b}}$ (and to $\mathbf{Ax} = \mathbf{b}$). QED

Observe that case (i) of Theorem 6-5 is possible only if $m \geq n$, since otherwise rank $(\mathbf{A}) \leq m < n$.

EXAMPLE 6-3 Consider the system whose augmented matrix is

$$[\mathbf{A}, \mathbf{b}] = \left[\begin{array}{rrrr:r} 1 & 0 & -1 & 2 & -1 \\ 1 & 1 & 1 & -1 & 2 \\ 0 & -1 & -2 & 3 & -3 \\ 5 & 2 & -1 & 4 & 1 \\ -1 & 2 & 5 & -8 & 7 \end{array}\right]$$

Adding appropriate multiples of row 1 to rows 2, 3, 4, and 5 yields

$$\left[\begin{array}{rrrr:r} 1 & 0 & -1 & 2 & -1 \\ 0 & 1 & 2 & -3 & 3 \\ 0 & -1 & -2 & 3 & -3 \\ 0 & 2 & 4 & -6 & 6 \\ 0 & 2 & 4 & -6 & 6 \end{array}\right]$$

Now, adding appropriate multiples of row 2 to rows 3, 4, and 5 yields

$$[\hat{\mathbf{A}}, \hat{\mathbf{b}}] = \left[\begin{array}{rrrr:r} 1 & 0 & -1 & 2 & -1 \\ 0 & 1 & 2 & -3 & 3 \\ \hdashline 0 & 0 & 0 & 0 & 0 \\ 0 & 0 & 0 & 0 & 0 \\ 0 & 0 & 0 & 0 & 0 \end{array}\right]$$

Hence, rank $(\mathbf{A})$ = rank $[\mathbf{A}, \mathbf{b}] = 2$, and

$$\mathbf{C} = \begin{bmatrix} -1 & 2 \\ 2 & -3 \end{bmatrix} \qquad \mathbf{f} = \begin{bmatrix} -1 \\ 3 \end{bmatrix}$$

From Eq. (6-7),

$$\begin{aligned}
\begin{bmatrix} \hat{x}_1 \\ \hat{x}_2 \end{bmatrix} &= \mathbf{f} - \mathbf{C}\begin{bmatrix} \hat{x}_3 \\ \hat{x}_4 \end{bmatrix} \\
&= \begin{bmatrix} -1 \\ 3 \end{bmatrix} - \begin{bmatrix} -1 & 2 \\ 2 & -3 \end{bmatrix}\begin{bmatrix} \hat{x}_3 \\ \hat{x}_4 \end{bmatrix} \\
&= \begin{bmatrix} -1 \\ 3 \end{bmatrix} - \begin{bmatrix} (-\hat{x}_3 + 2\hat{x}_4) \\ (\ 2\hat{x}_3 - 3\hat{x}_4) \end{bmatrix}
\end{aligned}$$

Letting $\hat{x}_3 = \alpha_1$, $\hat{x}_4 = \alpha_2$, and noting in this example $\hat{\mathbf{x}} = \mathbf{x}$ (since no column interchanges were performed), we can now state the complete solution to $\mathbf{Ax} = \mathbf{b}$ as follows:

$$\begin{aligned}
x_1 &= -1 + \alpha_1 - 2\alpha_2 \\
x_2 &= 3 - 2\alpha_1 + 3\alpha_2 \\
x_3 &= \alpha_1 \\
x_4 &= \alpha_2
\end{aligned}$$

where α_1 and α_2 are arbitrary scalars. ////

The proof of Theorem 6-5 also illustrates that the number of arbitrary scalars in the solution to a system of equations $\mathbf{Ax} = \mathbf{b}$ is $(n - r)$, where the system has n variables and $r = \text{rank}(\mathbf{A}) = \text{rank}[\mathbf{A}, \mathbf{b}]$. If $n = r$, the system has no arbitrary scalars and hence the solution is unique.

We note also that the columns of $\mathbf{A}$ which correspond to the columns of $\mathbf{I}_r$ in Eq. (6-5) are not necessarily unique. In any consistent system of equations $\mathbf{Ax} = \mathbf{b}$ for which rank $(\mathbf{A}) = r$, we can choose any r columns of $\mathbf{A}$ from which a nonsingular submatrix of order r can be formed as the columns which will correspond to $\mathbf{I}_r$.

EXAMPLE 6-4 Suppose we interchange columns 2 and 3 of the system of Example 6-3, resulting in the augmented matrix

$$\left[\begin{array}{rrrr|r} 1 & -1 & 0 & 2 & -1 \\ 1 & 1 & 1 & -1 & 2 \\ 0 & -2 & -1 & 3 & -3 \\ 5 & -1 & 2 & 4 & 1 \\ -1 & 5 & 2 & -8 & 7 \end{array}\right]$$

with $\hat{\mathbf{x}} = [x_1 \quad x_3 \quad x_2 \quad x_4]^T$. Adding appropriate multiples of row 1 to rows 2, 3, 4, and 5 yields

$$\left[\begin{array}{cccc:c} 1 & -1 & 0 & 2 & -1 \\ 0 & 2 & 1 & -3 & 3 \\ 0 & -2 & -1 & 3 & -3 \\ 0 & 4 & 2 & -6 & 6 \\ 0 & 4 & 2 & -6 & -6 \end{array}\right]$$

Adding appropriate multiples of row 2 to rows 3, 4, and 5 yields

$$\left[\begin{array}{cccc:c} 1 & -1 & 0 & 2 & -1 \\ 0 & 2 & 1 & -3 & 3 \\ 0 & 0 & 0 & 0 & 0 \\ 0 & 0 & 0 & 0 & 0 \\ 0 & 0 & 0 & 0 & 0 \end{array}\right]$$

Now, we must perform a backward pass, by multiplying row 2 by $\frac{1}{2}$ and then adding the resulting row 2 to row 1, yielding the desired result:

$$[\hat{\mathbf{A}}, \hat{\mathbf{b}}] = \left[\begin{array}{cc:cc:c} 1 & 0 & \frac{1}{2} & \frac{1}{2} & \frac{1}{2} \\ 0 & 1 & \frac{1}{2} & -\frac{3}{2} & \frac{3}{2} \\ \hdashline 0 & 0 & 0 & 0 & 0 \\ 0 & 0 & 0 & 0 & 0 \\ 0 & 0 & 0 & 0 & 0 \end{array}\right]$$

Thus,

$$\begin{bmatrix} 1 & 0 \\ 0 & 1 \end{bmatrix} \begin{bmatrix} x_1 \\ x_3 \end{bmatrix} = \begin{bmatrix} \frac{1}{2} \\ \frac{3}{2} \end{bmatrix} - \begin{bmatrix} \frac{1}{2} & \frac{1}{2} \\ \frac{1}{2} & -\frac{3}{2} \end{bmatrix} \begin{bmatrix} x_2 \\ x_4 \end{bmatrix}$$

$$\begin{bmatrix} x_1 \\ x_3 \end{bmatrix} = \begin{bmatrix} (\frac{1}{2} - \frac{1}{2}x_2 - \frac{1}{2}x_4) \\ (\frac{3}{2} - \frac{1}{2}x_2 + \frac{3}{2}x_4) \end{bmatrix}$$

Setting $x_2 = \beta_1$, $x_4 = \beta_2$, we have

$$x_1 = \tfrac{1}{2} - \tfrac{1}{2}\beta_1 - \tfrac{1}{2}\beta_2$$
$$x_2 = \beta_1$$
$$x_3 = \tfrac{3}{2} - \tfrac{1}{2}\beta_1 + \tfrac{3}{2}\beta_2$$
$$x_4 = \beta_2$$

where β_1, β_2 are arbitrary scalars. ////

We conclude this section with a summary of important results, some of which the reader is asked to prove in the exercises.

SYSTEMS OF m EQUATIONS IN n VARIABLES, $\mathbf{Ax} = \mathbf{b}$

Relationship of m and n	Rank (A) $= r$ Rank [A, b] $= p$		Solutions to $\mathbf{Ax} = \mathbf{b}$	Source
$m < n$	$r < m < n$	$r \neq p$	No solution	Theorem 6-5
	$r < m < n$	$r = p$	Infinity of solutions	Theorem 6-5
	$r = m$	p *must* $= r$	Infinity of solutions	Exercise 16
$m = n$	$r < m$	$r \neq p$	No solution	Theorem 6-5
	$r < m$	$r = p$	Infinity of solutions	Theorem 6-5
	$r = m = n$	p *must* $= r$	Unique solution	Theorem 6-5 and Exercise 16
$m > n$	$r \leq n < m$	$r \neq p$	No solution	Theorem 6-5
	$r < n < m$	$r = p$	Infinity of solutions	Theorem 6-5
	$r = n$	$r = p$	Unique solution	Theorem 6-5

6-3 LINEAR INDEPENDENCE

We shall now consider the results of Secs. 6-1 and 6-2 from a slightly different perspective, one which is frequently quite helpful in gaining certain insights into the theory of linear equations and which also introduces some concepts which can be generalized in a meaningful way into other branches of mathematics. The latter will be considered in Chap. 7.

We begin by defining:

A set of m-component vectors $\mathbf{a}_1, \mathbf{a}_2, \ldots, \mathbf{a}_k$ is said to be *linearly independent* if and only if the only values for the set of scalars $\beta_1, \beta_2, \ldots, \beta_k$ satisfying

$$\beta_1 \mathbf{a}_1 + \beta_2 \mathbf{a}_2 + \cdots + \beta_k \mathbf{a}_k = \mathbf{0} \tag{6-9}$$

are $\beta_1 = \beta_2 = \cdots = \beta_k = 0$.

If there exist scalars $\beta_1, \beta_2, \ldots, \beta_k$, not all equal to zero, which satisfy Eq. (6-9), then the set of vectors $\mathbf{a}_1, \mathbf{a}_2, \ldots, \mathbf{a}_k$ is said to be *linearly dependent*.

EXAMPLE 6-5 Let us determine whether the vectors

$$\mathbf{a}_1 = \begin{bmatrix} 1 \\ 0 \\ -2 \\ 4 \end{bmatrix} \qquad \mathbf{a}_2 = \begin{bmatrix} 1 \\ 1 \\ 1 \\ -2 \end{bmatrix} \qquad \mathbf{a}_3 = \begin{bmatrix} 4 \\ 3 \\ 1 \\ -2 \end{bmatrix}$$

are linearly independent.

Setting $\beta_1\mathbf{a}_1 + \beta_2\mathbf{a}_2 + \beta_3\mathbf{a}_3 = \mathbf{0}$ yields

$$\beta_1\begin{bmatrix}1\\0\\-2\\4\end{bmatrix} + \beta_2\begin{bmatrix}1\\1\\1\\-2\end{bmatrix} + \beta_3\begin{bmatrix}4\\3\\1\\-2\end{bmatrix} = \begin{bmatrix}0\\0\\0\\0\end{bmatrix}$$

Or

$$\begin{bmatrix}1 & 1 & 4\\0 & 1 & 3\\-2 & 1 & 1\\4 & -2 & -2\end{bmatrix}\begin{bmatrix}\beta_1\\\beta_2\\\beta_3\end{bmatrix} = \begin{bmatrix}0\\0\\0\\0\end{bmatrix}$$

Thus, we have a system of four equations in three variables, whose augmented matrix is

$$[\mathbf{A}, \mathbf{0}] = \left[\begin{array}{rrr:r}1 & 1 & 4 & 0\\0 & 1 & 3 & 0\\-2 & 1 & 1 & 0\\4 & -2 & -2 & 0\end{array}\right]$$

Adding appropriate multiples of row 1 to rows 2, 3, and 4 yields:

$$\left[\begin{array}{rrr:r}1 & 1 & 4 & 0\\0 & 1 & 3 & 0\\0 & 3 & 9 & 0\\0 & -6 & -18 & 0\end{array}\right]$$

Adding appropriate multiples of row 2 to rows† 1, 3, and 4 yields:

$$\left[\begin{array}{rr:r:r}1 & 0 & 1 & 0\\0 & 1 & 3 & 0\\\hdashline 0 & 0 & 0 & 0\\0 & 0 & 0 & 0\end{array}\right]$$

Hence, we obtain

$$\beta_1 = -\beta_3$$
$$\beta_2 = -3\beta_3$$

Therefore, since β_3 is arbitrary, we may set $\beta_3 = 1$ and we find that the vectors $\mathbf{a}_1$, $\mathbf{a}_2$, $\mathbf{a}_3$ are linearly dependent. ////

† Note that we are using a Gauss-Jordan type of procedure here. Wherever a Gaussian elimination type of procedure is applicable, so is a Gauss-Jordan elimination (although as we have seen in Chap. 4, the latter may require more arithmetic operations).

EXAMPLE 6-6 Consider the vectors $\mathbf{a}_1$, $\mathbf{a}_2$ of Example 6-5 and the vector

$$\mathbf{a}_4 = \begin{bmatrix} 4 \\ 3 \\ 1 \\ 1 \end{bmatrix}$$

These three vectors are linearly independent, since the only solution to $\beta_1 \mathbf{a}_1 + \beta_2 \mathbf{a}_2 + \beta_4 \mathbf{a}_4 = \mathbf{0}$ is $\beta_1 = \beta_2 = \beta_4 = 0$:

$$\left[\begin{array}{ccc|c} 1 & 1 & 4 & 0 \\ 0 & 1 & 3 & 0 \\ -2 & 1 & 1 & 0 \\ 4 & -2 & 1 & 0 \end{array}\right] \to \left[\begin{array}{ccc|c} 1 & 1 & 4 & 0 \\ 0 & 1 & 3 & 0 \\ 0 & 3 & 9 & 0 \\ 0 & -6 & -15 & 0 \end{array}\right]$$

$$\left[\begin{array}{ccc|c} 1 & 0 & 1 & 0 \\ 0 & 1 & 3 & 0 \\ 0 & 0 & 0 & 0 \\ 0 & 0 & 3 & 0 \end{array}\right] \to \left[\begin{array}{ccc|c} 1 & 0 & 1 & 0 \\ 0 & 1 & 3 & 0 \\ 0 & 0 & 3 & 0 \\ 0 & 0 & 0 & 0 \end{array}\right]$$

$$\to \left[\begin{array}{ccc|c} 1 & 0 & 0 & 0 \\ 0 & 1 & 0 & 0 \\ 0 & 0 & 1 & 0 \\ 0 & 0 & 0 & 0 \end{array}\right]$$

Or

$$I_3 \boldsymbol{\beta} = \mathbf{0}$$

where $\boldsymbol{\beta} = [\beta_1 \quad \beta_2 \quad \beta_4]^T$. ////

The two examples above illustrate that the problem of determining whether a given set of vectors $\mathbf{a}_1, \mathbf{a}_2, \ldots, \mathbf{a}_k$ is linearly independent may be accomplished by solving the system of equations

$$\mathbf{A}\boldsymbol{\beta} = \mathbf{0} \tag{6-10}$$

where $\mathbf{A} = [\mathbf{a}_1 \quad \mathbf{a}_2 \quad \cdots \quad \mathbf{a}_k]$, $\boldsymbol{\beta} = [\beta_1 \quad \beta_2 \quad \cdots \quad \beta_k]^T$. Moreover, since $\boldsymbol{\beta} = \mathbf{0}$ is always a solution of (6-10), the vectors $\mathbf{a}_1, \mathbf{a}_2, \ldots, \mathbf{a}_k$ will be linearly independent if and only if $\boldsymbol{\beta} = \mathbf{0}$ is the unique solution. By Theorem 6-5, Eq. (6-10) has a unique solution only if rank $(\mathbf{A}) = k$. Hence, we have:

Theorem 6-6 The vectors $\mathbf{a}_1, \mathbf{a}_2, \ldots, \mathbf{a}_k$ are linearly independent if and only if rank $(\mathbf{A}) = k$, where $\mathbf{A} = [\mathbf{a}_1 \quad \mathbf{a}_2 \quad \cdots \quad \mathbf{a}_k]$.

As a special case of Theorem 6-6 we have:

Theorem 6-7 A square matrix $\mathbf{A}$ is nonsingular if and only if its columns are linearly independent.

PROOF If $\mathbf{A} = [\mathbf{a}_1 \quad \mathbf{a}_2 \quad \cdots \quad \mathbf{a}_n]$ and $\mathbf{A}$ is of order n, then $\mathbf{A}$ is nonsingular if and only if rank $(\mathbf{A}) = n$, since if rank $(\mathbf{A}) = n$, the largest nonsingular square submatrix of $\mathbf{A}$ is $\mathbf{A}$ itself and if rank $(\mathbf{A}) < n$, then det $(\mathbf{A})$ is obviously zero.

Thus, if rank $(\mathbf{A}) = n$, then $\mathbf{A}$ is nonsingular and by Theorem 6-6, the column vectors $\mathbf{a}_1, \mathbf{a}_2, \ldots, \mathbf{a}_n$ are linearly independent. On the other hand, if the columns $\mathbf{a}_1, \mathbf{a}_2, \ldots, \mathbf{a}_n$ are linearly independent, then by Theorem 6-6, rank $(\mathbf{A}) = n$ and therefore $\mathbf{A}$ is nonsingular. QED

In Sec. 6-2 we noted that if the coefficient matrix of a consistent system of equations $\mathbf{Ax} = \mathbf{b}$ is of rank r, then the columns of $\mathbf{A}$ (and hence, the corresponding variables associated with each column) which correspond to the $\mathbf{I}_r$ of Eq. (6-7) are not necessarily uniquely determined. This point was illustrated in Examples 6-3 and 6-4. From Theorem 6-7 we now see that we can choose any r linearly independent columns of $\mathbf{A}$ as those which will correspond to $\mathbf{I}_r$. This means that we can solve for any r variables whose corresponding columns are linearly independent in terms of the remaining $(n - r)$ variables, where $\mathbf{A}$ is of order $m \times n$; these latter $(n - r)$ variables are then the $(n - r)$ arbitrary variables of the solution. As we shall see in Chap. 8, we may desire to investigate the set of solutions of a given system of equations $\mathbf{Ax} = \mathbf{b}$ by examining different combinations of the r variables which appear on the left-hand side of Eq. (6-7). Toward this end, we prove:

Theorem 6-8 If $\mathbf{a}_1, \mathbf{a}_2, \ldots, \mathbf{a}_k$ are linearly independent vectors and if $\mathbf{b}$ is a vector such that

$$\alpha_1 \mathbf{a}_1 + \alpha_2 \mathbf{a}_2 + \cdots + \alpha_k \mathbf{a}_k = \mathbf{b} \tag{6-11}$$

then the set of vectors $\mathbf{b}, \mathbf{a}_2, \mathbf{a}_3, \ldots, \mathbf{a}_k$ is linearly independent if and only if $\alpha_1 \neq 0$.

PROOF We first prove that if $\mathbf{b}, \mathbf{a}_2, \mathbf{a}_3, \ldots, \mathbf{a}_k$ are linearly independent, then $\alpha_1 \neq 0$. If α_1 were equal to zero, then we could write Eq. (6-11) as follows:

$$(-1)\mathbf{b} + \alpha_2 \mathbf{a}_2 + \alpha_3 \mathbf{a}_3 + \cdots + \alpha_k \mathbf{a}_k = \mathbf{0} \tag{6-12}$$

But Eq. (6-12) states that there exist scalars $\lambda_1, \lambda_2, \ldots, \lambda_k$ not all zero such that

$$\lambda_1 \mathbf{b} + \lambda_2 \mathbf{a}_2 + \cdots + \lambda_k \mathbf{a}_k = \mathbf{0}$$

(i.e., let $\lambda_1 = -1, \lambda_2 = \alpha_2, \ldots, \lambda_k = \alpha_k$) which means that $\mathbf{b}, \mathbf{a}_2, \mathbf{a}_3, \ldots, \mathbf{a}_k$ are linearly dependent; this contradicts our hypothesis that $\mathbf{b}$, $\mathbf{a}_2$, $\mathbf{a}_3, \ldots, \mathbf{a}_k$ are linearly independent. Therefore, α_1 cannot equal zero. This proves the "if" part of the theorem.

Now, we must prove that if $\alpha_1 \neq 0$, then $\mathbf{b}$, $\mathbf{a}_2$, $\mathbf{a}_3$, ..., $\mathbf{a}_k$ are linearly independent:

Suppose we write

$$\lambda_1 \mathbf{b} + \lambda_2 \mathbf{a}_2 + \cdots + \lambda_k \mathbf{a}_k = \mathbf{0} \tag{6-13}$$

Then, the vectors $\mathbf{b}$, $\mathbf{a}_2$, ..., $\mathbf{a}_k$ are linearly independent if and only if $\lambda_1 = \lambda_2 = \cdots = \lambda_k = 0$. Let us show that $\lambda_1 = \lambda_2 = \cdots \lambda_k = 0$. Substituting Eq. (6-11) into Eq. (6-13) yields

$$\lambda_1(\alpha_1 \mathbf{a}_1 + \alpha_2 \mathbf{a}_2 + \cdots + \alpha_k \mathbf{a}_k) + \lambda_2 \mathbf{a}_2 + \cdots + \lambda_k \mathbf{a}_k = \mathbf{0} \tag{6-14}$$

Upon rearranging the terms of Eq. (6-14), we find that

$$(\lambda_1\alpha_1)\mathbf{a}_1 + (\lambda_1\alpha_2 + \lambda_2)\mathbf{a}_2 + (\lambda_1\alpha_3 + \lambda_3)\mathbf{a}_3 + \cdots + (\lambda_1\alpha_k + \lambda_k)\mathbf{a}_k = \mathbf{0}$$

Since $\mathbf{a}_1$, $\mathbf{a}_2$, ..., $\mathbf{a}_k$ are linearly independent, it must be true that

$$\begin{aligned} \lambda_1\alpha_1 &= 0 \\ \lambda_1\alpha_2 + \lambda_2 &= 0 \\ \lambda_1\alpha_3 + \lambda_3 &= 0 \\ &\vdots \\ \lambda_1\alpha_k + \lambda_k &= 0 \end{aligned} \tag{6-15}$$

Moreover, since $\alpha_1 \neq 0$, the first equation of (6-15) will be satisfied if and only if $\lambda_1 = 0$. Substitution of $\lambda_1 = 0$ into the remaining equations of (6-15) yields the desired result: $\lambda_1 = \lambda_2 = \lambda_3 = \cdots = \lambda_k = 0$.

QED

The left-hand side of Eq. (6-11) is called a *linear combination* of the vectors $\mathbf{a}_1$, $\mathbf{a}_2$, ..., $\mathbf{a}_k$; the vector $\mathbf{b}$ is said to be a linear combination of $\mathbf{a}_1, \mathbf{a}_2, \ldots, \mathbf{a}_k$.

There are many other results pertaining to the concept of linear independence which we have deferred to the Exercises. Where such results are important and useful for general application, we have labeled them as "theorems" so that both the reader and the author may more readily reference them in subsequent chapters.

6-4 HOMOGENEOUS SYSTEMS OF EQUATIONS

We shall now devote several paragraphs to consideration of a special case of the prior results of this chapter, which we shall then apply in Chaps. 7 and 9 to facilitate some of the theory with which we shall be concerned in those chapters.

Specifically, let us investigate the consequences of Theorem 6-5 on the system of equations $\mathbf{Ax} = \mathbf{0}$. Such a system, in which the right-hand side $\mathbf{b}$ is $\mathbf{0}$, is called a *homogeneous system of equations.* The vector $\mathbf{x} = \mathbf{0}$ is always a solution to a homogeneous system of equations; $\mathbf{x} = \mathbf{0}$ is called the trivial solution. Frequently, we shall be interested in whether a particular homogeneous system of equations has solutions other than the trivial solution, that is, solutions in which at least one variable is nonzero. Such solutions are called *nontrivial* solutions.

In Sec. 6-3 we learned that the problem of determining whether a given set of vectors $\mathbf{a}_1, \mathbf{a}_2, \ldots, \mathbf{a}_k$ is linearly dependent or linearly independent is equivalent to determining whether the homogeneous system $\mathbf{Ax} = \mathbf{0}$ has nontrivial solutions or not, respectively, where $\mathbf{A} = [\mathbf{a}_1 \quad \mathbf{a}_2 \quad \cdots \quad \mathbf{a}_k]$.

Thus, the following are essentially restatements of previous theorems:

Theorem 6-9 If $\mathbf{A}$ is of order $m \times n$, then the homogeneous system of equations $\mathbf{Ax} = \mathbf{0}$ possesses nontrivial solutions if and only if rank $(\mathbf{A}) < n$. If $m < n$, then $\mathbf{Ax} = \mathbf{0}$ always possesses nontrivial solutions.

Theorem 6-10 If $\mathbf{A}$ is a square matrix of order n, then $\mathbf{Ax} = \mathbf{0}$ possesses nontrivial solutions if and only if $\mathbf{A}$ is singular (i.e., det $(\mathbf{A}) = 0$).

Another useful result is:

Theorem 6-11 If $\mathbf{x}_1$ and $\mathbf{x}_2$ are any two solutions to the homogeneous system of equations $\mathbf{Ax} = \mathbf{0}$, then any linear combination of $\mathbf{x}_1$ and $\mathbf{x}_2$ is also a solution.

PROOF Let $\mathbf{y} = \alpha_1\mathbf{x}_1 + \alpha_2\mathbf{x}_2$ be any linear combination of $\mathbf{x}_1, \mathbf{x}_2$. Then we wish to show that $\mathbf{Ay} = \mathbf{0}$. But

$$\begin{aligned}\mathbf{Ay} &= \mathbf{A}(\alpha_1\mathbf{x}_1 + \alpha_2\mathbf{x}_2)\\ &= \alpha_1(\mathbf{Ax}_1) + \alpha_2(\mathbf{Ax}_2)\\ &= \alpha_1\mathbf{0} + \alpha_2\mathbf{0}\\ &= \mathbf{0}\end{aligned}$$

QED

EXERCISES

For each of the Exercises 1 to 4, determine the rank of the given matrix:

1 $\begin{bmatrix} 1 & 0 & -1 & 2 \\ 2 & 3 & 1 & -1 \\ 0 & 2 & 2 & 1 \\ -3 & 1 & 4 & 1 \end{bmatrix}$ *2* $\begin{bmatrix} 1 & 1 & -1 & 2 \\ 4 & 2 & 1 & -1 \\ 2 & 4 & 6 & 0 \\ 9 & 0 & -6 & -3 \\ -2 & 7 & 0 & 4 \end{bmatrix}$

3 $\begin{bmatrix} 0 & 1 & 1 & 1 & -2 \\ 1 & 0 & -1 & 1 & 1 \\ 1 & 1 & 0 & 2 & -1 \\ 1 & -2 & -3 & -1 & 5 \end{bmatrix}$ *4* $\begin{bmatrix} 1 & -2 & 1 & 2 & 1 \\ -2 & 4 & 0 & -3 & 1 \\ 3 & -6 & 0 & 5 & -2 \\ -1 & 2 & 0 & 0 & 1 \end{bmatrix}$

5 Prove that if $\mathbf{A}$ is of order $m \times n$, then $\mathbf{A}$ has a right inverse if and only if rank $(\mathbf{A}) = m$. [*Hint:* Consider the equation $\mathbf{AX} = \mathbf{I}_m$, where $\mathbf{X}$ is of order $n \times m$ and $\mathbf{X} = [\mathbf{x}_1 \quad \mathbf{x}_2 \quad \cdots \quad \mathbf{x}_m]$. Then $\mathbf{AX} = \mathbf{I}$ may be thought of as m systems of equations $\mathbf{Ax}_j = \mathbf{e}_j$, $j = 1, 2, \ldots, m$, where $\mathbf{e}_j$ is the jth column of $\mathbf{I}_m$. Show that each of these systems has solutions if and only if rank $(\mathbf{A}) = m$.]

6 Prove that if $\mathbf{A}$ is of order $m \times n$, then $\mathbf{A}$ has a left inverse if and only if rank $(\mathbf{A}) = n$.

7 Combine the results of Exercises 5 and 6 to prove that a nonsquare matrix cannot have both a right and a left inverse.

8 Combine the results of Exercises 5 and 6 to prove that every square matrix has either no inverse or both a right and a left inverse.

9 Use Theorems 3-13 and 6-1 to prove:

Theorem 6-12: If $\mathbf{B}$ is any nonsingular square matrix of order n and $\mathbf{A}$ is any $m \times n$ matrix, then rank $(\mathbf{A}) =$ rank $(\mathbf{AB})$; if $\mathbf{C}$ is any nonsingular matrix of order m, then rank $(\mathbf{CA}) =$ rank $(\mathbf{A})$.

10 Prove Theorem 6-3.

11 Write down a step-by-step detailed description of how to apply the results of Theorem 6-5 to determine whether a given system of equations $\mathbf{Ax} = \mathbf{b}$ is consistent, and if so, to find its solutions. Part of your description should include either a Gaussian or Gauss-Jordan type of elimination procedure.

12 Prepare a flow chart for the method you have described in Exercise 11.

13 Prove **Theorem 6-13:** A set of m component vectors $\mathbf{a}_1, \mathbf{a}_2, \ldots, \mathbf{a}_k$ is necessarily linearly dependent if $k > m$.

14 Prove **Theorem 6-14:** If any subset of a set of vectors $\mathbf{a}_1, \mathbf{a}_2, \ldots, \mathbf{a}_k$ is linearly dependent, then the entire set is also linearly dependent. [*Hint:* For simplicity, assume the first p vectors $\mathbf{a}_1, \mathbf{a}_2, \ldots, \mathbf{a}_p$ are linearly dependent, where $p < k$. Then, by definition of linear dependence, there exists a set of scalars $\beta_1, \beta_2, \ldots, \beta_p$, not all zero, such that $\beta_1\mathbf{a}_1 + \beta_2\mathbf{a}_2 + \cdots + \beta_p\mathbf{a}_p = \mathbf{0}$. Now, consider the equation

$$\beta_1\mathbf{a}_1 + \beta_2\mathbf{a}_2 + \cdots + \beta_p\mathbf{a}_p + \beta_{p+1}\mathbf{a}_{p+1} + \cdots + \beta_k\mathbf{a}_k = \mathbf{0}$$

where $\beta_{p+1} = \beta_{p+2} = \cdots = \beta_k = 0$.]

15 Use Theorem 6-14 to prove:

Theorem 6-15: If a set of vectors $\mathbf{a}_1, \mathbf{a}_2, \ldots, \mathbf{a}_k$ is linearly independent, then every subset of $\mathbf{a}_1, \mathbf{a}_2, \ldots, \mathbf{a}_k$ is also linearly independent. [*Hint:* Show that the assumption that a subset is linearly dependent leads to a contradiction by applying Theorem 6-14.]

16 Given a system of m equations in n variables, $\mathbf{Ax} = \mathbf{b}$, prove: If rank $(\mathbf{A}) = m$, then the system is consistent; moreover, the solution is unique only if $m = n$.

17 Determine whether each of the following sets of vectors is linearly independent or linearly dependent:

(*a*) $\begin{bmatrix} 0 \\ 1 \\ -1 \\ 2 \end{bmatrix} \quad \begin{bmatrix} 1 \\ 1 \\ 1 \\ 0 \end{bmatrix} \quad \begin{bmatrix} 5 \\ 1 \\ 9 \\ -8 \end{bmatrix}$ (*b*) $\begin{bmatrix} 1 \\ 0 \\ 3 \end{bmatrix} \quad \begin{bmatrix} 0 \\ 1 \\ 0 \end{bmatrix} \quad \begin{bmatrix} 1 \\ 1 \\ 1 \end{bmatrix}$

(*c*) $\begin{bmatrix} 1 \\ 0 \\ 3 \end{bmatrix} \quad \begin{bmatrix} 0 \\ 1 \\ 0 \end{bmatrix} \quad \begin{bmatrix} 1 \\ 1 \\ 1 \end{bmatrix} \quad \begin{bmatrix} 7 \\ -5 \\ 4 \end{bmatrix} \quad \begin{bmatrix} 5 \\ 0 \\ 4 \end{bmatrix}$

(*d*) $\begin{bmatrix} 1 \\ -1 \\ 2 \\ 0 \end{bmatrix} \quad \begin{bmatrix} 0 \\ 1 \\ 4 \\ 1 \end{bmatrix} \quad \begin{bmatrix} 0 \\ 1 \\ 1 \\ 0 \end{bmatrix}$

18 Find all solutions for each of the following systems or show that the system is inconsistent:

(*a*) $[\mathbf{A}, \mathbf{b}] = \left[\begin{array}{rrrr:r} 1 & 0 & -1 & 3 & 4 \\ 2 & 2 & 0 & 1 & -3 \\ -3 & -2 & 1 & 4 & -1 \\ -1 & 2 & 3 & 16 & -15 \end{array}\right]$

(*b*) $[\mathbf{A}, \mathbf{b}] = \left[\begin{array}{rrrr:r} 2 & -3 & 0 & 5 & 7 \\ 0 & 1 & 3 & 4 & 5 \\ 4 & -6 & 1 & 9 & 9 \end{array}\right]$

(*c*) $[\mathbf{A}, \mathbf{b}] = \left[\begin{array}{rrr:r} 1 & 0 & 3 & 4 \\ 0 & 1 & -2 & -1 \\ 1 & 2 & 4 & 7 \\ 2 & 2 & 2 & 6 \end{array}\right]$

(*d*) $[\mathbf{A}, \mathbf{b}] = \left[\begin{array}{rrr:r} 0 & 1 & -2 & -1 \\ 1 & 2 & 4 & 7 \\ 2 & 2 & 2 & 7 \\ 1 & 0 & 3 & 4 \end{array}\right]$

19 Find all nontrivial solutions (or show that none exists) for each of the homogeneous systems whose coefficient matrices are:

(*a*) $\mathbf{A} = \begin{bmatrix} 1 & 2 & 0 & -3 \\ 0 & 3 & 2 & 1 \\ -1 & 1 & 2 & 4 \\ -3 & 1 & 4 & 1 \end{bmatrix}$ (*b*) $\mathbf{A} = \begin{bmatrix} 1 & 0 & -1 & -3 \\ 2 & 3 & 1 & 1 \\ 0 & 2 & 2 & 4 \\ -3 & 1 & 4 & 1 \end{bmatrix}$

(*c*) $\mathbf{A} = \begin{bmatrix} 0 & 1 & 1 & 1 & -2 \\ 0 & 1 & 1 & 1 & 1 \\ 1 & 1 & 0 & 2 & -1 \\ 1 & -2 & -3 & -1 & 5 \end{bmatrix}$ (*d*) $\mathbf{A} = \begin{bmatrix} 0 & 1 & 1 & 1 \\ 1 & 0 & 1 & -2 \\ 1 & -1 & 0 & -3 \\ 1 & 1 & 2 & -1 \\ -2 & 1 & -1 & 5 \end{bmatrix}$

(*e*) $\mathbf{A} = \begin{bmatrix} 0 & 1 & -2 \\ 1 & 2 & 4 \\ 2 & 2 & 2 \\ 1 & 0 & 3 \end{bmatrix}$ (*f*) $\mathbf{A} = \begin{bmatrix} 0 & 1 & 2 & 1 \\ 1 & 2 & 2 & 0 \\ -2 & 4 & 2 & 3 \end{bmatrix}$

20 Geometrically, the equation $ax_1 + bx_2 + cx_3 = d$ describes a plane in three-dimensional space, with coordinate axes x_1, x_2, x_3. Thus, the solutions to a system of m equations in three variables may be considered as the intersection of the m planes described by each of the equations.

What is the geometrical interpretation of the different situations which result when the rank of the coefficient matrix of such a system is 1, 2, and 3, respectively? Be sure to consider separately the cases of consistent and inconsistent systems.

7

VECTOR SPACES

7-1 INTRODUCTION; DEFINITION

Consider the set S of all solutions for a given homogeneous system of equations $\mathbf{Ax} = \mathbf{0}$. According to Theorem 6-11, if $\mathbf{x}_1$ and $\mathbf{x}_2$ are any two solutions of $\mathbf{Ax} = \mathbf{0}$, then:

1 $\alpha\mathbf{x}_1$ is also a solution, for any scalar α.
2 And $(\mathbf{x}_1 + \mathbf{x}_2)$ is also a solution.

In the terminology of sets, we say that if $\mathbf{x}_1$ and $\mathbf{x}_2$ are any two elements of the set S, then $\alpha\mathbf{x}_1$ and $(\mathbf{x}_1 + \mathbf{x}_2)$ are also elements of S. Any set of vectors to which the previous statement applies is called a *vector space*. The reader may easily verify that the sets S_1, S_2, and S_3 are examples of vector spaces:

$$S_1 = \left\{ \begin{bmatrix} x_1 \\ x_2 \end{bmatrix} \middle| \; x_1 = 0 \right\}^\dagger \tag{7-1}$$

$$S_2 = \left\{ \begin{bmatrix} x_1 \\ x_2 \end{bmatrix} \middle| \; 3x_1 - 4x_2 = 0 \right\}$$

$$S_3 = \left\{ \begin{bmatrix} x_1 \\ x_2 \\ x_3 \end{bmatrix} \middle| \; \begin{matrix} x_2 = 0 \\ x_1 = 2x_3 \end{matrix} \right\}$$

† Equation (7-1) is read, "S_1 is the set of all two-component vectors $\begin{bmatrix} x_1 \\ x_2 \end{bmatrix}$ such that $x_1 = 0$."

Below are several examples of sets of vectors which are *not* vector spaces:

$$S_4 = \left\{ \begin{bmatrix} x_1 \\ x_2 \end{bmatrix} \middle| \; x_1 = -1 \right\}$$

$$S_5 = \left\{ \begin{bmatrix} x_1 \\ x_2 \end{bmatrix} \middle| \; x_1 = {x_2}^2 \right\}$$

S_4 is not a vector space because although the vectors $\begin{bmatrix} -1 \\ 0 \end{bmatrix}$, $\begin{bmatrix} -1 \\ 2 \end{bmatrix}$ are elements of S_4, the sum $\begin{bmatrix} -1 \\ 0 \end{bmatrix} + \begin{bmatrix} -1 \\ 2 \end{bmatrix} = \begin{bmatrix} -2 \\ 2 \end{bmatrix}$ is clearly not an element of S_4, its first component not being equal to (-1). Similarly, $\begin{bmatrix} 1 \\ 1 \end{bmatrix}$, $\begin{bmatrix} 1 \\ -1 \end{bmatrix}$ are both elements of S_5, whereas $\begin{bmatrix} 1 \\ 1 \end{bmatrix} + \begin{bmatrix} 1 \\ -1 \end{bmatrix} = \begin{bmatrix} 2 \\ 0 \end{bmatrix}$ is obviously not in S_5, and therefore S_5 is not a vector space.

We have already seen that the set of all solutions to a given homogeneous system of equations is a vector space. Another typical set of vectors which turns out to be a vector space is the following: Given a set of vectors $\mathbf{a}_1, \mathbf{a}_2, \ldots, \mathbf{a}_n$, let C be the set of *all* linear combinations of $\mathbf{a}_1, \mathbf{a}_2, \ldots, \mathbf{a}_n$:

$$C = \{\mathbf{b} \,|\, \mathbf{b} = \alpha_1 \mathbf{a}_1 + \alpha_2 \mathbf{a}_2 + \cdots + \alpha_n \mathbf{a}_n\} \qquad (7\text{-}2)$$

It is easy to show that C is a vector space: If $\mathbf{b}_1$ and $\mathbf{b}_2$ are elements of C, then each is some linear combination of $\mathbf{a}_1, \mathbf{a}_2, \ldots, \mathbf{a}_n$. Thus, we can write

$$\mathbf{b}_1 = \lambda_1 \mathbf{a}_1 + \lambda_2 \mathbf{a}_2 + \cdots + \lambda_n \mathbf{a}_n$$

$$\mathbf{b}_2 = \beta_1 \mathbf{a}_1 + \beta_2 \mathbf{a}_2 + \cdots + \beta_n \mathbf{a}_n$$

Now, for any scalar α, it is clear that $\alpha \mathbf{b}_1$ is also a linear combination of $\mathbf{a}_1, \mathbf{a}_2, \ldots, \mathbf{a}_n$, since

$$\alpha \mathbf{b}_1 = (\alpha\lambda_1)\mathbf{a}_1 + (\alpha\lambda_2)\mathbf{a}_2 + \cdots + (\alpha\lambda_n)\mathbf{a}_n$$

Moreover,

$$\mathbf{b}_1 + \mathbf{b}_2 = (\lambda_1 + \beta_1)\mathbf{a}_1 + (\lambda_2 + \beta_2)\mathbf{a}_2 + \cdots + (\lambda_n + \beta_n)\mathbf{a}_n$$

is also a linear combination of $\mathbf{a}_1, \mathbf{a}_2, \ldots, \mathbf{a}_n$.

Theorem 7-1 The set of all linear combinations of a given finite set of vectors is a vector space.

We can combine the properties 1 and 2 of our definition of a vector space to prove:

Theorem 7-2 A set of vectors V is a vector space if and only if, for every pair of vectors $\mathbf{a}_1$, $\mathbf{a}_2$ in V, the linear combination $(\beta_1\mathbf{a}_1 + \beta_2\mathbf{a}_2)$ is also in V, for any scalars β_1, β_2.

We leave the proof of Theorem 7-2 and also of Theorem 7-3 below as exercises for the reader.

Theorem 7-3 If $\mathbf{a}_1, \mathbf{a}_2, \ldots, \mathbf{a}_k$ are vectors from a vector space V, then every linear combination of $\mathbf{a}_1, \mathbf{a}_2, \ldots, \mathbf{a}_k$ is also in V.

EXAMPLE 7-1 Consider the vector space

$$V = \left\{ \begin{bmatrix} x_1 \\ x_2 \\ x_3 \end{bmatrix} \middle| \; x_2 = 2x_3 \right\}$$

The vectors

$$\mathbf{a}_1 = \begin{bmatrix} 1 \\ 2 \\ 1 \end{bmatrix} \quad \text{and} \quad \mathbf{a}_2 = \begin{bmatrix} 0 \\ -4 \\ -2 \end{bmatrix}$$

are clearly elements of V. We now show that the set of all linear combinations of $\mathbf{a}_1$ and $\mathbf{a}_2$ is V itself. By Theorem 7-3, every linear combination of $\mathbf{a}_1$, $\mathbf{a}_2$ is in V; on the other hand, we must show that every element of V is a linear combination of $\mathbf{a}_1, \mathbf{a}_2$. First, we note that every vector in V is of the form

$$\mathbf{b} = \begin{bmatrix} \lambda_1 \\ 2\lambda_2 \\ \lambda_2 \end{bmatrix}$$

Therefore, we must show that there exist scalars β_1, β_2 such that

$$\mathbf{b} = \beta_1\mathbf{a}_1 + \beta_2\mathbf{a}_2$$

In other words, we must determine whether the following system of equations is consistent:

$$[\mathbf{a}_1 \; \mathbf{a}_2]\begin{bmatrix} \beta_1 \\ \beta_2 \end{bmatrix} = \mathbf{b}$$

$$\begin{bmatrix} 1 & 0 \\ 2 & -4 \\ 1 & -2 \end{bmatrix}\begin{bmatrix} \beta_1 \\ \beta_2 \end{bmatrix} = \begin{bmatrix} \lambda_1 \\ 2\lambda_2 \\ \lambda_2 \end{bmatrix}$$

The augmented matrix for the above system is

$$\left[\begin{array}{rr:c} 1 & 0 & \lambda_1 \\ 2 & -4 & 2\lambda_2 \\ 1 & -2 & \lambda_2 \end{array}\right]$$

Using elementary row operations yields

$$\left[\begin{array}{rr:c} 1 & 0 & \lambda_1 \\ 0 & -4 & 2\lambda_2 - 2\lambda_1 \\ 0 & -2 & \lambda_2 - \lambda_1 \end{array}\right] \rightarrow \left[\begin{array}{rr:c} 1 & 0 & \lambda_1 \\ 0 & 1 & \frac{1}{2}(\lambda_1 - \lambda_2) \\ \hdashline 0 & 0 & 0 \end{array}\right]$$

Hence, the system is consistent, and has the unique solution $\beta_1 = \lambda_1$, $\beta_2 = \frac{1}{2}(\lambda_1 - \lambda_2)$. ////

Example 7-1 suggests that there is a direct relationship between vector spaces and systems of equations. Let us pursue this idea further. We shall eventually show that:

1 Every vector space can be described by a homogeneous system of equations.
2 In every vector space V there exists a finite set of vectors S such that every vector in V can be expressed as a linear combination of the vectors in S.

In the vector space V of Example 7-1 we found such a set S; namely, $S = \{\mathbf{a}_1, \mathbf{a}_2\}$. Thus, although we have not yet proved statement 2 above, we have shown by example that in some vector spaces, such a set S does in fact exist. The set S is called a *spanning set* of the vector space V; the vectors in S are said to *span* V.

EXAMPLE 7-2 Consider the vector space

$$V = \left\{ \begin{bmatrix} x_1 \\ x_2 \\ x_3 \\ x_4 \end{bmatrix} \middle| \begin{array}{l} x_1 = 0 \\ x_2 + x_3 - 2x_4 = 0 \end{array} \right\}$$

and the set

$$S = \left\{ \begin{bmatrix} 0 \\ 1 \\ 1 \\ 1 \end{bmatrix}, \begin{bmatrix} 0 \\ 0 \\ 2 \\ 1 \end{bmatrix}, \begin{bmatrix} 0 \\ 2 \\ 0 \\ 1 \end{bmatrix} \right\}$$

Let us show that S is indeed a spanning set of V. Again, we begin by characterizing the vectors of V:

Observe that $\mathbf{b}$ is in V if and only if

$$\mathbf{b} = \begin{bmatrix} 0 \\ \lambda_1 \\ \lambda_2 \\ \frac{1}{2}(\lambda_1 + \lambda_2) \end{bmatrix}$$

Hence we must determine whether $\mathbf{b}$ can be expressed as a linear combination of the vectors in S. Setting

$$\alpha_1 \begin{bmatrix} 0 \\ 1 \\ 1 \\ 1 \end{bmatrix} + \alpha_2 \begin{bmatrix} 0 \\ 0 \\ 2 \\ 1 \end{bmatrix} + \alpha_3 \begin{bmatrix} 0 \\ 2 \\ 0 \\ 1 \end{bmatrix} = \begin{bmatrix} 0 \\ \lambda_1 \\ \lambda_2 \\ \frac{1}{2}(\lambda_1 + \lambda_2) \end{bmatrix}$$

yields the system of equations

$$\begin{bmatrix} 0 & 0 & 0 \\ 1 & 0 & 2 \\ 1 & 2 & 0 \\ 1 & 1 & 1 \end{bmatrix} \begin{bmatrix} \alpha_1 \\ \alpha_2 \\ \alpha_3 \end{bmatrix} = \begin{bmatrix} 0 \\ \lambda_1 \\ \lambda_2 \\ \frac{1}{2}(\lambda_1 + \lambda_2) \end{bmatrix}$$

from which we obtain

$$\left[\begin{array}{ccc|c} 0 & 0 & 0 & 0 \\ 1 & 0 & 2 & \lambda_1 \\ 1 & 2 & 0 & \lambda_2 \\ 1 & 1 & 1 & \frac{1}{2}(\lambda_1 + \lambda_2) \end{array}\right] \rightarrow \left[\begin{array}{ccc|c} 1 & 1 & 1 & \frac{1}{2}(\lambda_1 + \lambda_2) \\ 1 & 0 & 2 & \lambda_1 \\ 1 & 2 & 0 & \lambda_2 \\ 0 & 0 & 0 & 0 \end{array}\right]$$

$$\rightarrow \left[\begin{array}{ccc|c} 1 & 1 & 1 & \frac{1}{2}(\lambda_1 + \lambda_2) \\ 0 & -1 & 1 & \frac{1}{2}(\lambda_1 - \lambda_2) \\ 0 & 1 & -1 & \frac{1}{2}(\lambda_2 - \lambda_1) \\ 0 & 0 & 0 & 0 \end{array}\right] \rightarrow \left[\begin{array}{ccc|c} 1 & 1 & 1 & \frac{1}{2}(\lambda_1 + \lambda_2) \\ 0 & 1 & -1 & \frac{1}{2}(\lambda_2 - \lambda_1) \\ 0 & 0 & 0 & 0 \\ 0 & 0 & 0 & 0 \end{array}\right]$$

$$\rightarrow \left[\begin{array}{cc|c|c} 1 & 0 & 2 & \lambda_1 \\ 0 & 1 & -1 & \frac{1}{2}(\lambda_2 - \lambda_1) \\ \hline 0 & 0 & 0 & 0 \\ 0 & 0 & 0 & 0 \end{array}\right]$$

Hence,

$$\alpha_1 = \lambda_1 - 2\alpha_3$$
$$\alpha_2 = \tfrac{1}{2}(\lambda_2 - \lambda_1) + \alpha_3$$
$$\alpha_3 = \alpha_3$$

and the system is in fact consistent, with one arbitrary variable. If we set $\alpha_3 = 0$, we find that we can express **b** as a linear combination of the first two vectors of S:

$$\mathbf{b} = \lambda_1 \begin{bmatrix} 0 \\ 1 \\ 1 \\ 1 \end{bmatrix} + \tfrac{1}{2}(\lambda_2 - \lambda_1) \begin{bmatrix} 0 \\ 0 \\ 2 \\ 1 \end{bmatrix} = \begin{bmatrix} 0 \\ \lambda_1 \\ \lambda_2 \\ \tfrac{1}{2}(\lambda_1 + \lambda_2) \end{bmatrix}$$

Thus, not only have we determined that S is a spanning set of V, but in addition we have found that the first two vectors of S alone also form a spanning set of V. The reader may have observed that these latter two vectors are linearly independent, whereas the three vectors of S are linearly dependent. ////

We now turn to the problem of proving statement 2. To do so, we need some preliminary results:

Theorem 7-4 A set of vectors is linearly dependent if and only if at least one of the vectors is a linear combination of the others.

PROOF (*a*) Let $\mathbf{a}_1, \mathbf{a}_2, \ldots, \mathbf{a}_k$ be linearly dependent. Then, by definition, there exist scalars $\beta_1, \beta_2, \ldots, \beta_k$ not all zero, such that

$$\beta_1 \mathbf{a}_1 + \beta_2 \mathbf{a}_2 + \cdots + \beta_k \mathbf{a}_k = 0$$

Suppose $\beta_1 \neq 0$. Then we can write

$$\mathbf{a}_1 = -\frac{\beta_2}{\beta_1}\mathbf{a}_2 + -\frac{\beta_3}{\beta_1}\mathbf{a}_3 + \cdots + -\frac{\beta_k}{\beta_1}\mathbf{a}_k$$

Hence, $\mathbf{a}_1$ is a linear combination of $\mathbf{a}_2, \mathbf{a}_3, \ldots, \mathbf{a}_k$.

(*b*) If $\mathbf{a}_1$ is a linear combination of $\mathbf{a}_2, \mathbf{a}_3, \ldots, \mathbf{a}_k$, then

$$\mathbf{a}_1 = \lambda_2 \mathbf{a}_2 + \lambda_3 \mathbf{a}_3 + \cdots + \lambda_k \mathbf{a}_k$$

and

$$(-1)\mathbf{a}_1 + \lambda_2 \mathbf{a}_2 + \cdots + \lambda_k \mathbf{a}_k = \mathbf{0}$$

Hence, the vectors are linearly dependent. QED

Theorem 7-5 Every vector space, whose elements are m component vectors, has at most m linearly independent vectors.

PROOF This theorem is a direct consequence of Theorem 6-13 (see Exercise 13, Chap. 6). QED

If we combine the results of Theorems 7-4 and 7-5, we have:

Theorem 7-6 Given any vector space V, there exists a finite subset S of vectors in V such that (a) the vectors in S are linearly independent, and (b) every vector in V is a linear combination of the vectors in S.

PROOF Suppose V consists of m-component vectors. According to Theorem 7-5, the particular vector space V has at most p linearly independent vectors, where $p \leq m$. Then, let $S = \{\mathbf{a}_1, \mathbf{a}_2, \ldots, \mathbf{a}_p\}$ be any such set of p linearly independent vectors from V, and let $\mathbf{b}$ be *any* vector in V. The $p + 1$ vectors $\mathbf{a}_1, \mathbf{a}_2, \ldots, \mathbf{a}_p, \mathbf{b}$ are linearly dependent (since by assumption, p is the maximum number of linearly independent vectors in V) and by Theorem 7-4 at least one of $\mathbf{a}_1, \mathbf{a}_2, \ldots, \mathbf{a}_p, \mathbf{b}$ must be a linear combination of the others. If we can show that $\mathbf{b}$ must be a linear combination of $\mathbf{a}_1, \mathbf{a}_2, \ldots, \mathbf{a}_p$, then we will have proved the theorem.

Consider the linear combination

$$\lambda_1 \mathbf{a}_1 + \lambda_2 \mathbf{a}_2 + \cdots + \lambda_p \mathbf{a}_p + \lambda_{p+1} \mathbf{b} = \mathbf{0} \tag{7-3}$$

Since the vectors $\mathbf{a}_1, \mathbf{a}_2, \ldots, \mathbf{a}_p, \mathbf{b}$ are linearly dependent, at least one of $\lambda_1, \lambda_2, \ldots, \lambda_p, \lambda_{p+1}$ is nonzero. We would of course like to show that $\lambda_{p+1} \neq 0$. However, if λ_{p+1} is equal to zero, then Eq. (7-3) becomes

$$\lambda_1 \mathbf{a}_1 + \lambda_2 \mathbf{a}_2 + \cdots + \lambda_p \mathbf{a}_p = \mathbf{0}$$

and since $\mathbf{a}_1, \mathbf{a}_2, \ldots, \mathbf{a}_p$ are linearly independent, it must also be true that $\lambda_1 = \lambda_2 = \cdots = \lambda_p = 0$. Thus, the assumption that $\lambda_{p+1} = 0$ leads to the conclusion that all the λ_j's of Eq. (7-3) are zero. Since we know that at least one λ_j must be nonzero, our assumption that $\lambda_{p+1} = 0$ must be false. Therefore, $\lambda_{p+1} \neq 0$ and by the reasoning employed in the proof of Theorem 7-4, $\mathbf{b}$ must be a linear combination of $\mathbf{a}_1, \mathbf{a}_2, \ldots, \mathbf{a}_p$.

QED

Theorem 7-6 states that every vector space not only has a spanning set, but has a spanning set whose vectors are linearly independent. In the next section we shall show that for a given vector space, every such spanning set of linearly independent vectors contains the same number of vectors.

7-2 DIMENSION AND BASIS

We have seen that any vector space V of m-component vectors has at most m linearly independent vectors, perhaps fewer than m. If a particular vector space V has at most p linearly independent vectors (and has at least one such set of vectors), then *any* set of p linearly independent vectors of V is called a *basis* for V. Moreover, p is called the *dimension* of V.

In the proof of Theorem 7-6, we showed that any basis of V is also a spanning set. Let us state this result separately:

Theorem 7-7 If $S = \{\mathbf{a}_1, \mathbf{a}_2, \ldots, \mathbf{a}_p\}$ is a basis for a vector space V, then S also spans V.

Thus, according to Theorem 7-7, if V is a p-dimensional vector space and $\mathbf{a}_1, \mathbf{a}_2, \ldots, \mathbf{a}_p$ are linearly independent vectors in V, then $\mathbf{a}_1, \mathbf{a}_2, \ldots, \mathbf{a}_p$ span V; i.e., V is precisely the set of all linear combinations of $\mathbf{a}_1, \mathbf{a}_2, \ldots, \mathbf{a}_p$:

$$V = \{\mathbf{b} \mid \mathbf{b} = \alpha_1 \mathbf{a}_1 + \alpha_2 \mathbf{a}_2 + \cdots + \alpha_p \mathbf{a}_p\}$$

If the vectors $\mathbf{a}_j$ above are m-component vectors, and if we let $\mathbf{a}_j = [a_{1j} \quad a_{2j} \quad \cdots \quad a_{mj}]^T$, $j = 1, 2, \ldots, p$, then V is the set of *all* vectors $\mathbf{b} = [b_1 \quad b_2 \quad \cdots \quad b_m]^T$ such that

$$\alpha_1 \mathbf{a}_1 + \alpha_2 \mathbf{a}_2 + \cdots + \alpha_p \mathbf{a}_p = \mathbf{b}$$

$$\alpha_1 \begin{bmatrix} a_{11} \\ a_{21} \\ \vdots \\ a_{m1} \end{bmatrix} + \alpha_2 \begin{bmatrix} a_{12} \\ a_{22} \\ \vdots \\ a_{m2} \end{bmatrix} + \cdots + \alpha_p \begin{bmatrix} a_{1p} \\ a_{2p} \\ \vdots \\ a_{mp} \end{bmatrix} = \begin{bmatrix} b_1 \\ b_2 \\ \vdots \\ b_m \end{bmatrix}$$

Or

$$\begin{bmatrix} a_{11} & a_{12} & \cdots & a_{1p} \\ a_{21} & a_{22} & \cdots & a_{2p} \\ \cdot & \cdot & \cdots & \cdot \\ \cdot & \cdot & \cdots & \cdot \\ \cdot & \cdot & \cdots & \cdot \\ a_{m1} & a_{m2} & \cdots & a_{mp} \end{bmatrix} \begin{bmatrix} \alpha_1 \\ \alpha_2 \\ \vdots \\ \alpha_p \end{bmatrix} = \begin{bmatrix} b_1 \\ b_2 \\ \cdot \\ \cdot \\ \cdot \\ b_m \end{bmatrix} \tag{7-4}$$

Letting $\mathbf{A} = [\mathbf{a}_1 \quad \mathbf{a}_2 \quad \cdots \quad \mathbf{a}_p]$ we have that the augmented matrix for the system of equations (7-4) is $[\mathbf{A}, \mathbf{b}]$. Moreover, the vector space V is the set of all vectors $\mathbf{b}$ such that the system (7-4) is consistent. We can determine what restrictions this places on the components of $\mathbf{b}$ by reducing the augmented matrix $[\mathbf{A}, \mathbf{b}]$, using a Gaussian elimination type of method, to the following form:

$$\left[\begin{array}{c|c} \mathbf{I}_p & \mathbf{f} \\ \hline \mathbf{O}_{(m-p)\times p} & \mathbf{g} \end{array}\right] \tag{7-5}$$

Since the columns of **A** are linearly independent, we must be able to obtain the above form. Thus, the system is consistent if and only if $\mathbf{g} = \mathbf{0}$. However, each component of **g** is merely a linear combination of $b_1, b_2, \ldots, b_p$, which was obtained in the course of the elimination procedure employed above (i.e., by adding multiples of one row to another, etc.). Perhaps the best way to illustrate this result is with an example:

EXAMPLE 7-3 Consider a vector space V which has a basis consisting of

$$\mathbf{a}_1 = \begin{bmatrix} 0 \\ 1 \\ 1 \\ 1 \end{bmatrix} \qquad \mathbf{a}_2 = \begin{bmatrix} 0 \\ 0 \\ 2 \\ 1 \end{bmatrix}$$

(These are the same $\mathbf{a}_1$, $\mathbf{a}_2$ of Example 7-2.)

Then, V is the set of all linear combinations of $\mathbf{a}_1$, $\mathbf{a}_2$; that is, the set of all $\mathbf{b} = [b_1 \quad b_2 \quad b_3 \quad b_4]^T$ such that

$$\alpha_1 \begin{bmatrix} 0 \\ 1 \\ 1 \\ 1 \end{bmatrix} + \alpha_2 \begin{bmatrix} 0 \\ 0 \\ 2 \\ 1 \end{bmatrix} = \begin{bmatrix} b_1 \\ b_2 \\ b_3 \\ b_4 \end{bmatrix}$$

The augmented matrix for this system is

$$\left[\begin{array}{cc|c} 0 & 0 & b_1 \\ 1 & 0 & b_2 \\ 1 & 2 & b_3 \\ 1 & 1 & b_4 \end{array}\right]$$

from which we obtain

$$\left[\begin{array}{cc|c} 1 & 0 & b_2 \\ 0 & 0 & b_1 \\ 1 & 2 & b_3 \\ 1 & 1 & b_4 \end{array}\right] \rightarrow \left[\begin{array}{cc|c} 1 & 0 & b_2 \\ 0 & 0 & b_1 \\ 0 & 2 & b_3 - b_2 \\ 0 & 1 & b_4 - b_2 \end{array}\right] \rightarrow \left[\begin{array}{cc|c} 1 & 0 & b_2 \\ 0 & 1 & \frac{1}{2}(b_3 - b_2) \\ \hline 0 & 0 & b_1 \\ 0 & 0 & b_4 - \frac{1}{2}b_2 - \frac{1}{2}b_3 \end{array}\right]$$

Hence, in this example, the **g** of (7-5) is

$$\mathbf{g} = \begin{bmatrix} b_1 \\ b_4 - \frac{1}{2}b_2 - \frac{1}{2}b_3 \end{bmatrix}$$

and $\mathbf{b}$ is in V if and only if $\mathbf{g} = \mathbf{0}$:

$$\begin{aligned} b_1 &= 0 \\ b_4 - \tfrac{1}{2}b_2 - \tfrac{1}{2}b_3 &= 0 \end{aligned} \qquad (7\text{-}6)$$

Since, the second of these two equations is equivalent to $b_2 + b_3 - 2b_4 = 0$, the reader will note that V is precisely the same vector space as that of Example 7-2, which we found was spanned by $\mathbf{a}_1$ and $\mathbf{a}_2$.

Note also that the system (7-6) is homogeneous. ////

Example 7-3 and the discussion preceding it lead us directly to:

Theorem 7-8 Every vector space V may be characterized by a homogeneous system of equations. In particular, for every vector space V there exists a homogeneous system of equations† $\mathbf{Bb} = \mathbf{0}$ such that a vector $\hat{\mathbf{b}}$ is in $\mathbf{V}$ if and only if $\hat{\mathbf{b}}$ is a solution of $\mathbf{Bb} = \mathbf{0}$.

Theorems 7-1, 7-6, and 7-8 tell us that any vector space V can be specified by either (*a*) providing a basis for V or (*b*) providing the homogeneous system of equations $\mathbf{Bb} = \mathbf{0}$ referred to in Theorem 7-8.

In Example 7-3 we have seen how to determine the homogeneous system $\mathbf{Bb} = \mathbf{0}$, given a basis for a vector space. Now, let us investigate the problem of determining a basis for the vector space consisting of the set of all solutions of a given homogeneous system of equations. This latter problem is much more important, and we shall require a knowledge of how to solve it in Chap. 9.

Recall that a basis for a vector space V consists of any set of p linearly independent vectors of V, where p is the dimension of V. Thus, in order to find a basis, we must first know p. We resolve the latter problem in:

Theorem 7-9 The set of all solutions to a homogeneous system of equations $\mathbf{Bb} = \mathbf{0}$ is a vector space of dimension $p = n - \text{rank}\,(\mathbf{B})$, where $\mathbf{B}$ is of order $m \times n$.

PROOF Let $r = \text{rank}\,(\mathbf{B})$. Then by Theorem 6-3 there exists a sequence of elementary row operations (with, possibly, column interchanges) which reduces $\mathbf{B}$ to the form

$$\left[\begin{array}{c:c} \mathbf{I}_r & \mathbf{C}_{r \times (n-r)} \\ \hdashline \mathbf{O}_{(m-r) \times r} & \mathbf{O}_{(m-r) \times (n-r)} \end{array}\right]$$

† If V is the set of all m-component vectors, then the homogeneous system of equations is just the trivial set: $\mathbf{Ob} = \mathbf{0}$.

Then, the above matrix is the coefficient matrix for an equivalent homogeneous system, and we see that any solution $\mathbf{b}$ may be expressed as follows:

Let

$$\mathbf{b} = \begin{bmatrix} \mathbf{y}_1 \\ \mathbf{y}_2 \end{bmatrix} \qquad \mathbf{y}_1 = \begin{bmatrix} b_1 \\ b_2 \\ \vdots \\ b_r \end{bmatrix} \qquad \mathbf{y}_2 = \begin{bmatrix} b_{r+1} \\ b_{r+2} \\ \vdots \\ b_n \end{bmatrix}$$

Then

$$\mathbf{I}_r \mathbf{y}_1 + \mathbf{C}\mathbf{y}_2 = \mathbf{0}$$

$$\mathbf{y}_1 = -\mathbf{C}\mathbf{y}_2 \tag{7-7}$$

As we have discussed in Chap. 6, the $(n - r)$ components of $\mathbf{y}_2$ may be arbitrarily assigned values. If we denote these values by $\beta_1, \beta_2, \ldots, \beta_{n-r}$, and if we also denote the $(n - r)$ columns of $\mathbf{C}$ by $\mathbf{c}_1, \mathbf{c}_2, \ldots, \mathbf{c}_{n-r}$, then Eq. (7-7) becomes

$$\mathbf{y}_1 = -\beta_1\mathbf{c}_1 - \beta_2\mathbf{c}_2 - \cdots - \beta_{n-r}\mathbf{c}_{n-r} \tag{7-8}$$

Moreover, an obvious identity is

$$\mathbf{y}_2 = \begin{bmatrix} \beta_1 \\ \beta_2 \\ \vdots \\ \beta_{n-r} \end{bmatrix} = \beta_1 \begin{bmatrix} 1 \\ 0 \\ 0 \\ \vdots \\ 0 \end{bmatrix} + \beta_2 \begin{bmatrix} 0 \\ 1 \\ 0 \\ \vdots \\ 0 \end{bmatrix} + \cdots + \beta_{n-r} \begin{bmatrix} 0 \\ 0 \\ 0 \\ \vdots \\ 1 \end{bmatrix} \tag{7-9}$$

If we define $\mathbf{e}_j$ as a vector whose jth component is 1 and whose remaining components are zero, then Eq. (7-9) becomes

$$\mathbf{y}_2 = \beta_1\mathbf{e}_1 + \beta_2\mathbf{e}_2 + \cdots + \beta_{n-r}\mathbf{e}_{n-r} \tag{7-10}$$

Since the above discussion has been completely general, we see that *any* solution $\mathbf{b}$ can be expressed as

$$\mathbf{b} = \begin{bmatrix} \mathbf{y}_1 \\ \mathbf{y}_2 \end{bmatrix} = \beta_1 \begin{bmatrix} -\mathbf{c}_1 \\ \mathbf{e}_1 \end{bmatrix} + \beta_2 \begin{bmatrix} -\mathbf{c}_2 \\ \mathbf{e}_2 \end{bmatrix} + \cdots + \beta_{n-r} \begin{bmatrix} -\mathbf{c}_{n-r} \\ \mathbf{e}_{n-r} \end{bmatrix} \tag{7-11}$$

where the above equation is obtained by combining Eqs. (7-8) and (7-10).

Moreover, it is easy to see that the $p = n - r$ vectors

$$\begin{bmatrix} -\mathbf{c}_1 \\ \mathbf{e}_1 \end{bmatrix} \quad \begin{bmatrix} -\mathbf{c}_2 \\ \mathbf{e}_2 \end{bmatrix} \quad \cdots \quad \begin{bmatrix} -\mathbf{c}_{n-r} \\ \mathbf{e}_{n-r} \end{bmatrix}$$

are linearly independent, by direct application of Theorem 6-6.

It now remains to show that no set of $(p+1)$ vectors in the vector space can be linearly independent. But any vector in this space is, by definition, a solution of $\mathbf{Bb} = \mathbf{0}$ and hence may be expressed as a linear combination of the vectors

$$\mathbf{v}_j = \begin{bmatrix} -\mathbf{c}_j \\ \mathbf{e}_j \end{bmatrix} \qquad j = 1, 2, \ldots, n - r \qquad \text{(7-12)}$$

by Eq. (7-11). Thus, if $\mathbf{b}_1, \mathbf{b}_2, \ldots, \mathbf{b}_p, \mathbf{b}_{p+1}$ are any such solutions, then we can write

$$\begin{aligned} \mathbf{b}_1 &= \beta_{11}\mathbf{v}_1 + \beta_{12}\mathbf{v}_2 + \cdots + \beta_{1p}\mathbf{v}_p \\ \mathbf{b}_2 &= \beta_{21}\mathbf{v}_1 + \beta_{22}\mathbf{v}_2 + \cdots + \beta_{2p}\mathbf{v}_p \\ &\vdots \\ \mathbf{b}_p &= \beta_{p1}\mathbf{v}_1 + \beta_{p2}\mathbf{v}_2 + \cdots + \beta_{pp}\mathbf{v}_p \\ \mathbf{b}_{p+1} &= \beta_{p+1,1}\mathbf{v}_1 + \beta_{p+1,2}\mathbf{v}_2 + \cdots + \beta_{p+1,p}\mathbf{v}_p \end{aligned} \qquad \text{(7-13)}$$

Let us now show that $\mathbf{b}_1, \mathbf{b}_2, \ldots, \mathbf{b}_{p+1}$ must be linearly dependent. To do so, we form the linear combination

$$\alpha_1\mathbf{b}_1 + \alpha_2\mathbf{b}_2 + \cdots + \alpha_{p+1}\mathbf{b}_{p+1} = \mathbf{0} \qquad \text{(7-14)}$$

and show that there exist nontrivial values of $\alpha_1, \alpha_2, \ldots, \alpha_{p+1}$ which satisfy (7-14):

$$\begin{aligned} \mathbf{0} &= \alpha_1\mathbf{b}_1 + \alpha_2\mathbf{b}_2 + \cdots + \alpha_{p+1}\mathbf{b}_{p+1} \\ &= \sum_{j=1}^{p+1} \alpha_j \mathbf{b}_j \\ &= \sum_{j=1}^{p+1} \alpha_j \{\beta_{j1}\mathbf{v}_1 + \beta_{j2}\mathbf{v}_2 + \cdots + \beta_{jp}\mathbf{v}_p\} \\ &= \sum_{j=1}^{p+1} \alpha_j \left\{ \sum_{i=1}^{p} \beta_{ji}\mathbf{v}_i \right\} \\ &= \sum_{i=1}^{p} \left\{ \sum_{j=1}^{p+1} \alpha_j \beta_{ji} \right\} \mathbf{v}_i = \mathbf{0} \end{aligned}$$

Since $\mathbf{v}_1, \mathbf{v}_2, \ldots, \mathbf{v}_p$ are linearly independent, the coefficients $\{\sum_{j=1}^{p+1} \alpha_j \beta_{ji}\}$ must be all zero:

$$\sum_{j=1}^{p+1} \alpha_j \beta_{ji} = 0 \qquad i = 1, 2, \ldots, p \qquad \text{(7-15)}$$

Equation (7-15) represents a homogeneous system of p equations in the $(p+1)$ variables $\alpha_1, \alpha_2, \ldots, \alpha_{p+1}$. By Theorem 6-9, this system possesses nontrivial solutions since the rank of the coefficient matrix is at most p.

To summarize this proof, we have shown that:

Given a homogeneous system of equations $\mathbf{Bb} = \mathbf{0}$, where $\mathbf{B}$ is $m \times n$ and rank $(\mathbf{B}) = r$, then (*a*) the set of all solutions to $\mathbf{Bb} = \mathbf{0}$ is a vector space; (*b*) if $p = n - r$, then the p vectors $\mathbf{v}_1, \mathbf{v}_2, \ldots, \mathbf{v}_p$ defined by (7-12) are linearly independent; (*c*) every set of $(p + 1)$ vectors in this vector space is linearly dependent.

Combining (*b*) and (*c*), we see that the dimension of the vector space is p, since this is the maximum number of linearly independent vectors in the space. QED

In the course of proving the above theorem, we have not only shown the dimension of the vector space to be p, but we have actually found a basis for it as well; namely, the p linearly independent vectors $\mathbf{v}_1, \mathbf{v}_2, \ldots, \mathbf{v}_p$:

Theorem 7-10 The vectors $\mathbf{v}_1, \mathbf{v}_2, \ldots, \mathbf{v}_p$, defined by Eqs. (7-7) through (7-12), constitute a basis for the vector space consisting of the set of all solutions to $\mathbf{Bb} = \mathbf{0}$, where $\mathbf{B}$ is an $m \times n$ matrix of rank r (and where $p = n - r$).

As the reader has no doubt observed from the discussions throughout this chapter, the phrase "the vector space consisting of the set of all solutions to the homogeneous system of equations $\mathbf{Bb} = \mathbf{0}$" is rather awkward, to say the least. Moreover, this phrase has occurred quite frequently in our discussions. Therefore, we shall assign a special name to the above vector space: henceforth, it shall be called the *null space* of $\mathbf{B}$.

EXAMPLE 7-4 Let us find the dimension of, and a basis for, the null space of

$$\mathbf{B} = \begin{bmatrix} 1 & 2 & 2 & 0 \\ 0 & 1 & 2 & 1 \\ -2 & -1 & 2 & 3 \end{bmatrix}$$

Performing elementary row operations on $\mathbf{B}$ yields

$$\rightarrow \begin{bmatrix} 1 & 2 & 2 & 0 \\ 0 & 1 & 2 & 1 \\ 0 & 3 & 6 & 3 \end{bmatrix} \rightarrow \left[\begin{array}{cc|cc} 1 & 0 & -2 & -2 \\ 0 & 1 & 2 & 1 \\ \hline 0 & 0 & 0 & 0 \end{array}\right]$$

Hence, rank $(\mathbf{B}) = 2$ and, by Theorem 7-9, the dimension of the null space of $\mathbf{B}$ is $n - r = 4 - 2 = 2$.

Since

$$\mathbf{C} = \begin{bmatrix} -2 & -2 \\ 2 & 1 \end{bmatrix}$$

and

$$\mathbf{c}_1 = \begin{bmatrix} -2 \\ 2 \end{bmatrix} \qquad \mathbf{c}_2 = \begin{bmatrix} -2 \\ 1 \end{bmatrix}$$

we see, by Theorem 7-10, that

$$\mathbf{v}_1 = \begin{bmatrix} -\mathbf{c}_1 \\ \mathbf{e}_1 \end{bmatrix} = \begin{bmatrix} 2 \\ -2 \\ 1 \\ 0 \end{bmatrix} \qquad \mathbf{v}_2 = \begin{bmatrix} -\mathbf{c}_2 \\ \mathbf{e}_2 \end{bmatrix} = \begin{bmatrix} 2 \\ -1 \\ 0 \\ 1 \end{bmatrix}$$

The reader should verify that $\mathbf{v}_1$ and $\mathbf{v}_2$ are indeed linearly independent and that they are both members of the null space of $\mathbf{B}$ (i.e., that $\mathbf{B}\mathbf{v}_1 = \mathbf{0}$ and $\mathbf{B}\mathbf{v}_2 = \mathbf{0}$). ////

We can also use some of the methodology employed in the proof of Theorem 7-9 to prove several additional extremely important results:

Theorem 7-11 If a linearly independent set of p vectors $\mathbf{a}_1, \mathbf{a}_2, \ldots, \mathbf{a}_p$ spans a vector space V, then the dimension of V is equal to p.

PROOF Since the p vectors $\mathbf{a}_1, \mathbf{a}_2, \ldots, \mathbf{a}_p$ span V, then every vector in V is a linear combination of these vectors. Thus, if $\mathbf{b}_1, \mathbf{b}_2, \ldots, \mathbf{b}_p, \mathbf{b}_{p+1}$ are any $(p + 1)$ vectors in V, then we can write

$$\mathbf{b}_i = \sum_{j=1}^{p} \alpha_{ij} \mathbf{a}_j \qquad i = 1, 2, \ldots, p + 1$$

Let us now show that the $(p + 1)$ vectors $\mathbf{b}_i$, $i = 1, 2, \ldots, p + 1$, are linearly dependent. Let

$$\sum_{i=1}^{p+1} \beta_i \mathbf{b}_i = \mathbf{0}$$

$$\sum_{i=1}^{p+1} \beta_i \left(\sum_{j=1}^{p} \alpha_{ij} \mathbf{a}_j \right) = \mathbf{0}$$

$$\sum_{j=1}^{p} \left(\sum_{i=1}^{p+1} \alpha_{ij} \beta_i \right) \mathbf{a}_j = \mathbf{0}$$

Since $\mathbf{a}_1, \mathbf{a}_2, \ldots, \mathbf{a}_p$ are linearly independent, it must be true that

$$\sum_{i=1}^{p+1} \alpha_{ij} \beta_i = 0 \qquad j = 1, 2, \ldots, p$$

Thus, we have a homogeneous system of p equations in $(p + 1)$ variables $\beta_1, \beta_2, \ldots, \beta_{p+1}$, which always possesses nontrivial solutions. Hence, $\mathbf{b}_1, \mathbf{b}_2, \ldots, \mathbf{b}_{p+1}$ must be linearly dependent, and the maximum number of linearly independent vectors in V is therefore equal to p.

QED

Theorem 7-12 If V is a vector space of dimension p, then every linearly independent spanning set of V contains exactly p vectors.

PROOF Suppose $\mathbf{a}_1, \mathbf{a}_2, \ldots, \mathbf{a}_k$ is any linearly independent set of vectors which spans V. Then, by Theorem 7-11, the dimension of V must be k. However, by hypothesis, the dimension of $V = p$. Therefore, $k = p$, and we have shown that every linearly independent spanning set of V must contain exactly p vectors. QED

If we compare Theorem 7-12 with our definition of "basis," we immediately obtain:

Theorem 7-13 Given a vector space V of dimension p, then every basis of V is also a linearly independent spanning set; and conversely.

7-3 ABSTRACT VECTOR SPACES

At the beginning of this chapter we stated a definition of a vector space which contained two "axioms," or properties. Namely, our definition stated that:

A set of vectors V is called a *vector space* if for any two vectors $\mathbf{x}_1$ and $\mathbf{x}_2$ in V and any scalar α:

1 $(\mathbf{x}_1 + \mathbf{x}_2)$ is in V.
2 $\alpha\mathbf{x}_1$ is in V.

However, we wish to point out at this time that the above definition applies only to "vectors" which are in fact either row or column vectors, i.e., matrices with either one row or one column. Of course, these are the only kinds of vectors we have defined in this text. But, the term *vector* has a much broader definition in mathematics, and we shall now discuss the latter briefly, as the reader is apt to encounter it in many linear algebra texts.

Specifically, the most general definition of a vector space is the following: A set of "objects" (or "elements") V is called a *vector space* if, for any elements a, b, and c of V, and any scalars α, β:

1 $(a + b)$ is in V.

2 $a + b = b + a$

3 $a + (b + c) = (a + b) + c$

4 For any pair of elements a, b, there exists an element e in V such that $b - e = a$

5 There exists a unique element "0" in V such that $a + 0 = a$

6 αa is in V.

7 $\alpha(\beta a) = (\alpha\beta)a$

8 $(\alpha + \beta)a = \alpha a + \beta a$

9 $1a = a$

The elements of any such vector space are called *vectors*. Thus, our concept of a vector is merely a special case of the above. Note that our earlier definition of a vector space consists of properties 1 and 6 and that the other seven properties of the more general definition are automatically satisfied by column or row vectors.

Below are several examples of vector spaces whose elements are not column or row vectors:

(*a*) The set V_1 of all polynomials of degree less than or equal to 3:

$$V_1 = \{p(x) | p(x) = a_0 + a_1 x + a_2 x^2 + a_3 x^3\}$$

(*b*) The set V_2 of all functions of a single variable t which are continuous over the interval $a \le t \le b$:

$$V_2 = \{f(t) | f(t) \text{ is continuous for } a \le t \le b\}$$

(*Note:* The sum of two continuous functions is itself a continuous function.)

(*c*) The set V_3 of all functions $f(t)$ which are continuous over the interval $a \le t \le b$ and which satisfy $f(a) = f(b)$.

The reader should verify for himself that each of the above is indeed a vector space.

Vector spaces of the above types are of great importance in many areas of engineering and physics and in the field of numerical analysis.

Most† of our earlier theorems and concepts concerning vector spaces can be generalized to these abstract vector spaces:

† Two exceptions are Theorems 7-6 and 7-8.

Given any vector space V, and elements (i.e., vectors) $a_1, a_2, \ldots, a_k$ of V, then $a_1, a_2, \ldots, a_k$ are linearly independent if and only if the only set of scalars satisfying

$$\alpha_1 a_1 + \alpha_2 a_2 + \cdots + \alpha_k a_k = 0$$

is the set $\alpha_1 = \alpha_2 = \cdots = \alpha_k = 0$. A vector space V is of dimension p if p is the maximum number of linearly independent vectors in V. However, it is possible that a vector space will not have a finite dimension, in which case it is called an infinite dimensional vector space; an example of the latter is the vector space V_2 of all continuous functions.

EXAMPLE 7-5 Consider the vector space

$$V_4 = \{p(x) \mid p(x) = a_0 + a_1 x + a_2 x^2\}$$

We shall show that V_4 is of dimension 3 and that a basis for V_4 is the following set of vectors:

$$p_1(x) = 1 \qquad p_2(x) = x \qquad p_3(x) = x^2$$

First of all, it is obvious that every vector in V_4 can be expressed as a linear combination of $p_1(x)$, $p_2(x)$, and $p_3(x)$. Hence, these vectors span V_4. To show that $p_1(x)$, $p_2(x)$, and $p_3(x)$ are indeed linearly independent, we write

$$\alpha_1 p_1(x) + \alpha_2 p_2(x) + \alpha_3 p_3(x) = 0$$
$$\alpha_1 1 + \alpha_2 x + \alpha_3 x^2 = 0$$

In other words, the function $\alpha_1 + \alpha_2 x + \alpha_3 x^2$ must equal zero for *all values* of x. This can occur if and only if $\alpha_1 = \alpha_2 = \alpha_3 = 0$. ////

EXERCISES

1 Determine whether each of the following sets is a vector space. For each set which you decide is not a vector space, provide a counterexample to illustrate your decision.

(*a*) $S_1 = \left\{ \begin{bmatrix} x_1 \\ x_2 \\ x_3 \\ x_4 \end{bmatrix} \middle| \; x_1 - 2x_2 + 3x_3 = x_4 \right\}$

(*b*) $S_2 = \left\{ \begin{bmatrix} x_1 \\ x_2 \\ x_3 \\ x_4 \end{bmatrix} \middle| \; x_1 x_2 = 0 \right\}$

(*c*) $S_3 = \left\{ \begin{bmatrix} x_1 \\ x_2 \\ x_3 \end{bmatrix} \middle| \; x_1 - x_2 \geq 0 \right\}$

(*d*) $S_4 = \left\{ \begin{bmatrix} x_1 \\ x_2 \\ x_3 \\ x_4 \end{bmatrix} \middle| \; \begin{matrix} x_1 = 2x_2 \\ x_3 - x_4 = 0 \end{matrix} \right\}$

(*e*) $S_5 = \left\{ \begin{bmatrix} x_1 \\ x_2 \\ x_3 \end{bmatrix} \middle| \; x_1 - 2x_2 = x_3 + 4 \right\}$

2 Prove Theorem 7-2.

3 Prove Theorem 7-3.

4 Show that the set

$$S = \left\{ \begin{bmatrix} 0 \\ 0 \\ 1 \\ 3 \end{bmatrix} \begin{bmatrix} 1 \\ 0 \\ 0 \\ 1 \end{bmatrix} \begin{bmatrix} 2 \\ 1 \\ 0 \\ 0 \end{bmatrix} \begin{bmatrix} 0 \\ 3 \\ 2 \\ 0 \end{bmatrix} \right\}$$

is a spanning set for the vector space

$$V = \left\{ \begin{bmatrix} x_1 \\ x_2 \\ x_3 \\ x_4 \end{bmatrix} \middle| \; x_1 - 2x_2 + 3x_3 = x_4 \right\}$$

Are the vectors in S linearly independent?

5 Determine whether the set

$$T = \left\{ \begin{bmatrix} 1 \\ 3 \\ 2 \\ 1 \end{bmatrix} \begin{bmatrix} -1 \\ 2 \\ 2 \\ 1 \end{bmatrix} \begin{bmatrix} 0 \\ 1 \\ 1 \\ 1 \end{bmatrix} \right\}$$

is a spanning set for the vector space V of Exercise 4.

6 Determine whether the set

$$U = \left\{ \begin{bmatrix} 1 \\ 3 \\ 2 \\ 1 \end{bmatrix} \begin{bmatrix} -1 \\ 2 \\ 2 \\ 1 \end{bmatrix} \begin{bmatrix} 5 \\ 5 \\ 2 \\ 1 \end{bmatrix} \right\}$$

is a spanning set for the vector space V of Exercise 4.

7 Determine the dimension of, and a basis for, the vector space of Exercise 4.

8 Determine the dimension of, and a basis for, the vector space

$$V = \left\{ \begin{bmatrix} x_1 \\ x_2 \\ x_3 \\ x_4 \end{bmatrix} \middle| \begin{array}{l} x_1 - 2x_3 = 0 \\ x_2 + x_3 - x_4 = 0 \end{array} \right\}$$

9 Determine the dimension of, and a basis for, the vector space which is the set of all linear combinations of the following vectors:

$$\begin{bmatrix} 1 \\ 1 \\ 1 \\ 2 \end{bmatrix} \quad \begin{bmatrix} 0 \\ 1 \\ 0 \\ -2 \end{bmatrix} \quad \begin{bmatrix} -1 \\ 1 \\ 1 \\ 0 \end{bmatrix} \quad \begin{bmatrix} 0 \\ -2 \\ -1 \\ 1 \end{bmatrix}$$

10 Determine the dimension of, and a basis for, the vector space which is the set of all linear combinations of the following vectors:

$$\begin{bmatrix} 0 \\ 1 \\ 1 \\ 1 \end{bmatrix} \quad \begin{bmatrix} 1 \\ 0 \\ 1 \\ 1 \end{bmatrix} \quad \begin{bmatrix} 1 \\ 1 \\ 0 \\ 1 \end{bmatrix} \quad \begin{bmatrix} 1 \\ 1 \\ 1 \\ 0 \end{bmatrix} \quad \begin{bmatrix} 2 \\ -3 \\ 4 \\ 1 \end{bmatrix}$$

11 Find a matrix whose null space is the vector space of Exercise 4.

12 Find a matrix whose null space is the vector space of Exercise 9.

13 Find the dimension of, and a basis for, the null space of

$$A = \begin{bmatrix} 1 & 2 & 0 & -3 \\ 0 & 3 & 2 & 1 \\ -1 & 1 & 2 & 4 \\ -3 & 1 & 4 & 1 \end{bmatrix}$$

14 Find the dimension of, and a basis for, the null space of

$$B = \begin{bmatrix} 0 & 1 & 1 & 1 & -2 \\ 0 & 1 & 1 & 1 & 1 \\ 1 & 1 & 0 & 2 & -1 \\ 1 & -2 & -3 & -1 & 5 \end{bmatrix}$$

15 Find the dimension of, and a basis for, the null space of

$$C = \begin{bmatrix} 2 & -3 & 4 & 1 \\ 1 & 1 & 1 & 0 \\ 1 & 1 & 0 & 1 \\ 1 & 0 & 1 & 1 \\ 0 & 1 & 1 & 1 \end{bmatrix}$$

16 Show that the following set is an abstract vector space, as defined in Sec. 7-3:

$$S = \{p(x) \mid p(x) = a_0 + a_1 x + a_2 x^2 \quad \text{and} \quad a_0 = 2a_1\}$$

What is the dimension of S? Find a basis for it.

8

LINEAR PROGRAMMING

8-1 INTRODUCTION

We have seen in the past few chapters that, in general, a system of m equations in n variables can be expected to possess an infinity of solutions if $m < n$. This is particularly so if the system of equations in question describes an actually existing physical system or process, in which the variables may represent, for example, production levels of different items. Typically, in such a system, the equations are not inconsistent, and furthermore, there is more than one set (and therefore, an infinity) of feasible values for these variables.

In this chapter we shall study such systems from a perspective different from that employed in the preceding chapters. Instead of attempting to find—or characterize—the set of *all* solutions to such a system, we shall be concerned with determining *one* solution which "optimizes" the system, in a sense to be made more precise below.

Consider the following situation. A soup manufacturer has a line of p varieties of soup. The manufacturing process for each variety requires the use of m different pieces of machinery (e.g., food graters, mixers, kettles, and canning and packaging equipment). The following data are known:

1 Machine i may be used up to b_i minutes per month, $i = 1, 2, \ldots, m$. (This amount varies from machine to machine owing to differing maintenance requirements.)

2 Each case of soup variety j requires a_{ij} minutes on machine i, $i = 1, 2, \ldots, m$ and $j = 1, 2, \ldots, p$.

3 Each case of soup variety j yields a net profit to the company of c_j dollars, $j = 1, 2, \ldots, p$.

The company wishes to determine how many cases of each variety to produce per month, so that the total net profit is as large as possible. If we define the variables

$$x_j = \text{number of cases of variety } j \text{ to be produced} \qquad j = 1, 2, \ldots, p$$

then the total net profit z can be expressed as

$$z = c_1 x_1 + c_2 x_2 + \cdots + c_p x_p \tag{8-1}$$

There are restrictions on how many cases of each variety j can be produced, which are specified by the time limitations on the machinery. Thus, the total amount of time required of machine i is equal to $(a_{i1}x_1 + a_{i2}x_2 + \cdots + a_{ip}x_p)$; this quantity cannot exceed b_i minutes:

$$a_{i1}x_1 + a_{i2}x_2 + \cdots + a_{ip}x_p \leq b_i \qquad i = 1, 2, \ldots, m \tag{8-2}$$

Furthermore, it is clear that we cannot allow the variables x_j to have negative values. Hence, we have the nonnegativity restrictions:

$$x_j \geq 0 \qquad j = 1, 2, \ldots, p \tag{8-3}$$

We can now precisely state the manufacturer's problem in quantitative terms: we wish to find a set of values for $x_1, x_2, \ldots, x_p$ which satisfy the inequalities (8-2) and the nonnegativity restrictions (8-3), and which make the net profit z as large as possible. That is, we wish to find $x_1, x_2, \ldots, x_p$ which

$$\begin{aligned} &\text{Maximize } z = \sum_{j=1}^{p} c_j x_j \\ &\text{such that: } \sum_{j=1}^{p} a_{ij} x_j \leq b_i \qquad i = 1, 2, \ldots, m \\ &\qquad\qquad x_j \geq 0 \qquad j = 1, 2, \ldots, p \end{aligned} \tag{8-4}$$

Any problem which can be formulated in the above manner is called a *linear programming problem.* Such a problem has three main components:

1 A linear *objective function*, the quantity to be optimized (in some cases we may wish to minimize instead of maximize):

$$z = \sum_{j=1}^{p} c_j x_j$$

2 A linear *constraint set:* (8-2)

3 Nonnegativity restrictions on the variables: (8-3)

Since all our theory relating to linear systems deals with linear equations rather than inequalities, it is desirable to convert the inequalities (8-2) into equations. We can do so by introducing one new variable for each of the m constraint inequalities (8-2) as follows: for inequality i, we write

$$\sum_{j=1}^{p} a_{ij} x_{ij} + x_{p+i} = b_i \qquad i = 1, 2, \ldots, m \tag{8-5}$$

The variable x_{p+i} may be interpreted in our example above as the number of minutes that machine i is unused in the production of the p varieties of soup. Thus, x_{p+i} must also be a nonnegative quantity. Such a variable is frequently called a *slack variable.* Since we must add a slack variable x_{p+i} to each of the m constraint inequalities, we now have a total of $(p + m)$ variables, $x_1, x_2, \ldots, x_p, x_{p+1}, \ldots, x_{p+m}$. Letting $p + m = n$, we can express the constraint set as a system of m linear equations in n variables and can use our conventional notation for such a system: $\mathbf{Ax} = \mathbf{b}$. However, we must also remember to modify the nonnegativity restrictions (8-3) to include the slack variables:

$$x_j \geq 0 \qquad j = 1, 2, \ldots, p, \ldots, n \tag{8-6}$$

Thus, we have shown how to convert a set of linear inequalities into a set of linear equations (plus nonnegativity restrictions on the variables). Henceforth, we shall call any problem with a linear objective function and linear constraints (inequalities and/or equations) a linear programming problem.

Once the constraints of a linear programming problem have been converted into equations, we can express the problem in matrix notation, as follows:

Let

$$\mathbf{c}^T = [c_1 \quad c_2 \quad \cdots \quad c_n]$$
$$\mathbf{b}^T = [b_1 \quad b_2 \quad \cdots \quad b_m]$$
$$\mathbf{x}^T = [x_1 \quad x_2 \quad \cdots \quad x_n]$$
$$\mathbf{A} = [a_{ij}]_{m \times n}$$

Then, the linear programming problem becomes

$$\begin{aligned} &\text{Maximize } z = \mathbf{c}^T\mathbf{x} \\ &\text{subject to:} \quad \mathbf{Ax} = \mathbf{b} \\ &\qquad\qquad\quad x_j \geq 0 \qquad j = 1, 2, \ldots, n \end{aligned} \tag{8-7}$$

(*Note:* Since the coefficients c_j in the objective function refer to unit profits, the value of c_j corresponding to any slack variable will be zero.)

Let us now consider several examples of linear programming problems.

EXAMPLE 8-1 Suppose our soup company has some surplus time available on its equipment and wishes to utilize this spare time by introducing two new varieties of soup, shrimp gumbo and clam chowder. Suppose further that only three processes are involved, preparation of ingredients, cooking time, and canning, and that 1200 hours, 1290 hours, and 900 hours will be available during the next year for each of these three processes, respectively (that is, the equipment required for preparing the ingredients will be available for 1200 hours, etc.). Below is a table summarizing the number of hours required per case of soup by each process, for each of the two varieties:

Process	Number of hours per case	
	Shrimp gumbo	Clam chowder
I. Preparation of ingredients	4	1
II. Cooking time	1	3
III. Canning, labeling, and packing	2	2

The company's marketing and sales department has determined that the net profit of these soups will be \$3 per case of shrimp gumbo and \$2 per case of clam chowder. How many cases of each kind of soup should the company produce in order to maximize its net profit?

To formulate this problem as a linear programming problem, we must first define the variables:

Let x_1 = number of cases of shrimp gumbo to be produced

x_2 = number of cases of clam chowder to be produced

Then, the objective function is:

$$\text{Maximize } z = 3x_1 + 2x_2$$

That is, z is the total net profit realized from x_1 cases of shrimp gumbo and x_2 cases of clam chowder.

From the table above, we see that process I (preparation of ingredients) requires 4 hours per case of shrimp gumbo and 1 hour per case of clam chowder. Thus, if x_1 and x_2 cases of these varieties are produced, respectively, then the total time required to complete process I is $(4x_1 + 1x_2)$ hours. This quantity cannot exceed 1200 hours, the total available time for preparation of ingredients. Hence,

$$4x_1 + 1x_2 \leq 1200$$

In a similar manner, we obtain the two constraints corresponding to process II and process III, respectively:

Process II: $1x_1 + 3x_2 \leq 1290$

Process III: $2x_1 + 2x_2 \leq 900$

In addition, we have the nonnegativity restrictions, $x_1 \geq 0$, $x_2 \geq 0$.

Thus, the complete linear programming problem is:

$$\begin{aligned} \text{Maximize } z &= 3x_1 + 2x_2 \\ \text{subject to: } & 4x_1 + 1x_2 \leq 1200 \\ & 1x_1 + 3x_2 \leq 1290 \\ & 2x_1 + 2x_2 \leq 900 \\ & x_1, x_2 \geq 0 \end{aligned} \tag{8-8}$$

////

EXAMPLE 8-2 An automobile manufacturer has two assembly plants serving a region in which six dealerships are located. Every month the six dealers place an order for cars with the manufacturer, and he must then decide how many cars to ship from each plant to each dealer, so that every order is filled. Moreover, the manufacturer would like to minimize the total shipping cost involved in filling these orders. Below is a table giving the cost of shipping each car from a given plant to a given dealer.

SHIPPING COST PER CAR (IN DOLLARS)

To / From plant	Dealer					
	1	2	3	4	5	6
A	15	10	25	20	15	20
B	20	20	15	25	20	15

Plant *A* has a manufacturing capacity of 1000 cars per month, and plant *B* can produce 1500 cars a month. The current orders from the six dealers are as follows:

Dealer	1	2	3	4	5	6
Number of cars ordered	300	340	250	330	280	310

To formulate the above as a linear programming problem, let us define

x_{Aj} = number of cars to be shipped from plant A to dealer j

x_{Bj} = number of cars to be shipped from plant B to dealer j

where $j = 1, 2, 3, 4, 5, 6$.
The objective function is

$$\text{Minimize } z = 15x_{A1} + 10x_{A2} + 25x_{A3} + 20x_{A4} + 15x_{A5} + 20x_{A6} + 20x_{B1} + 20x_{B2} + 15x_{B3} + 25x_{B4} + 20x_{B5} + 15x_{B6}$$

Since plant A can ship no more than 1000 cars, and since the total quantity shipped from plant A is $(x_{A1} + x_{A2} + x_{A3} + x_{A4} + x_{A5} + x_{A6})$, we have the constraint

$$x_{A1} + x_{A2} + x_{A3} + x_{A4} + x_{A5} + x_{A6} \leq 1000$$

In a similar manner we obtain a constraint for plant B:

$$x_{B1} + x_{B2} + x_{B3} + x_{B4} + x_{B5} + x_{B6} \leq 1500$$

Moreover, for each of the six dealerships, we must ensure that the order will be filled. Since the total number of cars shipped to dealer j is $(x_{Aj} + x_{Bj})$, we have:

$$x_{A1} + x_{B1} = 300$$
$$x_{A2} + x_{B2} = 340$$
$$x_{A3} + x_{B3} = 250$$
$$x_{A4} + x_{B4} = 330$$
$$x_{A5} + x_{B5} = 280$$
$$x_{A6} + x_{B6} = 310$$

Upon adding the nonnegativity restrictions we have then completed the linear programming formulation of this problem:

$$x_{Aj}, x_{Bj} \geq 0 \qquad j = 1, 2, 3, 4, 5, 6$$ ////

EXAMPLE 8-3† A cement plant produces 2.5 million barrels of cement a year. The particulate pollution emission rate of the plant is currently 2 pounds of particulates per barrel of cement produced. A new law requires the plant to reduce its emissions to 0.8 million pounds per year.

† This example is due to R. E. Kohn [7].

A control device (method 1) costing \$0.14 per barrel more than the current production method to operate would yield emissions of 0.5 pound of particulates per barrel, and another type of control device (method 2) costing \$0.18 extra per barrel would reduce emissions to 0.2 pound of particulates per barrel of cement produced.

The company wishes to determine how many barrels of cement to produce with each method (including the current method) in order to minimize its cost increase, while at the same time satisfying the new law.

Thus, the variables are:

x_1 = number of barrels of cement produced by method 1

x_2 = number of barrels of cement produced by method 2

x_3 = number of barrels of cement produced by the current method

The objective function is

$$\text{Minimize } z = 0.14x_1 + 0.18x_2$$

We need to concern ourselves here only with the extra cost of the two new methods. Hence, the current method has a cost of zero in the objective function.

Since the company produces 2.5 million barrels of cement annually,

$$x_1 + x_2 + x_3 = 2.5 \text{ million}$$

The total amount of particulates produced is $(0.5x_1 + 0.2x_2 + 2x_3)$ pounds. By the new law, this quantity cannot exceed 0.8 million pounds. Hence,

$$0.5x_1 + 0.2x_2 + 2x_3 \leq 0.8 \text{ million}$$

Also, of course, we require that

$$x_1 \geq 0 \qquad x_2 \geq 0 \qquad x_3 \geq 0$$ ////

8-2 GRAPHICAL SOLUTION OF A LINEAR PROGRAMMING PROBLEM

Now that we have defined a linear programming problem and presented several examples of such problems, we turn to the question of solving one. Our eventual goal is to develop a solution technique that takes advantage of the theory we have developed concerning systems of equations and is computationally efficient as well. Before doing so, however, it will be instructive to use a geometric approach to study the properties of linear programming problems.

Let us consider the problem of Example 8-1, defined by (8-8):

$$\begin{aligned} &\text{Maximize } z = 3x_1 + 2x_2 \\ &\text{subject to:} \quad 4x_1 + x_2 \leq 1200 \\ &\qquad x_1 + 3x_2 \leq 1290 \\ &\qquad 2x_1 + 2x_2 \leq 900 \\ &\qquad x_1, x_2 \geq 0 \end{aligned} \tag{8-8}$$

Because of the nonnegativity restrictions, we need to consider only the first quadrant of the x_1x_2 plane. In this quadrant, let us graph the first constraint of (8-8), namely, $4x_1 + x_2 \leq 1200$. The shaded region below represents the set of all points which satisfy the constraint $4x_1 + x_2 \leq 1200$ *and* the nonnegativity restrictions.

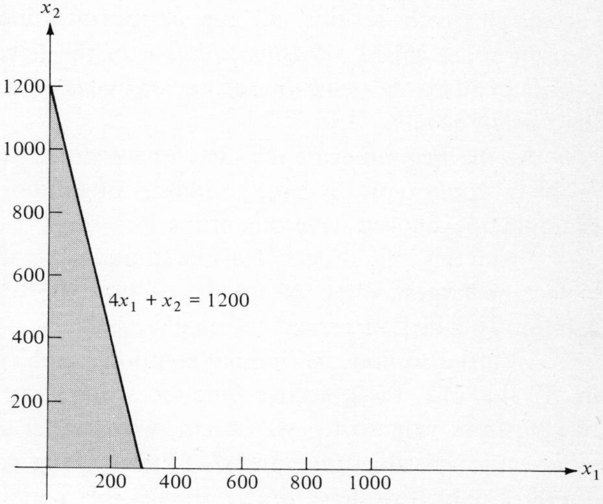

Now, let us add the second constraint, $x_1 + 3x_2 \leq 1290$; in the figure on page 184, the heavily shaded region is the set of all points satisfying both the first two constraints, as well as the nonnegativity restrictions:

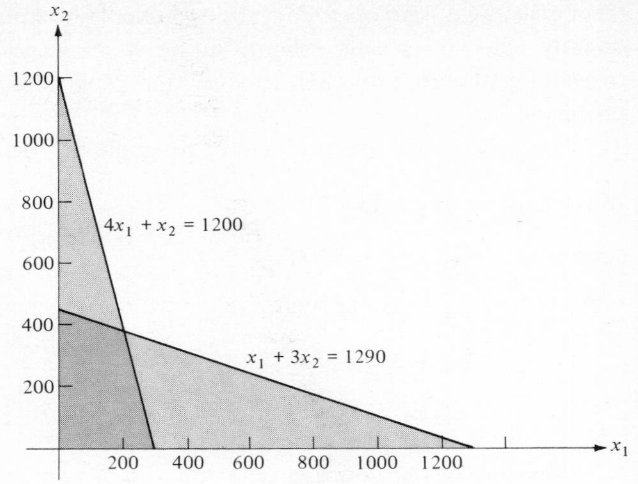

Finally, upon adding the third constraint, $2x_1 + 2x_2 \leq 900$, we obtain the shaded region as shown in Fig. 8-1. Any point in the shaded region in Fig. 8-1 satisfies the constraints of (8-8); the set of all points which satisfy these constraints is precisely this shaded region.

For any linear programming problem, a point (or set of values for the variables) which satisfies all the constraints, including the nonnegativity restrictions, is called a *feasible solution* of the linear programming problem. The set of all feasible solutions of a given linear programming problem is called the *feasible region.*

At the beginning of the chapter we noted that a linear programming problem would typically have an infinity of solutions; we see that our current example does indeed have this property.

From this infinite set of feasible solutions, we would like to find one which yields the largest value of $z = 3x_1 + 2x_2$, the objective function. Such a solution is called an *optimal solution.*

In order to find the optimal solution graphically, we observe that for a fixed value of z, the objective function is just a straight line. Moreover, for two different values of z, we obtain two parallel straight lines. Thus, if we choose a set of values for z and plot the resulting parallel straight lines on the graph on which we have drawn the feasible region, some of these lines will contain points which lie within the feasible region, while others will lie entirely outside the feasible region. We desire to find the line corresponding to the largest value of z which has at least one point lying within the feasible region. In Fig. 8-2 we have drawn some lines corresponding to various different values of z to illustrate this discussion.

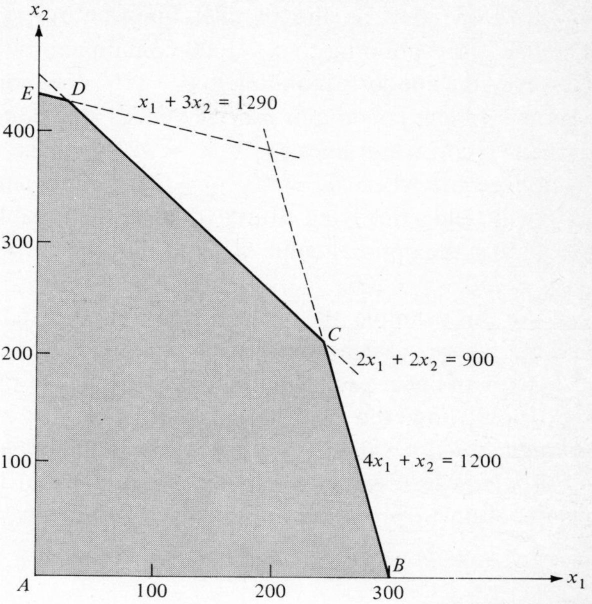

FIGURE 8-1

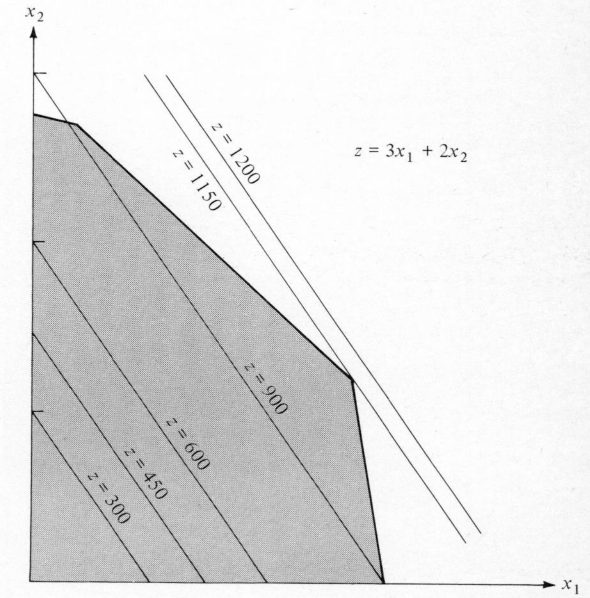

FIGURE 8-2

We see that as the parallel lines "move to the right," z increases. The line corresponding to $z = 1200$ contains no points in the feasible region. However, the line corresponding to $z = 1150$ does contain a point in the feasible region. In fact, it contains exactly one feasible point, namely, the intersection of the two constraint lines $4x_1 + x_2 = 1200$ and $2x_1 + 2x_2 = 900$. This intersection occurs when $x_1 = 250$, $x_2 = 200$. Since any value of z greater than 1150 will yield a line lying entirely outside the feasible region, we conclude that $z = 1150$ is the optimal value of z and that the corresponding optimal solution is $x_1 = 250$, $x_2 = 200$.

In this example, the optimal solution occurred at a "corner point" of the feasible region. Such points are frequently called *extreme points.*

Not all linear programming problems have optimal solutions, as we shall see below; however, *if a linear programming problem does have an optimal solution, at least one of its extreme points will be optimal.* This statement can be proven rigorously (see references [2], [5]) and is one of the fundamental results of linear programming theory. Intuitively, we can see why it is true:

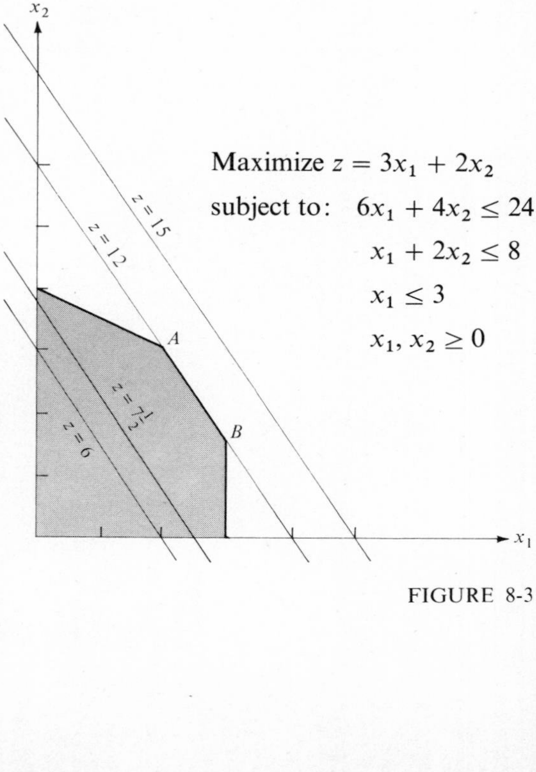

FIGURE 8-3

Starting with an objective function line which passes through the feasible region if we continue drawing parallel straight lines with increasing values of z, we will eventually draw a line which either touches the feasible region at exactly one point or is coincident with one of the boundary lines of the feasible region. The next parallel line we draw will then contain no points in the feasible region. The case mentioned above in which the last objective function line is coincident with one of the boundary lines is illustrated in Fig. 8-3. Note that two of the extreme points in this example are optimal (labeled A and B). Also, any point on the line between A and B is optimal, as well.

Earlier in this section we mentioned that not every linear programming problem has an optimal solution. There are basically two situations in which this can occur, and these are best described by means of several examples.

EXAMPLE 8-4 Consider the linear programming problem

$$\text{Maximize } z = 3x_1 + 2x_2$$
$$\text{subject to: } 2x_1 - x_2 \geq 6$$
$$x_1 + x_2 \leq 2$$
$$x_1, x_2 \geq 0$$

This example is illustrated in Fig. 8-4. There are two shaded regions, one representing the set of all points satisfying the constraint $x_1 + x_2 \leq 2$ and the nonnegativity restrictions (labeled A), the other representing the set of all points satisfying the constraint $2x_1 - x_2 \geq 6$ and the nonnegativity restrictions

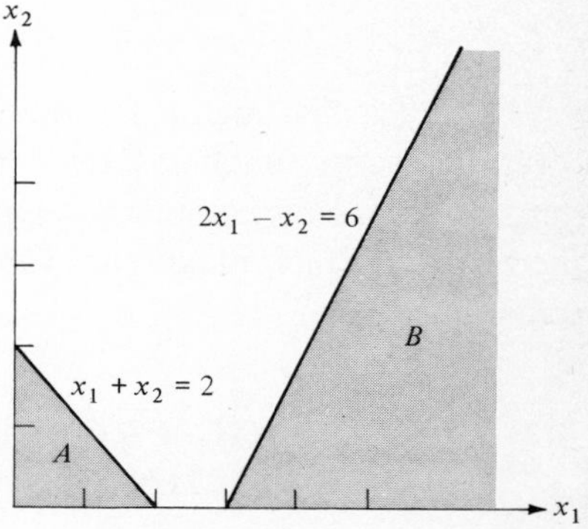

FIGURE 8-4

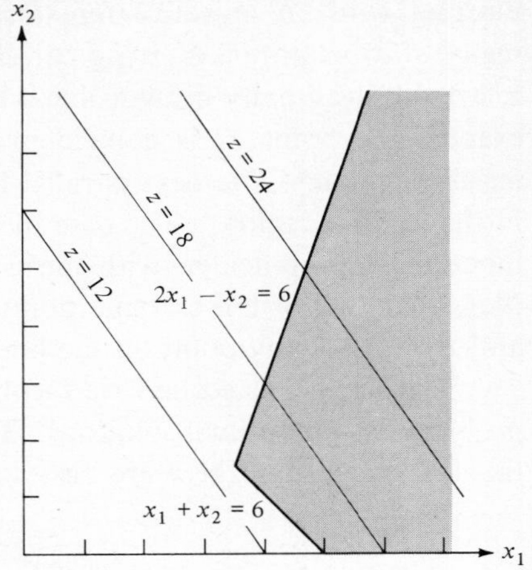

FIGURE 8-5

(labeled B). Observe that there is no point which satisfies all the constraints (including the nonnegativity restrictions).

Hence, this linear programming problem has no feasible solution, and, therefore, no optimal solution. ////

EXAMPLE 8-5 Consider the linear programming problem

$$\text{Maximize } z = 3x_1 + 2x_2$$

$$\begin{aligned} \text{subject to:} \quad 2x_1 - x_2 &\geq 6 \\ x_1 + x_2 &\geq 6 \\ x_1, x_2 &\geq 0 \end{aligned}$$

The feasible region for this problem is shown in Fig. 8-5, along with several objective function lines. Note that we can continue drawing objective function lines, corresponding to increasing values of z, indefinitely. That is, there is no finite maximum value of z. Such a problem is said to be unbounded. ////

8-3 ALGEBRAIC PROPERTIES OF LINEAR PROGRAMMING PROBLEMS

In the previous section we examined a linear programming problem from a geometric viewpoint. This enabled us to gain an intuitive insight into the general shape of the feasible region for a linear programming problem and some characteristics of an optimal solution. However, we obviously cannot apply a graphical solution procedure to any linear programming problem which has more than three variables; hence, we must turn to an algebraic approach. To do so, let us try and determine how to characterize an extreme point algebraically, by returning to Example 8-1 and examining the extreme points of this problem.

As we observed earlier, it is desirable (algebraically) to work with a system of equations, rather than inequalities. Hence, we shall begin by converting the constraints of Example 8-1 to equations, by adding three nonnegative slack variables x_3, x_4, x_5, one to each of the constraint inequalities given in (8-8). This yields the following three constraint equations:

$$\begin{aligned} 4x_1 + x_2 + x_3 \qquad\qquad &= 1200 \\ x_1 + 3x_2 \qquad + x_4 \qquad &= 1290 \\ 2x_1 + 2x_2 \qquad\qquad + x_5 &= 900 \end{aligned} \tag{8-9}$$

From Fig. 8-1 we observe that there are five extreme points, labeled *A*, *B*, *C*, *D*, and *E*. Extreme point *A* is the origin, and $x_1 = x_2 = 0$. Thus, by Eqs. (8-9), $x_3 = 1200$, $x_4 = 1290$, and $x_5 = 900$ at extreme point *A*. Extreme point *B* is the intersection of the x_1 axis with the equation $4x_1 + x_2 = 1200$. Hence, $x_2 = 0$ and $x_1 = 300$. Substitution of these values into Eqs. (8-9) yields $x_3 = 0$, $x_4 = 990$, $x_5 = 300$.

Extreme point *C* occurs at the intersection of the two equations $4x_1 + x_2 = 1200$ and $2x_1 + 2x_2 = 900$. As we have previously noted, this intersection occurs when $x_1 = 250$, $x_2 = 200$. Substitution of these values into Eqs. (8-9) yields $x_3 = 0$, $x_4 = 440$, $x_5 = 0$.

In a similar manner we can obtain values for all the variables for each of the remaining two extreme points. Table 8-1 summarizes these calculations.

Observe that each of the above extreme points has exactly two variables equal to zero and three nonzero variables. Observe also that in each case the

Table 8-1 EXTREME POINTS OF EXAMPLE 8-1

Extreme point	x_1	x_2	x_3	x_4	x_5
A	0	0	1200	1290	900
B	300	0	0	990	300
C	250	200	0	440	0
D	30	420	660	0	0
E	0	430	770	0	40

values for these nonzero variables are the values which would be obtained by *setting* the two zero variables to zero and solving the remaining system of three equations in three variables. Consider, for example, extreme point D. If we set $x_4 = x_5 = 0$, Eqs. (8-9) become:

$$\begin{aligned} 4x_1 + x_2 + x_3 &= 1200 \\ x_1 + 3x_2 \qquad &= 1290 \\ 2x_1 + 2x_2 \qquad &= 900 \end{aligned}$$

The augmented matrix for the above system is

$$\left[\begin{array}{ccc|c} 4 & 1 & 1 & 1200 \\ 1 & 3 & 0 & 1290 \\ 2 & 2 & 0 & 900 \end{array}\right]$$

from which we obtain

$$\left[\begin{array}{ccc|c} 4 & 1 & 1 & 1200 \\ 0 & \frac{11}{4} & -\frac{1}{4} & 990 \\ 0 & \frac{3}{2} & -\frac{1}{2} & 300 \end{array}\right] \rightarrow \left[\begin{array}{ccc|c} 4 & 1 & 1 & 1200 \\ 0 & \frac{11}{4} & -\frac{1}{4} & 990 \\ 0 & 0 & -\frac{4}{11} & -240 \end{array}\right]$$

$$\rightarrow \left[\begin{array}{ccc|c} 4 & 1 & 0 & 540 \\ 0 & \frac{11}{4} & 0 & 1155 \\ 0 & 0 & 1 & 660 \end{array}\right] \rightarrow \left[\begin{array}{ccc|c} 4 & 0 & 0 & 120 \\ 0 & 1 & 0 & 420 \\ 0 & 0 & 1 & 660 \end{array}\right]$$

$$\rightarrow \left[\begin{array}{ccc|c} 1 & 0 & 0 & 30 \\ 0 & 1 & 0 & 420 \\ 0 & 0 & 1 & 660 \end{array}\right]$$

$$\begin{aligned} x_1 &= 30 \\ x_2 &= 420 \\ x_3 &= 660 \end{aligned}$$

Thus, finding the values of the variables at any extreme point involves two steps:

1 Determine which two variables should be set equal to zero.
2 Solve the resulting system of equations for the values of the remaining variables.

Step 2 cannot be accomplished unless the resulting square coefficient matrix is nonsingular; if it is not, then we will not be able to uniquely determine the values of the variables (see Theorem 6-5). Moreover, this coefficient matrix is nonsingular if and only if its columns are linearly independent (Theorem 6-7).

Note also that the number of variables involved in step 2 is equal to the number of constraints (excluding the nonnegativity restrictions), which is three in our current example.

These facts suggest the following definition:

Basic solution†: Given a system of m equations in n variables $(n > m)$ $\mathbf{Ax} = \mathbf{b}$, if $(n - m)$ variables are set to zero and if the columns of $\mathbf{A}$ corresponding to the remaining m variables are linearly independent, then the uniquely determined values of these variables along with the $(n - m)$ zero variables is called a *basic solution* of $\mathbf{Ax} = \mathbf{b}$.

The term *basic* is used here because the m linearly independent columns are a basis for the set of all m-component vectors.

A basic solution in which all the variables are nonnegative is called a *basic feasible solution*, since such a solution is a feasible solution for a linear programming problem whose constraints are: $\mathbf{Ax} = \mathbf{b}$, $x_j \geq 0$ $(j = 1, 2, \ldots, n)$.

Not every basic solution is also feasible. For example, suppose we set $x_1 = x_3 = 0$; Eqs. (8-9) then become

$$\begin{Bmatrix} x_2 & & & = 1200 \\ 3x_2 + x_4 & & & = 1290 \\ 2x_2 + & & + x_5 & = 900 \end{Bmatrix}$$

The three columns of the coefficient matrix above are linearly independent, since this system has the unique solution $x_2 = 1200$, $x_4 = -2310$, $x_5 = -1500$. However, this solution is obviously not a feasible solution.

As we have noted previously, the values of the variables for each of the extreme points in our example can be obtained by the process described in the definition of a basic solution; our experience thus far suggests that a basic feasible solution is nothing more than an extreme point, and vice versa. This is in fact true, and we shall state the result as a theorem, for future reference:

Theorem 8-1 In any linear programming problem, for every extreme point there is a corresponding basic feasible solution; and for every basic feasible solution there is a corresponding extreme point.

For a proof of Theorem 8-1 the reader should consult references [2], [4], or [5].

† We assume here and throughout this chapter that rank $(\mathbf{A}) = m =$ number of equations; if a given system of equations has rank less than the number of equations in it, then one or more of the equations is redundant and can be removed, yielding a reduced system with rank equal to the number of equations.

We have already concluded that, in order to find an optimal solution to a linear programming problem, we need only to investigate its extreme points. Theorem 8-1 tells us that, alternatively, we need only to investigate basic feasible solutions.

Hence, we now have an algebraic procedure for finding an optimal solution for any linear programming problem:

1 *Compute all basic feasible solutions*, by setting all combinations of $(n - m)$ variables to zero and solving the resulting system of m equations in m variables. If the system doesn't have a unique solution, it is not basic and should be discarded; if the system does have a unique solution but not all the variables are nonnegative, the solution is not feasible and should be discarded.

2 Compute the value of the objective function for each basic feasible solution found in step 1. The largest of these corresponds to the optimal solution.

We described step 1 in detail above to indicate that there is a finite number of basic feasible solutions (even though there is an infinity of feasible solutions) for any linear programming problem. However, the possible number of basic feasible solutions can be extremely large, even for relatively small values of m and n. From step 1 we see that the possible number of basic solutions is equal to the number of ways of choosing $(n - m)$ variables (to set to zero) out of n total variables; i.e., there are $\binom{n}{n-m} = n!/[m!(n-m)!]$ possible basic solutions. Below is a table giving the value of $\binom{n}{n-m}$ for several different values of m and n to indicate how rapidly this number grows.

n \ m	5	10	20
5	1		
10	252	1	
15	3,003	3,003	
20	15,504	184,756	1
30	142,506	30,045,015	30,045,015
40	658,006	847,660,528	137,846,528,820
50	2,118,760	10,272,278,170	47,129,212,243,960

Obviously, the algebraic procedure described above is impractical except for very small linear programming problems. Unfortunately, most practical applications of linear programming involve problems with hundreds—or even thousands—of variables. For example, if the automobile manufacturer of

Example 8-2 had 10 assembly plants and 250 dealerships (certainly not unrealistic numbers), then the number of variables in the resulting linear programming problem would be 2500, not counting slack variables, and there would be 260 constraints.

In the following section we shall present a method for finding the optimal solution of a linear programming problem which does not enumerate all the basic solutions but is nevertheless able to determine an optimal basic feasible solution. This method, called the simplex method, has proven to be extremely effective in solving linear programming problems, including very large ones. Computer codes that have been written for the simplex method are capable of handling problems which have tens of thousands of variables and several hundred constraints. In fact, the author is aware of some recent variations of the simplex method which have been used to solve certain linear programming problems with as many as 280,000 variables and over 50,000 constraints!

This capability of the simplex method to solve very large problems has made linear programming one of the most widely used techniques in management science and systems analysis.

8-4 THE SIMPLEX METHOD

The simplex method for solving linear programming problems was originally developed by George Dantzig in 1947. Essentially, the simplex method is an algorithm which generates a sequence of basic feasible solutions in such a way that each new solution generated has a value of the objective function z which is at least as large† as the value of z for the most recently computed solution. In this manner, the sequence of solutions has a corresponding sequence of increasing (or at least nondecreasing) values of the objective function. In other words, once we have computed one basic feasible solution and its corresponding value of z, the simplex method will automatically exclude from consideration all basic feasible solutions with smaller values of z.

Moreover, the simplex method is also able to determine when it has computed an optimal basic feasible solution, so that it is not necessary to attempt to find better solutions when none in fact exists.

Let us investigate these characteristics in more detail. To begin with, we consider the computational aspects of finding a basic feasible solution. From its definition, we see that a basic solution may be found by solving a

† We are assuming that the linear programming problem under discussion is a "maximization" problem.

system of m equations in m variables; if in the resulting solution all variables have nonnegative values, the solution is a basic feasible solution. However, we desire to compute a sequence of basic feasible solutions, and it is computationally quite inefficient to solve an entirely different set of equations for each new basic solution, for two reasons:

1 In computing several different basic solutions, many of the columns in the resulting systems will be the same, and hence some computational savings may be achievable by taking advantage of this fact.

2 If a basic solution is computed merely by choosing which variables to set to zero, there is no way of telling whether the resulting solution will be feasible, until a great deal of the computation is completed.

In order to eliminate these difficulties, we introduce the concept of "adjacent" basic feasible solutions: Two basic feasible solutions are said to be *adjacent* if $(m - 1)$ of the m basic variables in each solution are common to both; the *basic variables* are the variables not *set* to zero in a basic solution. For example, the basic feasible solutions corresponding to extreme points A and B of Table 8-1 are adjacent basic feasible solutions, since both have the two basic variables x_4 and x_5 (out of the three basic variables in each solution). Geometrically, adjacent basic feasible solutions lie along the same edge of the feasible region, as can be seen from Fig. 8-1.

Algebraically, it is a simple matter to compute an adjacent basic feasible solution from a given basic feasible solution. To see how to do so, suppose we reorder the columns of the coefficient matrix of our constraints, $\mathbf{Ax} = \mathbf{b}$, so that the first m columns correspond to the m basic variables and the last $(n - m)$ columns correspond to the variables we will set to zero (called the *nonbasic variables*). Thus, we let

$$\mathbf{B}_{m \times m} = \text{first } m \text{ columns of } \mathbf{A}$$

$$\mathbf{N}_{m \times (n-m)} = \text{last } (n - m) \text{ columns of } \mathbf{A}$$

$$\mathbf{x}_B = m\text{-component vector, containing the basic variables}$$

$$\mathbf{x}_N = (n - m)\text{-component vector, containing the nonbasic variables}$$

We now have

$$\mathbf{Ax} = \mathbf{b}$$

$$[\mathbf{B}, \mathbf{N}]\begin{bmatrix}\mathbf{x}_B \\ \mathbf{x}_N\end{bmatrix} = \mathbf{b}$$

$$\mathbf{Bx}_B + \mathbf{Nx}_N = \mathbf{b}$$

$$\mathbf{x}_B + (\mathbf{B}^{-1}\mathbf{N})\mathbf{x}_N = \mathbf{B}^{-1}\mathbf{b} \qquad (8\text{-}10)$$

(We know that $\mathbf{B}^{-1}$ exists by the discussion preceding the definition of a basic solution in Sec. 8-3.) In Eq. (8-10) we have solved for the basic variables $\mathbf{x}_B$ in terms of the nonbasic variables $\mathbf{x}_N$; that is, given any set of values of the $(n - m)$ nonbasic variables, we can immediately obtain the values of the basic variables from Eq. (8-10). In particular, in a basic solution, we set the nonbasic variables to zero and obtain $\mathbf{x}_B = \mathbf{B}^{-1}\mathbf{b}$.

Suppose now that we wish to find an adjacent basic solution. In such a solution, $(m - 1)$ of the basic variables are the same, while one of the nonbasic variables becomes a basic variable and one of the basic variables becomes nonbasic. Let $\mathbf{Y} = [\mathbf{y}_1 \quad \mathbf{y}_2 \quad \cdots \quad \mathbf{y}_n]$ be the coefficient matrix of (8-10), and let $\mathbf{v} = \mathbf{B}^{-1}\mathbf{b}$. Then, the first m columns of $\mathbf{Y}$ are equal to the corresponding m columns of the identity matrix $\mathbf{I}_m$:

$$\mathbf{y}_j = \mathbf{e}_j \qquad j = 1, 2, \ldots, m$$

The last $(n - m)$ $\mathbf{y}_j$'s correspond to the columns of $\mathbf{B}^{-1}\mathbf{N}$. If we let $\mathbf{x}_B = [x_{B1} \quad x_{B2} \quad \cdots \quad x_{Bm}]^T$ and $\mathbf{x}_N = [x_{N1} \quad x_{N2} \quad \cdots \quad x_{Np}]^T$, $p = n - m$, then we can write the augmented matrix for (8-10) as follows:

$$\begin{array}{ccccccccc} x_{B1} & x_{B2} & \cdots & x_{Bm} & x_{N1} & x_{N2} & \cdots & x_{Np} & \\ [\mathbf{e}_1 & \mathbf{e}_2 & \cdots & \mathbf{e}_m & \mathbf{y}_{m+1} & \mathbf{y}_{m+2} & \cdots & \mathbf{y}_n & | \ \mathbf{v}] \end{array}$$

We have written each variable over its corresponding column for clarity. Suppose we are interested in computing the adjacent basic solution in which we replace the rth basic variable x_{Br} with the kth nonbasic variable x_{Nk}. Let us interchange the columns corresponding to these two variables in the above augmented matrix. We obtain the augmented matrix

$$\begin{array}{cccccccccccccc} 1 & x_{B2} & \cdots & x_{B,r-1} & x_{Nk} & x_{B,r+1} & \cdots & x_{Bm} & x_{N1} & \cdots & x_{N,k-1} & x_{Br} & x_{N,k+1} & \cdots & x_{Np} \\ & \mathbf{e}_2 & \cdots & \mathbf{e}_{r-1} & \mathbf{y}_{m+k} & \mathbf{e}_{r+1} & \cdots & \mathbf{e}_m & \mathbf{y}_{m+1} & \cdots & \mathbf{y}_{m+k-1} & \mathbf{e}_r & \mathbf{y}_{m+k+1} & \cdots & \mathbf{y}_n \ | \end{array}$$

Now, we wish to solve for $x_{B1}, x_{B2}, \ldots, x_{B,r-1}, x_{Nk}, x_{B,r+1}, \ldots, x_{Bm}$ in terms of the remaining variables. However, in the above augmented matrix we already have all but x_{Nk} solved for. Hence, we need only to solve for x_{Nk}, and this is easily accomplished by performing one step of a Gauss-Jordan elimination; namely, add multiples of row r of the augmented matrix to each of the other $(n - 1)$ rows, so that the elements in column r of these $(n - 1)$ rows become zero; then, divide row r by the pivot element so that the element in row r, column r, is a 1. Note that the elements in row r and columns $1, 2, \ldots, r - 1, r + 1, \ldots, m$ are all zeros, so that the first m

columns remain unchanged, with the exception of course of column r. After performing this Gauss-Jordan step, the augmented matrix is now

$$\begin{matrix} \hat{x}_{B1} & \hat{x}_{B2} & \cdots & \hat{x}_{B,r-1} & \hat{x}_{Br} & \hat{x}_{B,r+1} & \cdots & \hat{x}_{Bm} & \hat{x}_{N1} & \hat{x}_{N2} & \cdots & \hat{x}_{Np} & \\ [\mathbf{e}_1 & \mathbf{e}_2 & \cdots & \mathbf{e}_{r-1} & \mathbf{e}_r & \mathbf{e}_{r+1} & \cdots & \mathbf{e}_m & \hat{\mathbf{y}}_{m+1} & \hat{\mathbf{y}}_{m+2} & \cdots & \hat{\mathbf{y}}_n & | \ \hat{\mathbf{v}}] \end{matrix}$$

EXAMPLE 8-6 The constraints of Example 8-1, given by Eq. (8-9) are:

$$\begin{aligned} 4x_1 + x_2 + x_3 \qquad\qquad &= 1200 \\ x_1 + 3x_2 \qquad + x_4 \qquad &= 1290 \\ 2x_1 + 2x_2 \qquad\qquad + x_5 &= 900 \end{aligned}$$

The basic feasible solution corresponding to extreme point A has basic variables x_3, x_4, x_5 and nonbasic variables x_1, x_2. Hence, let $x_{B1} = x_3$, $x_{B2} = x_4$, $x_{B3} = x_5$; $x_{N1} = x_1$, $x_{N2} = x_2$. Equation (8-10) becomes:

$$\begin{bmatrix} x_{B1} \\ x_{B2} \\ x_{B3} \end{bmatrix} + \begin{bmatrix} 4 & 1 \\ 1 & 3 \\ 2 & 2 \end{bmatrix} \begin{bmatrix} x_{N1} \\ x_{N2} \end{bmatrix} = \begin{bmatrix} 1200 \\ 1290 \\ 900 \end{bmatrix}$$

The augmented matrix is

$$\begin{matrix} x_{B1} & x_{B2} & x_{B3} & x_{N1} & x_{N2} & \\ \end{matrix}$$
$$\left[\begin{array}{ccc:cc:c} 1 & 0 & 0 & 4 & 1 & 1200 \\ 0 & 1 & 0 & 1 & 3 & 1290 \\ 0 & 0 & 1 & 2 & 2 & 900 \end{array}\right]$$

Suppose we wish to replace the basic variable x_{B2} with the nonbasic variable x_{N1}. Then, according to the preceding discussion, we first interchange the columns corresponding to these two variables:

$$\begin{matrix} x_{B1} & x_{N1} & x_{B3} & x_{B2} & x_{N2} & \\ \end{matrix}$$
$$\left[\begin{array}{ccc:cc:c} 1 & 4 & 0 & 0 & 1 & 1200 \\ 0 & 1 & 0 & 1 & 3 & 1290 \\ 0 & 2 & 1 & 0 & 2 & 900 \end{array}\right]$$

Next, we add multiples of row 2 to rows 1 and 3, so that the elements in row 1, column 2, and in row 3, column 2, become zero:

$$\begin{matrix} \hat{x}_{B1} & \hat{x}_{B2} & \hat{x}_{B3} & \hat{x}_{N1} & \hat{x}_{N2} & \\ \end{matrix}$$
$$\left[\begin{array}{ccc:cc:c} 1 & 0 & 0 & -4 & -11 & -3960 \\ 0 & 1 & 0 & 1 & 3 & 1290 \\ 0 & 0 & 1 & -2 & -4 & -1680 \end{array}\right]$$

We have now obtained the desired form. Note that $\hat{x}_{B1}$, $\hat{x}_{B3}$ still correspond to x_3 and x_5, respectively, but that x_{B2} now corresponds to x_1. Also, $\hat{x}_{N1}$, $\hat{x}_{N2}$ now correspond to x_4 and x_2, respectively. Hence, the basic solution we have just computed is

$$x_3 = -3960$$
$$x_1 = 1290$$
$$x_5 = -1680$$

which is not a feasible solution. ////

Example 8-6 illustrates two points. First, although the notation may seem slightly complicated, the actual computation of an adjacent basic solution is quite simple. Secondly, the example illustrates that we still have not considered the problem of ensuring that the adjacent basic solution will be feasible. However, this latter difficulty is easily resolved. To do so, let us define the elements of the augmented matrix of (8-10) as follows:

$$\mathbf{y}_j = [y_{1j}, y_{2j}, \ldots, y_{mj}]^T \qquad j = 1, 2, \ldots, n$$
$$\mathbf{v} = [v_1, v_2, \ldots, v_m]^T$$

Then, we can discuss the Gauss-Jordan elimination step in terms of how each element in the augmented matrix is altered. Suppose we wish to replace the basic variable x_{Br} with the nonbasic variable which corresponds to column $\mathbf{y}_k$ (which is not necessarily the variable x_{Nk}, as before). Then, the Gauss-Jordan step can be described thus:

1 Add the multiple $-y_{ik}/y_{rk}$ of row r to row i, for each of the $(n-1)$ rows $i = 1, 2, \ldots, r-1, r+1, \ldots, m$.
2 Divide row r by y_{rk}.

Obviously, we require that $y_{rk} \neq 0$. The formulas for the new augmented matrix, which corresponds to the resulting adjacent basic solution, are:

$$\begin{aligned} i \neq r:&\quad \hat{y}_{ij} = y_{ij} - \frac{y_{ik}}{y_{rk}} y_{rj} \qquad j = 1, 2, \ldots, n \\ i = r:&\quad \hat{y}_{rj} = \frac{y_{rj}}{y_{rk}} \qquad j = 1, 2, \ldots, n \end{aligned} \tag{8-11}$$

$$\begin{aligned} i \neq r:&\quad \hat{x}_{Bi} \equiv \hat{v}_i = v_i - \frac{y_{ik}}{y_{rk}} v_r \\ &\quad \hat{x}_{Br} \equiv \hat{v}_r = \frac{v_r}{y_{rk}} \end{aligned} \tag{8-12}$$

These equations enable us to determine under what conditions the new basic solution will also be feasible (i.e., all variables nonnegative). From Eq. (8-12), we see that y_{rk} must be positive, or else $\hat{x}_{Br}$ will become negative (since all v_i are currently positive if our current solution is feasible, as we are assuming). Furthermore, we require that $\hat{x}_{Bi} \geq 0$, $i \neq r$, and hence

$$v_i - \frac{y_{ik}}{y_{rk}} v_r \geq 0 \qquad \text{all } i \neq r \qquad \text{(8-13)}$$

Since $y_{rk} > 0$ and $v_i \geq 0$, $i = 1, 2, \ldots, m$, it is clear that the inequality (8-13) will be satisfied for each y_{ik} such that $y_{ik} \leq 0$. But, for each $y_{ik} > 0$, we must require that (8-13) is satisfied. For $y_{ik} > 0$, we can rewrite (8-13) as

$$\frac{v_r}{y_{rk}} \leq \frac{v_i}{y_{ik}} \qquad \text{all } i \text{ for which } y_{ik} > 0 \qquad \text{(8-14)}$$

One way to ensure that (8-14) is satisfied is to compute the ratios v_i/y_{ik} for each $y_{ik} > 0$ and to choose r as the minimum of these ratios. That is, suppose a nonbasic variable has been somehow selected to become a basic variable (we shall see how to do this later), and that this nonbasic variable corresponds to column $\mathbf{y}_k$. Then, we can ensure that the next basic solution we compute will be feasible if we choose the basic variable x_{Br} to become nonbasic by choosing the r for which

$$\frac{v_r}{y_{rk}} = \min_{y_{ik} > 0} \left\{ \frac{v_i}{y_{ik}} \right\} \qquad \text{(8-15)}$$

Equation (8-15) is merely a restatement of the inequalities (8-14).

EXAMPLE 8-7 Again, consider the augmented matrix for the system of equations (8-9), as in Example 8-6. The basic variables are currently x_3, x_4, x_5. Suppose we have somehow decided that the nonbasic variable x_1 should become a basic variable. Since x_1 is in column 4 in the augmented matrix (below), we have that $k = 4$:

$$\begin{array}{ccc:cc:c} x_3 & x_4 & x_5 & x_1 & x_2 & \\ \end{array}$$
$$\left[\begin{array}{ccc:cc:c} 1 & 0 & 0 & 4 & 1 & 1200 \\ 0 & 1 & 0 & 1 & 3 & 1290 \\ 0 & 0 & 1 & 2 & 2 & 900 \end{array}\right]$$

The ratios v_i/y_{ik} are:

$$\frac{v_1}{y_{14}} = \frac{1200}{4} = 300$$

$$\frac{v_2}{y_{24}} = \frac{1290}{1} = 1290$$

$$\frac{v_3}{y_{34}} = \frac{900}{2} = 450$$

The minimum occurs when $i = 1$. Hence, the basic variable $x_{B1} = x_3$ is the variable which we should choose to become nonbasic.

Performing the Gauss-Jordan step (without first interchanging the columns) yields:

$$\begin{array}{ccccc} x_3 & x_4 & x_5 & x_1 & x_2 \end{array}$$
$$\left[\begin{array}{ccc:cc:c} \frac{1}{4} & 0 & 0 & 1 & \frac{1}{4} & 300 \\ 0 & 1 & 0 & 0 & \frac{11}{4} & 990 \\ 0 & 0 & 1 & 0 & \frac{3}{2} & 300 \end{array}\right]$$

The new basic variables are x_1, x_4, and x_5, with values 300, 990 and 300, respectively.

This basic solution is clearly feasible, and corresponds to extreme point B, in Table 8-1 and Fig. 8-1. ////

The above example illustrates that it is not necessary to actually interchange the columns before performing the Gauss-Jordan step; instead, we need only to perform this step by using Eqs. (8-11) and (8-12). After doing so, we will again have m of the $\mathbf{y}_j$ columns corresponding to the m columns of the identity matrix $\mathbf{I}_m$. These m columns therefore correspond to the m basic variables. Specifically, the column $\mathbf{e}_i$ (the ith column of $\mathbf{I}_m$) corresponds to the basic variable x_{Bi}, and the value of this variable is v_i. Thus, in Example 8-7 above, $\mathbf{e}_1$ corresponds to x_1, and hence x_{B1} is x_1, with $x_{B1} = x_1 = 300$; $\mathbf{e}_2$ corresponds to x_4 and so $x_{B2} = x_4 = 990$; and $\mathbf{e}_3$ corresponds to x_5, so $x_{B3} = x_5 = 300$.

In view of the above discussion, we will henceforth not perform any column interchanges, as these are completely unnecessary.

At the beginning of this section, we remarked that our goal was to develop an efficient method for generating a sequence of basic feasible solutions, in such a way that each succeeding solution had a corresponding value of the objective function at least as large as its predecessor. Thus far, we have seen how to compute efficiently one basic feasible solution from a given basic feasible solution; specifically, we compute an adjacent basic feasible solution. Now, let us consider the corresponding values of the objective function for these solutions, and see how we can choose an adjacent solution which does have a larger objective function value.

Suppose the objective function is

$$z = c_1x_1 + c_2x_2 + \cdots + c_nx_n$$

We can write this equation as

$$z - c_1x_1 - c_2x_2 - \cdots - c_nx_n = 0 \qquad (8\text{-}16)$$

Furthermore, since we have defined the coefficient matrix for (8-9) as $\mathbf{Y} = [y_{ij}]$, and the right-hand side of (8-9) by $\mathbf{v} = \mathbf{B}^{-1}\mathbf{b} = [v_i]$, we can now write (8-9) as follows:

$$\sum_{j=1}^{n} y_{ij}x_j = v_i \qquad i = 1, 2, \ldots, m \qquad (8\text{-}17)$$

Previously, we have assumed that the first m variables were the basic variables. It is not necessary to make this assumption, but the reader should keep in mind that each basic variable appears in exactly one of the m equations (8-17); namely, x_{B1} appears in the first equation only, with coefficient 1, x_{B2} appears in the second equation only, with coefficient 1, etc.

Now, let us add a multiple of each of the m equations (8-17) to the objective function equation (8-16), so that the coefficient of each basic variable becomes zero. The objective function equation will then be of the form

$$z + d_1x_1 + d_2x_2 + \cdots + d_nx_n = f \qquad (8\text{-}18)$$

If c_{Bi} is the c_j corresponding to the basic variable x_{Bi}, then

$$d_j = \begin{cases} \sum_{i=1}^{m} y_{ij}c_{Bi} - c_j & \text{if } x_j \text{ is nonbasic} \\ 0 & \text{if } x_j \text{ is basic} \end{cases} \qquad (8\text{-}19)$$

$$f = \sum_{i=1}^{m} c_{Bi}v_i = \sum_{i=1}^{n} c_{Bi}x_{Bi} \qquad (8\text{-}20)$$

Hence, f is the current value of the objective function. Equations (8-18) and (8-19) enable us to determine immediately whether the current basic feasible solution is optimal:

Theorem 8-2 A basic feasible solution is optimal if all $d_j \geq 0$, $j = 1, 2, \ldots, n$, where d_j is defined by Eq. (8-19).

PROOF The value of d_j for each basic variable is zero, by (8-19). All nonbasic variables are currently zero, by definition, and furthermore in any other basic feasible solution, at least one of these nonbasic variables will increase in value (become nonzero). However, if any such variable increases, and its corresponding d_j is positive, then by Eq. (8-18), the resulting objective function value must be smaller. Hence, if all d_j are greater than or equal to zero, the current solution is optimal. QED

The discussion in the above proof also leads us directly to:

Theorem 8-3 Given a basic feasible solution, if any $d_j < 0$, then an adjacent basic feasible solution in which the (currently) nonbasic variable x_j becomes basic will have an objective function value at least as great as the current objective function value.

Theorem 8-3 suggests how we may choose the nonbasic variable we wish to become basic, that is, the k of Eqs. (8-11) through (8-15) and the related discussion. *Any* nonbasic variable x_k whose corresponding d_k is negative will suffice. There are several ways we can choose k if more than one $d_j < 0$.

1 Choose the most negative d_j:

$$d_k = \min_j \{d_j\} \qquad (8\text{-}21)$$

2 Choose the first negative d_j discovered.

3 Choose k such that the resulting adjacent solution yields the largest possible increase in z.

The reader may suppose that method 3 is the preferred approach, since our goal is to maximize z. However, the most commonly used approach is method 1. There are several reasons for this, one of which is that method 1 requires substantially less computational effort than method 3 at each iteration (an iteration in this context being the computation of one basic feasible solution), while intuitively requiring fewer total iterations than method 2. But the author knows of no computational experience which shows that any one method is clearly superior to any other. In the interest of being consistent with common practice, however, we shall use method 1 hereinafter.

We are now ready to describe the simplex method for solving a linear programming problem:

The Simplex Method

1. Find an initial basic feasible solution. Compute the corresponding y_{ij}, v_i, d_j.

2. If all $d_j \geq 0$, the current solution is optimal; if any $d_j < 0$, compute

$$d_k = \min_j \{d_j\}$$

The nonbasic variable x_k is thus chosen to become basic.

3. Determine the basic variable x_{Br} which is to become nonbasic, by (8-15):

$$\frac{v_r}{y_{rk}} = \min_{y_{ik} > 0} \left\{\frac{v_i}{y_{ik}}\right\}$$

4. Compute the resulting adjacent basic feasible solution (in which x_{Br} is replaced by x_k) from Eqs. (8-11) and (8-12):

$$i \neq r: \begin{cases} \hat{y}_{ij} = y_{ij} - \dfrac{y_{ik}}{y_{rk}} y_{rj} & j = 1, 2, \ldots, n \\ \hat{x}_{Bi} = \hat{v}_i = v_i - \dfrac{y_{ik}}{y_{rk}} v_r \end{cases}$$

$$i = r: \begin{cases} \hat{y}_{rj} = \dfrac{y_{rj}}{y_{rk}} & j = 1, 2, \ldots, n \\ \hat{x}_{Br} = \hat{v}_r = \dfrac{v_r}{y_{rk}} \end{cases}$$

Also, compute the new d_j, from†

$$\hat{d}_j = d_j - \frac{v_r}{y_{rk}} d_k \qquad j = 1, 2, \ldots, n \qquad \text{(8-22)}$$

5. Return to step 2.

8-5 THE SIMPLEX TABLEAU; AN EXAMPLE

Each iteration of the simplex method involves essentially three steps: the selection of which nonbasic variable is to become basic; the selection of the basic variable which is to become nonbasic; the computation of the new values $\hat{y}_{ij}$, $\hat{v}_i$, $\hat{d}_j$ corresponding to the new basic feasible solution.

† The reader is asked in the Exercises to derive Eq. (8-22).

The first two steps require relatively little computational effort. The third step requires one step of Gauss-Jordan elimination, on an augmented matrix with $(m+1)$ rows and $(n+1)$ columns; the $(m+1)$st row corresponds to the objective function equation.

A convenient way to perform the latter, and to store the results, is by means of a tableau, whose format is indicated in Table 8-2; the tableau is called the simplex tableau.

The first column of the simplex tableau, labeled "basic variables," contains the name of each of the basic variables; that is, which variable corresponds to x_{B1}, which variable corresponds to x_{B2}, etc. The last column of the tableau contains the *values* of these basic variables in the current basic feasible solution (recall that the nonbasic variables are all zero), in rows 1 through m, and the current value of z in row $(m+1)$. The rest of row $(m+1)$ contains the current values of $d_1, d_2, \ldots, d_n$.

To illustrate the use of the simplex tableau and the simplex method itself, let us solve the linear programming problem of Example 8-1. After adding the slack variables to the constraints, the constraint set is given by Eq. (8-9):

$$\begin{aligned} 4x_1 + x_2 + x_3 \qquad\qquad &= 1200 \\ x_1 + 3x_2 \qquad + x_4 \qquad &= 1290 \\ 2x_1 + 2x_2 \qquad\qquad + x_5 &= 900 \end{aligned}$$

As we have seen in Example 8-6, the equations in the above form represent the basic feasible solution in which

$$\begin{aligned} x_{B1} &= x_3 = 1200 = v_1 \\ x_{B2} &= x_4 = 1290 = v_2 \\ x_{B3} &= x_5 = 900 = v_3 \end{aligned}$$

Table 8-2 FORMAT OF THE SIMPLEX TABLEAU

Basic variables	x_1	x_2	$\cdots$	x_n	$\mathbf{v}$
x_{B1}	y_{11}	y_{12}	$\cdots$	y_{1n}	v_1
x_{B2}	y_{21}	y_{22}	$\cdots$	y_{2n}	v_2
$\vdots$	$\vdots$	$\vdots$		$\vdots$	$\vdots$
x_{Bm}	y_{m1}	y_{m2}	$\cdots$	y_{mn}	v_m
z	d_1	d_2	$\cdots$	d_n	f

Also, we have that

$$\mathbf{y}_1 = \begin{bmatrix} y_{11} \\ y_{21} \\ y_{31} \end{bmatrix} = \begin{bmatrix} 4 \\ 1 \\ 2 \end{bmatrix} \qquad \mathbf{y}_2 = \begin{bmatrix} y_{12} \\ y_{22} \\ y_{32} \end{bmatrix} = \begin{bmatrix} 1 \\ 3 \\ 2 \end{bmatrix}$$

$$\mathbf{y}_3 = \begin{bmatrix} 1 \\ 0 \\ 0 \end{bmatrix} \qquad \mathbf{y}_4 = \begin{bmatrix} 0 \\ 1 \\ 0 \end{bmatrix} \qquad \mathbf{y}_5 = \begin{bmatrix} 0 \\ 0 \\ 1 \end{bmatrix}$$

Hence, the only additional information we need for the simplex tableau for this basic feasible solution is $d_1, d_2, \ldots, d_n$, and f.

The objective function of Example 8-1 is

$$z = 3x_1 + 2x_2 + 0x_3 + 0x_4 + 0x_5$$

or

$$z - 3x_1 - 2x_2 - 0x_3 - 0x_4 - 0x_5 = 0$$

Hence, the coefficients of the basic variables are 0, as we desire, so that $d_1 = -3, d_2 = -2, d_3 = d_4 = d_5 = 0, f = 0$. The reader should verify that the same results are obtained by using Eqs. (8-19) and (8-20).

Putting all these calculations into the simplex tableau yields the tableau given in Table 8-3.

We have thus completed step 1 of the simplex method, and proceed to step 2, the selection of which nonbasic variable to make basic:

$$\begin{aligned} d_k &= \min\{d_1, d_2\} \\ &= \min\{-3, -2\} \\ &= -3 = d_1 \end{aligned}$$

Hence, x_1 will become basic. Note that we excluded the d_j's corresponding to the basic variables in the above calculation; it would make no sense

Table 8-3 INITIAL SIMPLEX TABLEAU FOR EXAMPLE 8-1

Basic variables	x_1	x_2	x_3	x_4	x_5	$\mathbf{v}$
$x_{B1} = x_3$	4	1	1	0	0	1200
$x_{B2} = x_4$	1	3	0	1	0	1290
$x_{B3} = x_5$	2	2	0	0	1	900
z	-3	-2	0	0	0	0

to include these, as we are making a choice among nonbasic variables with negative d_j's. Furthermore, all d_j corresponding to basic variables are always zero (never negative).

We now go to step 3 and decide which basic variable will become nonbasic, by use of Eq. (8-15):

$$\frac{v_r}{y_{rk}} = \min_{y_{ik}>0} \left\{\frac{v_i}{y_{ik}}\right\}$$

$$= \min \left\{\frac{v_1}{y_{11}}, \frac{v_2}{y_{21}}, \frac{v_3}{y_{31}}\right\}$$

$$= \min \left\{\frac{1200}{4}, \frac{1290}{1}, \frac{900}{2}\right\}$$

$$= 300 = \frac{v_1}{y_{11}}$$

Thus, the variable corresponding to x_{B1} becomes nonbasic, which is x_3 in this case. The new x_{B1} will be x_1. To compute the next simplex tableau (step 4), we may use Eqs. (8-11), (8-12), and (8-22), directly; or, equivalently, we may perform one Gauss-Jordan elimination step, using the (r, k) element as the pivot element, and performing the calculations on all $(m + 1)$ rows; that is, add multiples of row $r = 1$ to each of the other rows so that the elements in column $k = 1$ of these rows become 0. The result is the simplex tableau corresponding to the adjacent feasible solution in which basic variable x_3 is replaced by x_1, and is given in Table 8-4.

Note that the objective function has increased from $z = 0$ to $z = 900$, but that this solution is still not optimal, since d_2 is negative. As d_2 is the only negative d_j, we choose x_2 to become basic and proceed to step 3:

Table 8-4 SECOND SIMPLEX TABLEAU FOR EXAMPLE 8-1

Basic variables	x_1	x_2	x_3	x_4	x_5	v
$x_{B1} = x_1$	1	$\frac{1}{4}$	$\frac{1}{4}$	0	0	300
$x_{B2} = x_4$	0	$\frac{11}{4}$	$-\frac{1}{4}$	1	0	990
$x_{B3} = x_5$	0	$\frac{3}{2}$	$-\frac{1}{2}$	0	1	300
z	0	$-\frac{5}{4}$	$\frac{3}{4}$	0	0	900

$$\begin{aligned}\frac{v_r}{y_{rk}} &= \min_{y_{ik}>0} \left\{\frac{v_i}{y_{ik}}\right\} \\ &= \min \left\{\frac{v_1}{y_{12}}, \frac{v_2}{y_{22}}, \frac{v_3}{y_{32}}\right\} \\ &= \min \left\{\frac{300}{1/4}, \frac{990}{11/4}, \frac{300}{3/2}\right\} \\ &= \min \{1200, 360, 200\} \\ &= 200 = \frac{v_3}{y_{32}}\end{aligned}$$

Thus, the variable corresponding to x_{B3} becomes nonbasic, and the new x_{B3} is x_2. The resulting simplex tableau is given in Table 8-5.

We now observe that all d_j are nonnegative, and hence the current solution is optimal, with

$$\begin{aligned} x_1 &= 250 \\ x_2 &= 200 \\ x_4 &= 440 \\ x_3 &= x_5 = 0 \end{aligned}$$

and $z = 1150$. This result concurs with the solution we obtained graphically in Sec. 8-2.

8-6 THE SIMPLEX METHOD: FURTHER DISCUSSION

There are several computational aspects of the simplex method which we have not yet discussed; these involve how to find an initial basic feasible solution; how to determine whether the constraint set has no feasible solution

Table 8-5 FINAL SIMPLEX TABLEAU FOR EXAMPLE 8-1

Basic variables	x_1	x_2	x_3	x_4	x_5	**v**
$x_{B1} = x_1$	1	0	$\frac{1}{3}$	0	$-\frac{1}{6}$	250
$x_{B2} = x_4$	0	0	$\frac{2}{3}$	1	$-\frac{11}{6}$	440
$x_{B3} = x_2$	0	1	$-\frac{1}{3}$	0	$\frac{2}{3}$	200
z	0	0	$\frac{1}{3}$	0	$\frac{5}{6}$	1150

(as in Example 8-4); how to determine if the problem has an unbounded solution (as in Example 8-5).

Finding an Initial Basic Feasible Solution

Step 1 of the simplex method, as described at the end of Sec. 8-4, requires that we "find an initial basic feasible solution" and compute the corresponding y_{ij}, v_i, d_j. To do so, we need to determine m basic variables $\mathbf{x}_B$ and express these in the form given by Eq. (8-10):

$$\mathbf{x}_B + (\mathbf{B}^{-1}\mathbf{N})\mathbf{x}_N = \mathbf{B}^{-1}\mathbf{b} \qquad (8\text{-}10)$$

However, as we have remarked previously, there is no computationally efficient method for determining in advance whether such a basic solution will turn out to be feasible. We might have to try many different combinations of m basic variables before finding a feasible combination. To avoid this difficulty, we introduce the following method for finding an initial basic feasible solution.

As before, suppose we seek to solve the linear programming problem

$$\begin{aligned} &\text{Maximize } z = \sum_{j=1}^{n} c_j x_j \\ &\text{subject to:} \quad \mathbf{Ax} = \mathbf{b} \\ &\qquad\qquad\quad x_j \geq 0 \qquad j = 1, 2, \ldots, n \end{aligned} \qquad (8\text{-}23)$$

Let us assume that the constraints $\mathbf{Ax} = \mathbf{b}$ have been adjusted so that $b_i \geq 0$, $i = 1, 2, \ldots, m$. This is easily accomplished by multiplying any equation with a negative b_i by (-1).

We will now add m new variables $w_1, w_2, \ldots, w_m$ to the problem, creating a modified linear programming problem:

$$\begin{aligned} &\text{Maximize } z = \sum_{j=1}^{n} c_j x_j - M \sum_{i=1}^{m} w_i \\ &\mathbf{Ax} + \mathbf{I}_m \mathbf{w} = \mathbf{b} \\ &x_j \geq 0 \qquad j = 1, 2, \ldots, n \\ &w_i \geq 0 \qquad i = 1, 2, \ldots, m \end{aligned} \qquad (8\text{-}24)$$

In the modified problem (8-24), M represents a very large positive number, and $\mathbf{w} = [w_1 \quad w_2 \quad \cdots \quad w_m]^T$. The variables $w_1, w_2, \ldots, w_m$ are called *artificial variables*. We make two observations about this modified problem:

1 An initial basic feasible solution is $\mathbf{w} = \mathbf{b}$, $\mathbf{x} = \mathbf{0}$, since these values of $\mathbf{x}$, $\mathbf{w}$ obviously satisfy the constraints, are nonnegative, and consist of the

m basic variables $w_1, w_2, \ldots, w_m$, whose columns are linearly independent. Moreover, these columns are the m columns of the identity $\mathbf{I}_m$. Hence,

$$\mathbf{y}_j = \mathbf{a}_j \qquad j = 1, 2, \ldots, n \tag{8-25}$$

$$\mathbf{y}_{n+i} = \mathbf{e}_i \qquad i = 1, 2, \ldots, m \tag{8-26}$$

$$x_{Bi} = w_i = v_i = b_i \qquad i = 1, 2, \ldots, m \tag{8-27}$$

As usual, $\mathbf{a}_j$ denotes the jth column of $\mathbf{A}$ and $\mathbf{e}_i$ denotes the ith column of $\mathbf{I}_m$. Thus the elements of the first m rows of the initial simplex tableau are easily determined.

2 If M is chosen large enough, the optimal solution to the modified problem (8-24) will have $\mathbf{w} = \mathbf{0}$. Any feasible solution to (8-24) in which $\mathbf{w} = \mathbf{0}$ is also a feasible solution for the original linear programming problem (8-23), as is evident by inspection of the two constraint sets for these problems. Hence, if we solve the modified problem (8-24), we will also obtain the optimal solution of (8-23).

Thus, in view of the above two comments, we see that the introduction of the m artificial variables provides an extremely fast and sure method of avoiding the apparent difficulties involved in finding an initial basic feasible solution for (8-23). It only remains for us to show how to compute the initial values for $d_1, d_2, \ldots, d_n, d_{n+1}, \ldots, d_{n+m}$, and f, where $d_{n+1}, d_{n+2}, \ldots, d_{n+m}$ correspond to $w_1, w_2, \ldots, w_m$.

This is a simple matter, and it is left as an exercise for the reader to show that

$$d_j = \begin{cases} -M\sum_{i=1}^{m} a_{ij} - c_j & j = 1, 2, \ldots, n \\ 0 & j = n+1, n+2, \ldots, n+m \end{cases} \tag{8-28}$$

Once the initial tableau is constructed, the simplex method may be used as described and illustrated in the previous two sections.

An equally good method for determining an initial basic feasible solution for (8-23) is one known as the "two-phase method"; it is discussed in the Exercises.

Determination of No Feasible Solution

As we have illustrated in Sec. 8-2 (Example 8-4), it is possible that a linear programming problem will have no feasible solution. If one attempts to solve such a problem, with the procedure described above using artificial variables, the fact that no feasible solution exists for (8-23) will mean that the optimal solution obtained for (8-24) will contain at least one positive artificial variable w_i; for, if the optimal solution to (8-24) contains all $w_i = 0$, then as we remarked earlier, such a solution is in fact a feasible solution to (8-23).

Hence, if no feasible solution exists for the original problem (8-23), this fact will be automatically discovered when one solves the modified problem (8-24).

Determination of an Unbounded Solution

Example 8-5 illustrates that it is possible for a linear programming problem to have an unbounded solution (i.e., there is no finite maximum value of z). In the simplex method, this fact is discovered when a situation is reached in which a nonbasic variable x_k is chosen to become basic (implying that its corresponding $d_k < 0$) and step 3 cannot be implemented, because all $y_{ik} \leq 0$, $i = 1, 2, \ldots, m$. The reason that this situation indicates an unbounded solution is that a negative d_k means that increasing the value of x_k (currently equal to zero) will increase the objective function z; however, the restriction on *how much* we can increase x_k in order to arrive at an adjacent basic feasible solution depends only on positive y_{ik}. If all y_{ik} are less than or equal to zero, there is no restriction on how much we can increase x_k, and thus, the objective function can also increase indefinitely.

Geometrically, this situation corresponds to moving from an extreme point along a constraint edge which is infinitely long, as in Fig. 8-5.

8-7 OTHER METHODS

We have now discussed in some detail the simplex method for solving linear programming problems. As we have already remarked, the simplex method has proved to be extremely efficient and capable of solving very large problems. There are several variations of the simplex method which are also quite useful in certain circumstances. In this section we shall briefly mention some of these methods and their applications, although we shall not provide complete descriptions of them. Our intent here is merely to make the reader aware of the existence of these methods. The reader interested in more detailed treatments of these topics is urged to consult the references given at the end of the chapter.

The Revised Simplex Method

The revised simplex method is the name given to a computational procedure for performing the calculations of the simplex method in a manner which is more efficient for digital computers. The revised simplex method follows the same sequence of iterations (basic feasible solutions) as the simplex method, using the same decision rules for determining which variable becomes basic [Eq. (8-21)] and which nonbasic [Eq. (8-15)] at each iteration. However, in the revised simplex method the entire simplex tableau is not stored in the computer; instead, the matrix $\mathbf{B}^{-1}$ of Eq. (8-10) is stored, along with some of the quantities $d_1, d_2, \ldots, d_n$. Thus, at each iteration the elements of $\mathbf{B}^{-1}$ are modified, rather than the entire simplex tableau. The primary advantage of the revised simplex method is thus the smaller amount of computer storage required.

A secondary advantage of revised simplex is that it is easier to control roundoff errors than it is with the simplex method. Recall from Chap. 5 that it is desirable to use a pivoting strategy when solving a system of equations on a digital computer; we have ignored this fact in our discussion of the simplex method. Thus, if one attempts to implement the simplex method on a digital computer in the form described in the preceding three sections, one can expect that roundoff difficulties may be encountered. However, since the revised simplex method requires only a knowledge of $\mathbf{B}^{-1}$, roundoff errors can be reduced by periodically recomputing $\mathbf{B}^{-1}$ directly by Gaussian elimination with a pivoting strategy.

Most commercially developed linear programming computer codes use the revised simplex method, for these reasons.

The Dual Simplex Method

Under certain circumstances one may encounter a situation in which one has a basic solution for a linear programming problem which is not feasible (at least one basic variable is negative) but which is "optimal," in the sense that all $d_j \geq 0$, $j = 1, 2, \ldots, n$. For such a problem, a method known as the dual simplex method has been developed. In the dual simplex method, one first selects one of the negative basic variables to become nonbasic, and then selects which nonbasic variable will become basic, in such a way that the next basic solution remains optimal ($\hat{d}_j \geq 0, j = 1, 2, \ldots, n$). Thus, when a basic solution is found which is also feasible, it is the desired optimal solution as well.

The dual simplex method is particularly useful when one has solved a given linear programming problem and then desires to add a constraint or two to the original problem. If these added constraints are not satisfied by the current optimal solution, then one or more variables will be negative and the dual simplex method may be applied. This procedure is usually much more efficient than solving the new linear programming problem from the beginning.

Another variation of the simplex method is the *primal-dual algorithm*, in which one begins with a basic solution which is neither feasible nor "optimal" (in the sense used above, all $d_j \geq 0$) and chooses the variables to become basic and nonbasic so that ultimately both optimality and feasibility are achieved.

The primal-dual algorithm has some special applications for which it is useful, but neither the primal-dual algorithm nor the dual simplex method has proven to be computationally superior to the simplex method for solving general linear programming problems.

The Transportation Problem

Some linear programming problems which arise commonly in practice have a special structure, which can frequently be used to advantage by adapting the simplex method to it. Perhaps the most widely used of these specially structured problems is the transportation problem, which can be described as follows:

A company has m warehouses and n retail outlets, and wishes to ship a commodity from the warehouses ("origins") to the retail outlets ("destinations"), at minimal cost. It costs c_{ij} dollars to ship each unit of the commodity from origin i to destination j. Furthermore, origin i has a_i units of the commodity available, and destination j requires b_j units of the commodity.

If we let x_{ij} denote the number of the commodity to be shipped from origin i to destination j, then the resulting linear programming problem is:

$$\begin{aligned} &\text{Minimize } z = \sum_{i=1}^{m} \sum_{j=1}^{n} c_{ij} x_{ij} \\ &\text{subject to: } \sum_{j=1}^{n} x_{ij} \leq a_i \qquad i = 1, 2, \ldots, m \\ &\qquad\qquad \sum_{i=1}^{m} x_{ij} = b_j \qquad j = 1, 2, \ldots, n \\ &\qquad\qquad x_{ij} \geq 0 \qquad i = 1, 2, \ldots, m \\ &\qquad\qquad\qquad\qquad j = 1, 2, \ldots, n \end{aligned} \tag{8-29}$$

Any problem which can be formulated in the form of (8-29) is called a *transportation* problem. Example 8-2 is such a problem. The inequality constraints of (8-29) are sometimes replaced by equations, and the equations of (8-29) are sometimes replaced by " $\geq$ " type inequalities.

Transportation problems have the following very useful properties:

1 A feasible solution always exists provided only that $\sum a_i \geq \sum b_j$ (supply meets demand).

2 The problem never has an unbounded solution.

3 An initial basic feasible solution is trivial to obtain.

4 Because of the special structure of the coefficient matrix, all the coefficients in the simplex tableau (except the d_{ij}) are either 0, 1, or -1.

These facts have led to a specialization of the simplex method for solving transportation problems, in which a tableau containing $2mn$ elements is used, instead of the usual simplex tableau which would require $(mn + 1) \times (m + n)$ elements for problem (8-29), a considerable saving both in computer storage requirements and in computational effort.

One of the main applications of the primal-dual algorithm occurs in conjunction with the solution of transportation problems.

Integer Linear Programming

In many applications of linear programming, it is required that some or all of the variables be integer-valued. The simplex method does not guarantee that this will be the case, and, in fact, it is typically true that the optimal solution of a linear programming problem will contain noninteger values for some of the variables. Since a basic solution is nothing more than the solution of a system of equations, it would be pure luck if the solution turned out to consist of integer values for all the variables.

A linear programming problem in which one or more variables are required to be integers is called an *integer linear programming problem* or, sometimes, an *integer programming problem.*

It is usually not sufficient to merely round the optimal solution of a linear programming problem to obtain the optimal solution to the corresponding integer programming problem. First of all, there are many ways to round this solution, since each of the non-integer-valued variables may be rounded "up" or "down" to the nearest integer; secondly, such a rounded solution may not satisfy all the constraints; thirdly, such a rounded solution may not be optimal.

Many methods have been developed for solving integer programming problems. Some of these are discussed in references [1], [2], [4], [6] and [9]. However, none of these methods has yet proved to be in any way as efficient as the simplex method, and in general, none can be confidently used for problems with more than a hundred or so constraints and more than 75 to 100 integer-restricted variables.

REFERENCES

1. COOPER, LEON, and DAVID STEINBERG: "Introduction to Methods of Optimization," Saunders, Philadelphia, 1970.
2. COOPER, LEON, and DAVID I. STEINBERG: "Methods and Applications of Linear Programming, Saunders, Philadelphia, 1974.
3. DANTZIG, GEORGE B.: "Linear Programming and Extensions," Princeton, Princeton, N.J., 1963.
4. GASS, SAUL I.: "Linear Programming," 3rd ed., McGraw-Hill, New York, 1969.
5. HADLEY, G.: "Linear Programming," Addison-Wesley, Reading, Mass., 1962.
6. HILLIER, F. S., and G. J. LIEBERMAN: "Introduction to Operations Research," Holden-Day, San Francisco, 1967.
7. KOHN, ROBERT E.: A Mathematical Programming Model for Air Pollution Control, *School Science and Mathematics*, June 1969.
8. STRUM, JAY E.: "Introduction of Linear Programming," Holden-Day, San Francisco, 1972.
9. WAGNER, HARVEY M.: "Principles of Operations Research," Prentice-Hall, Englewood Cliffs, N.J., 1969.

EXERCISES

1 Solve the following linear programming problem graphically:

$$\text{Maximize } z = 2x + 3x$$

$$\begin{aligned} \text{subject to:} \quad x_1 + x_2 &\leq 7 \\ -x_1 + 3x_2 &\leq 9 \\ 2x_1 + x_2 &\leq 12 \\ x_1, \quad x_2 &\geq 0 \end{aligned}$$

2 Determine all the basic solutions for the linear programming problem of Exercise 1. Indicate which of these is also feasible, and find the optimal basic feasible solution by evaluating the objective function at each basic feasible solution.

3 Solve the linear programming problem of Exercise 1 by the simplex method.

4 Solve the following linear programming problem, (*a*) graphically; (*b*) by the simplex method:

$$\text{Maximize } z = 2x_1 + x_2$$
$$\text{subject to:} \quad 2x_1 + 3x_2 \geq 6$$
$$x_1 + x_2 \leq 6$$
$$-2x_1 + x_2 \leq 3$$
$$2x_1 - x_2 \leq 8$$
$$x_1, \quad x_2 \geq 0$$

5 The simplex method, as we have described it in this chapter, will find the *maximum* value of an objective function. In general, if we wish to *minimize* an objective function

$$z = \sum_{j=1}^{n} c_j x_j$$

we can do so by maximizing the related objective function

$$\hat{z} = -z = -\sum_{j=1}^{n} c_j x_j$$

Illustrate this fact by solving the following linear programming problem graphically and by the simplex method:

$$\text{Minimize } z = 2x_1 + x_2$$

subject to the same constraint set as the problem of Exercise 4.

6 A linear programming problem may have an unbounded constraint set and yet still have a finite optimal solution. Illustrate this fact by solving graphically:

$$\text{Minimize } z = 3x_1 + 2x_2$$
$$\text{subject to:} \quad 2x_1 - x_2 \geq 6$$
$$x_1 + x_2 \geq 6$$
$$x_1, \quad x_2 \geq 0$$

7 Solve the problem of Exercise 6 by the simplex method.

8 Consider the problem of finding an initial basic feasible solution for the linear programming problem:

$$\begin{aligned} &\text{Maximize } z = \sum_{j=1}^{n} c_j x_j \\ &\text{subject to:} \quad \mathbf{Ax} = \mathbf{b} \\ &\qquad x_j \geq 0 \qquad j = 1, 2, \ldots, n \end{aligned} \tag{8-23}$$

Suppose we create the following modified problem:

$$\text{Maximize } \hat{z} = -\sum_{i=1}^{m} w_i$$

$$\text{subject to:} \quad \mathbf{Ax} + \mathbf{I}_m \mathbf{w} = \mathbf{b} \tag{8-30}$$

$$x_j \geq 0 \qquad j = 1, 2, \ldots, n$$

$$w_i \geq 0 \qquad i = 1, 2, \ldots, m$$

where the w_i are the artificial variables as defined in Sec. 8-6. As is the case with the modified problem (8-24), an initial basic feasible solution to (8-30) is $\mathbf{w} = \mathbf{b}$, $\mathbf{x} = \mathbf{0}$, and the optimal solution to (8-30) requires that $\mathbf{w} = \mathbf{0}$, provided a feasible solution exists for (8-23). The solution of (8-30) is called phase I. Upon its completion, one has an initial basic feasible solution for (8-23) and can then proceed to phase II: Calculate the values of $d_1, d_2, \ldots, d_n$ for the current basic feasible solution, by means of Eq. (8-19), and continue solving the problem by the simplex method. The columns of the simplex tableau which correspond to the artificial variables can be discarded once phase II is begun.

The procedure described above is called the *two-phase method*, and it is completely equivalent to the method described in Sec. 8-6, in the sense that the same sequence of basic feasible solutions will be generated by both methods, on any given problem.

Illustrate this fact by using the two-phase method to solve the linear programming problem of Exercise 4.

9 Solve the problem of Exercise 6 by the two-phase method.

10 Illustrate the simplex method's test for an unbounded solution, by attempting to solve the problem of Example 8-5 by the simplex method.

11 Use the method described in Sec. 8-6 to show that the problem of Example 8-4 has no feasible solution.

12 Use the two-phase method to show that the problem of Example 8-4 has no feasible solution.

13 Derive Eq. (8-22).

14 Derive Eq. (8-28).

15 Graph the following problem and attempt to solve it by the simplex method:

$$\text{Maximize } z = x_1 + 2x_2$$

$$\text{subject to:} \quad x_1 - x_2 \geq 6$$

$$x_1 + x_2 \geq 12$$

$$3x_1 + x_2 \leq 24$$

$$x_1, \quad x_2 \geq 0$$

16 (a) Graph the following problem:

$$\begin{aligned} &\text{Maximize } z = 5x_1 + 3x_2 \\ &\text{subject to:} \quad x_1 + x_2 \le 6 \\ &\qquad |x_1 - 2| \le 3 \\ &\qquad x_2 \le 4 \\ &\qquad x_1, \quad x_2 \ge 0 \end{aligned}$$

(b) Show how this problem can be formulated as a linear programming problem.

17 (a) Graph the following problem:

$$\begin{aligned} &\text{Minimize } z = x_1 + x_2 \\ &\text{subject to:} \quad |x_1 + 2x_2 - 3| \le 1 \\ &\qquad x_1 - x_2 \le 1 \\ &\qquad x_1, \quad x_2 \ge 0 \end{aligned}$$

(b) Formulate the above problem as a linear programming problem.

18 Formulate as a linear programming problem:

A cattle rancher desires to prepare, at minimum cost, a cattle feed which is a mixture of corn, hay, and wheat germ. He wants each 100 pounds of the mixture to contain at least 8 pounds of protein, 5 pounds of fat, and 6 pounds of minerals and vitamins. Below is a table of relevant information:

Ingredient	Contents per 100 lb			Cost per 100 lb
	Protein (lb)	Fat (lb)	Vitamins & minerals (lb)	
Corn	3	9	2	\$1.25
Hay	6	5	2	\$1.45
Wheat germ	10	4	3	\$1.80

19 Formulate as a linear programming problem:

A candy company manufactures three different assortments of mixed chocolates, which contain various pieces of milk chocolate, caramel-filled, and cream-filled. The table below provides information on the company's standards for each type of assortment, along with the selling prices of each:

Assortment name	requirements	Selling price per pound
Special	Not more than 20% creams Not more than 35% caramels	\$1.89
Deluxe	Not less than 30% creams Not more than 50% milk chocolates	\$2.39
Supreme	Between 40 and 50% creams Not less than 30% milk chocolates	\$2.79

The costs per pound of the chocolates are: $0.45 for the milk chocolates, $0.40 for the caramel-filled, and $0.60 for the cream-filled chocolates.

Determine the optimal mixture for each assortment, so that the net profit (sales minus costs) is a maximum.

20 Formulate as a linear programming problem:

An old but successful company is planning a major renovation of its corporate headquarters, to begin in five years, for which it will need a large amount of capital at that time. The company wishes to invest its currently available capital of 2 million dollars in such a way that as much money as possible will be available five years hence. The company may choose from among the following investment opportunities:

Investment opportunity	Years available	Length of time required	Yield (per annum)
A	0, 1, 2, 3, 4	2 years	6%
B	0, 1, 2, 3, 4	3 years	7%
C	1, 2, 3, 4	4 years	6%
D	3, 4	1 year	5%

For each of the investments, the company receives the yield annually, but must keep its principle invested for the entire "length of time required." An initial investment may be made only at the beginning of any year in which the opportunity is available [e.g., investment A is available at the beginning of year 0 (now), 1, 2, 3, or 4].

9

EIGENVALUE PROBLEMS

9-1 INTRODUCTION AND BASIC THEORY

In this chapter we shall direct our attention to a special type of matrix problem—known as an *eigenvalue* problem—which has wide application in many types of practical problems which occur frequently in the fields of engineering, physics, and matrix algebra itself. We have already mentioned two applications of eigenvalue problems in connection with the convergence of certain iteration methods (Chap. 4) and with the condition number of a matrix (Chap. 5).

Specifically, the eigenvalue problem may be stated as follows:

Given a square matrix $\mathbf{A} = [a_{ij}]_{n \times n}$, find scalars λ and nontrivial vectors $\mathbf{x}$ (not all components equal to zero) such that

$$\mathbf{Ax} = \lambda \mathbf{x} \qquad (9\text{-}1)$$

For each scalar λ and nontrivial vector $\mathbf{x}$ which satisfy Eq. (9-1), λ is called an *eigenvalue* of $\mathbf{A}$ and $\mathbf{x}$ is called an *eigenvector*† of $\mathbf{A}$ corresponding to λ.

† Since $\mathbf{x} = \mathbf{0}$ satisfies Eq. (9-1) for all λ, we shall call $\mathbf{x}$ an eigenvector *only* if not all of its components are zero.

The terms *characteristic value* (or *characteristic root*) and *characteristic vector* are also frequently used for such λ and $\mathbf{x}$, respectively. We shall restrict our attention throughout this chapter to *square* matrices $\mathbf{A}$.

The question naturally arises: Does every square matrix $\mathbf{A}$ possess eigenvalues and eigenvectors? Let us resolve this question. We may rewrite Eq. (9-1) as follows:

$$\mathbf{Ax} - \lambda\mathbf{x} = \mathbf{0}$$

or

$$(\mathbf{A} - \lambda\mathbf{I})\mathbf{x} = \mathbf{0} \tag{9-2}$$

Now, Eq. (9-2) is in a form we have seen before: It represents a system of n homogeneous equations in n unknowns (the components of the vector $\mathbf{x}$), with coefficient matrix $(\mathbf{A} - \lambda\mathbf{I})$. Of course, technically the scalar λ is also unknown. However, we learned in Chap. 6 that a system of n homogeneous equations in n unknowns possesses a nontrivial solution if and only if the determinant of the coefficient matrix is equal to zero; that is,

$$\det(\mathbf{A} - \lambda\mathbf{I}) = 0 \tag{9-3}$$

In component form, Eq. (9-3) is

$$\begin{vmatrix} a_{11} - \lambda & a_{12} & a_{13} & \cdots & a_{1n} \\ a_{21} & a_{22} - \lambda & a_{23} & \cdots & a_{2n} \\ a_{31} & a_{32} & a_{33} - \lambda & \cdots & a_{3n} \\ \vdots & \vdots & \vdots & & \vdots \\ a_{n1} & a_{n2} & a_{n3} & \cdots & a_{nn} - \lambda \end{vmatrix} = 0 \tag{9-4}$$

If we were to compute this determinant, we would obtain an nth-degree polynomial in λ:

$$\det(\mathbf{A} - \lambda\mathbf{I}) = (-1)^n\lambda^n + \alpha_{n-1}\lambda^{n-1} + \alpha_{n-2}\lambda^{n-2} + \cdots + \alpha_1\lambda + \alpha_0$$

This polynomial is called the *characteristic polynomial* of $\mathbf{A}$. Each value of λ which satisfies Eq. (9-4) yields a system of homogeneous equations of the form (9-2). Thus, the problem of finding the values of λ for which (9-2) possesses nontrivial solutions is the same as finding the roots of the characteristic polynomial:

$$(-1)^n\lambda^n + \alpha_{n-1}\lambda^{n-1} + \alpha_{n-2}\lambda^{n-2} + \cdots + \alpha_1\lambda + \alpha_0 = 0 \tag{9-5}$$

The fundamental theorem of algebra assures us that Eq. (9-5) possesses exactly n (not necessarily distinct) roots, some of which may be complex numbers. Each root is therefore an eigenvalue of $\mathbf{A}$ (and hence every square matrix does indeed possess at least one eigenvalue).

Let us assume that **A** has p ($p \leq n$) distinct eigenvalues $\lambda_1, \lambda_2, \ldots, \lambda_p$ (i.e., by "distinct" we mean that $\lambda_j \neq \lambda_k$ if $j \neq k$). For each such λ_j, the set of corresponding eigenvectors is the null space† of the matrix $(\mathbf{A} - \lambda_j \mathbf{I})$. To summarize, we now state:

Theorem 9-1 Let $\lambda_1, \lambda_2, \ldots, \lambda_p$ denote the distinct eigenvalues of a matrix **A**. Then **x** is an eigenvector corresponding to λ_j if and only if $\mathbf{x} \neq \mathbf{0}$ and **x** is in the null space of $(\mathbf{A} - \lambda_j \mathbf{I})$.

Now, let us examine the dimensionality of each of these null spaces. Again, we shall denote by $\lambda_1, \lambda_2, \ldots, \lambda_p$ the distinct eigenvalues of a matrix **A**. Then, the characteristic polynomial of **A** can be written in factored form as follows:

$$\det(\mathbf{A} - \lambda \mathbf{I}) = (\lambda - \lambda_1)^{m_1}(\lambda - \lambda_2)^{m_2} \cdots (\lambda - \lambda_p)^{m_p}(-1)^n \tag{9-6}$$

where m_j denotes the multiplicity of the root λ_j. This multiplicity is also called the *algebraic* multiplicity of λ_j. Since there is a total of exactly n roots to Eq. (9-5), and hence to Eq. (9-6), it must be true that

$$m_1 + m_2 + \cdots + m_p = n$$

We would now like to show that the dimension of the null space‡ of $(\mathbf{A} - \lambda_j \mathbf{I})$ is less than or equal to m_j, $j = 1, 2, \ldots, p$. To do so, we first prove:

Theorem 9-2 If **P** is any nonsingular matrix (of order n), then the matrices $\mathbf{P}^{-1}\mathbf{A}\mathbf{P}$ and **A** have the same characteristic polynomial.

PROOF The characteristic polynomial of **A** is $\det(\mathbf{A} - \lambda \mathbf{I})$. But

$$\begin{aligned}
\det(\mathbf{A} - \lambda \mathbf{I}) &= \det(\mathbf{A} - \lambda \mathbf{I}) \det(\mathbf{I}) = \det(\mathbf{A} - \lambda \mathbf{I}) \det(\mathbf{P}^{-1}\mathbf{P}) \\
&= \det(\mathbf{A} - \lambda \mathbf{I}) \det(\mathbf{P}^{-1}) \det(\mathbf{P}) \\
&= \det(\mathbf{P}^{-1}) \det(\mathbf{A} - \lambda \mathbf{I}) \det(\mathbf{P}) \\
&= \det[\mathbf{P}^{-1}(\mathbf{A} - \lambda \mathbf{I})\mathbf{P}] = \det[(\mathbf{P}^{-1}\mathbf{A} - \lambda \mathbf{P}^{-1})\mathbf{P}] \\
&= \det(\mathbf{P}^{-1}\mathbf{A}\mathbf{P} - \lambda \mathbf{I})
\end{aligned}$$

which is the characteristic polynomial of $\mathbf{P}^{-1}\mathbf{A}\mathbf{P}$. QED

Recall now that the dimension of a vector space is the maximum number of linearly independent vectors in the vector space. Suppose that the maximum

† The reader will recall from Chap. 7 that the set of all solutions to a set of homogeneous equations $\mathbf{Bx} = \mathbf{0}$ forms a vector space, called the *null space* of **B**.

‡ We shall denote by $N(\mathbf{A} - \lambda_j \mathbf{I})$ the null space of $(\mathbf{A} - \lambda_j \mathbf{I})$.

number of linearly independent eigenvectors in $N(\mathbf{A} - \lambda_j\mathbf{I})$ is d_j [and therefore d_j is the dimension of $N(\mathbf{A} - \lambda_j\mathbf{I})$], and let $\mathbf{x}_1, \mathbf{x}_2, \ldots, \mathbf{x}_{d_j}$ be a basis for $N(\mathbf{A} - \lambda_j\mathbf{I})$. Clearly, $d_j \leq n$, since the rank of $(\mathbf{A} - \lambda_j\mathbf{I})$ cannot exceed n. We then define the $n \times d_j$ matrix $\mathbf{B}$ whose columns are $\mathbf{x}_1, \mathbf{x}_2, \ldots, \mathbf{x}_{d_j}$:

$$\mathbf{B} = [\mathbf{x}_1 \quad \mathbf{x}_2 \quad \cdots \quad \mathbf{x}_{d_j}] \tag{9-7}$$

We may now construct a nonsingular matrix $\mathbf{P}$:

$$\mathbf{P}_{n \times n} = [\mathbf{B}_{n \times d_j} \quad \mathbf{C}_{n \times (n - d_j)}]$$

Thus, we have formed $\mathbf{P}$ by adding any $(n - d_j)$ linearly independent column vectors to the d_j linearly independent eigenvectors $\mathbf{x}_1, \mathbf{x}_2, \ldots, \mathbf{x}_{d_j}$ so that the resulting set of vectors is linearly independent. Let us define $\mathbf{P}^{-1}$ in partitioned form as:

$$\mathbf{P}^{-1} = \begin{bmatrix} \mathbf{G}_{d_j \times n} \\ \mathbf{H}_{(n - d_j) \times n} \end{bmatrix}$$

Since $\mathbf{P}^{-1}\mathbf{P} = \mathbf{I}_n$, we have that

$$\mathbf{P}^{-1}\mathbf{P} = \begin{bmatrix} \mathbf{G} \\ \mathbf{H} \end{bmatrix} [\mathbf{B} \quad \mathbf{C}] = \begin{bmatrix} \mathbf{GB} & \mathbf{GC} \\ \mathbf{HB} & \mathbf{HC} \end{bmatrix} = \begin{bmatrix} \mathbf{I}_{d_j} & \mathbf{O} \\ \mathbf{O} & \mathbf{I}_{(n-d_j)} \end{bmatrix}$$

or

$$\begin{aligned} \mathbf{GB} &= \mathbf{I}_{d_j} \\ \mathbf{HB} &= \mathbf{O}_{(n-d_j) \times d_j} \\ \mathbf{GC} &= \mathbf{O}_{d_j \times (n-d_j)} \\ \mathbf{HC} &= \mathbf{I}_{(n-d_j)} \end{aligned} \tag{9-8}$$

Now, we calculate

$$\begin{aligned} \mathbf{P}^{-1}\mathbf{AP} &= \begin{bmatrix} \mathbf{G} \\ \mathbf{H} \end{bmatrix} \mathbf{A}[\mathbf{B} \quad \mathbf{C}] \\ &= \begin{bmatrix} \mathbf{G} \\ \mathbf{H} \end{bmatrix} [\mathbf{AB} \quad \mathbf{AC}] \\ &= \begin{bmatrix} \mathbf{GAB} & \mathbf{GAC} \\ \mathbf{HAB} & \mathbf{HAC} \end{bmatrix} \end{aligned} \tag{9-9}$$

But from (9-7) we see that

$$\begin{aligned} \mathbf{AB} &= \mathbf{A}[\mathbf{x}_1 \quad \mathbf{x}_2 \quad \cdots \quad \mathbf{x}_{d_j}] \\ &= [\mathbf{Ax}_1 \quad \mathbf{Ax}_2 \quad \cdots \quad \mathbf{Ax}_{d_j}] \\ &= [\lambda_j\mathbf{x}_1 \quad \lambda_j\mathbf{x}_2 \quad \cdots \quad \lambda_j\mathbf{x}_{d_j}] \\ &= \lambda_j[\mathbf{x}_1 \quad \mathbf{x}_2 \quad \cdots \quad \mathbf{x}_{d_j}] \\ &= \lambda_j\mathbf{B} \end{aligned} \tag{9-10}$$

Substituting (9-10) into (9-9) yields

$$\mathbf{P}^{-1}\mathbf{AP} = \begin{bmatrix} \lambda_j \mathbf{GB} & \mathbf{GAC} \\ \lambda_j \mathbf{HB} & \mathbf{HAC} \end{bmatrix} \tag{9-11}$$

From the first two equations of (9-8) we see that (9-11) becomes

$$\mathbf{P}^{-1}\mathbf{AP} = \begin{bmatrix} \lambda_j \mathbf{I}_{d_j} & \mathbf{GAC} \\ \mathbf{O} & \mathbf{HAC} \end{bmatrix} \tag{9-12}$$

Calculating the characteristic polynomial of $\mathbf{P}^{-1}\mathbf{AP}$ from (9-12) yields:

$$\det(\mathbf{P}^{-1}\mathbf{AP} - \lambda\mathbf{I}) = (\lambda_j - \lambda)^{d_j} \det(\mathbf{HAC} - \lambda\mathbf{I}) \tag{9-13}$$

If we let $\det(\mathbf{HAC} - \lambda\mathbf{I}) = f(\lambda)$, then Eq. (9-13) may be written

$$\det(\mathbf{P}^{-1}\mathbf{AP} - \lambda\mathbf{I}) = (\lambda_j - \lambda)^{d_j} f(\lambda) \tag{9-14}$$

where $f(\lambda)$ is a polynomial of degree $(n - d_j)$.

Let us compare Eq. (9-14) with the characteristic polynomial of **A**, given in Eq. (9-6). We may express the latter as

$$\det(\mathbf{A} - \lambda\mathbf{I}) = (\lambda_j - \lambda)^{m_j} g(\lambda) \tag{9-15}$$

where

$$g(\lambda) = (-1)^{n+m_j}(\lambda - \lambda_1)^{m_1} \cdots (\lambda - \lambda_{j-1})^{m_{j-1}}(\lambda - \lambda_{j+1})^{m_{j+1}} \cdots (\lambda - \lambda_p)^{m_p}$$

Thus, $g(\lambda_j) \neq 0$. That is, $g(\lambda)$ does not contain the factor $(\lambda_j - \lambda)$.

By Theorem 9-2, the right-hand sides of Eqs. (9-14) and (9-15) must be equal:

$$(\lambda_j - \lambda)^{d_j} f(\lambda) = (\lambda_j - \lambda)^{m_j} g(\lambda) \tag{9-16}$$

Therefore, since $g(\lambda)$ does not contain the factor $(\lambda_j - \lambda)$, but $f(\lambda)$ might, Eq. (9-16) tells us immediately that $d_j \leq m_j$. We have now proved:

Theorem 9-3 If λ_j is an eigenvalue of multiplicity m_j of a matrix **A**, then the dimension d_j of the null space of $(\mathbf{A} - \lambda_j\mathbf{I})$ satisfies

$$1 \leq d_j \leq m_j \tag{9-17}$$

(d_j is often called the *geometric multiplicity* of λ_j.)

EXAMPLE 9-1 Let

$$\mathbf{A} = \begin{bmatrix} 3 & 2 & 1 & 1 & 2 \\ 0 & 3 & 0 & 1 & 1 \\ 0 & 0 & 3 & 2 & 2 \\ 0 & 0 & 0 & 4 & 1 \\ 0 & 0 & 0 & 0 & 4 \end{bmatrix}$$

Then,

$$\det(\mathbf{A} - \lambda\mathbf{I}) = (3 - \lambda)^3(4 - \lambda)^2$$

Thus, **A** has two distinct eigenvalues, $\lambda_1 = 3$ and $\lambda_2 = 4$, with (algebraic) multiplicities $m_1 = 3$, $m_2 = 2$, respectively.

The null space of $(\mathbf{A} - \lambda_1\mathbf{I})$ is determined by finding all the solutions to $(\mathbf{A} - \lambda_1\mathbf{I})\mathbf{x} = 0$, or

$$\begin{bmatrix} 0 & 2 & 1 & 1 & 2 \\ 0 & 0 & 0 & 1 & 1 \\ 0 & 0 & 0 & 2 & 2 \\ 0 & 0 & 0 & 1 & 1 \\ 0 & 0 & 0 & 0 & 1 \end{bmatrix} \begin{bmatrix} x_1 \\ x_2 \\ x_3 \\ x_4 \\ x_5 \end{bmatrix} = \begin{bmatrix} 0 \\ 0 \\ 0 \\ 0 \\ 0 \end{bmatrix}$$

This set of equations yields:

$$x_5 = x_4 = 0 \qquad 2x_2 + x_3 = 0$$

Thus, there are two arbitrary parameters, x_1 and one of x_2 or x_3; therefore, the dimension of $N(\mathbf{A} - 3\mathbf{I})$ is two: $d_1 = 2$. A basis for $N(\mathbf{A} - 3\mathbf{I})$ consists of any two linearly independent eigenvectors; for example,

$$\begin{bmatrix} 1 \\ 0 \\ 0 \\ 0 \\ 0 \end{bmatrix} \quad \text{and} \quad \begin{bmatrix} 0 \\ 1 \\ -2 \\ 0 \\ 0 \end{bmatrix}$$

Similarly, we calculate the null space of $(\mathbf{A} - 4\mathbf{I})$:

$$\begin{bmatrix} -1 & 2 & 1 & 1 & 2 \\ 0 & -1 & 0 & 1 & 1 \\ 0 & 0 & -1 & 2 & 2 \\ 0 & 0 & 0 & 0 & 1 \\ 0 & 0 & 0 & 0 & 0 \end{bmatrix} \begin{bmatrix} x_2 \\ x_2 \\ x_3 \\ x_4 \\ x_5 \end{bmatrix} = \begin{bmatrix} 0 \\ 0 \\ 0 \\ 0 \\ 0 \end{bmatrix}$$

Upon solving this set of equations we obtain:

$$\begin{aligned} x_1 &= 5x_4 \\ x_2 &= x_4 \\ x_3 &= 2x_4 \\ x_5 &= 0 \end{aligned}$$

Thus, x_4 is the only arbitrary parameter, and $d_2 = 1$. An eigenvector is $\begin{bmatrix} 5 \\ 1 \\ 2 \\ 1 \\ 0 \end{bmatrix}$.

////

EXAMPLE 9-2 Let $\mathbf{A} = \begin{bmatrix} 3 & 4 \\ 4 & -3 \end{bmatrix}$

$$\det(\mathbf{A} - \lambda\mathbf{I}) = \begin{vmatrix} 3-\lambda & 4 \\ 4 & -3-\lambda \end{vmatrix} = -(3-\lambda)(3+\lambda) - 16$$
$$= \lambda^2 - 25 = (\lambda - 5)(\lambda + 5)$$

Thus, the eigenvalues are $\lambda_1 = 5$, $\lambda_2 = -5$, and $m_1 = m_2 = 1$. From (9-17) we see that $d_1 = d_2 = 1$ also.

The null space of λ_1 is found from:

$$\begin{bmatrix} -2 & 4 \\ 4 & -8 \end{bmatrix} \begin{bmatrix} x_1 \\ x_2 \end{bmatrix} = \begin{bmatrix} 0 \\ 0 \end{bmatrix}$$
$$x_1 = 2x_2$$

The eigenvectors corresponding to $\lambda_1 = 5$ are therefore all scalar multiples of $\mathbf{x} = \begin{bmatrix} 2 \\ 1 \end{bmatrix}$; similarly we find that the eigenvectors corresponding to $\lambda_2 = -5$ are scalar multiples of $\mathbf{y} = \begin{bmatrix} -1 \\ 2 \end{bmatrix}$.

////

We conclude this section with several very important theorems.

Theorem 9-4 If $\mathbf{x}_1$ and $\mathbf{x}_2$ are eigenvectors of a matrix $\mathbf{A}$, corresponding to eigenvalues λ_1, λ_2, respectively, and if $\lambda_1 \neq \lambda_2$, then $\mathbf{x}_1$ and $\mathbf{x}_2$ are linearly independent.

PROOF Let

$$\alpha_1 \mathbf{x}_1 + \alpha_2 \mathbf{x}_2 = \mathbf{0} \qquad (9\text{-}18)$$

Premultiplying both sides of Eq. (9-18) by $\mathbf{A}$ yields:

$$\mathbf{A}(\alpha_1 \mathbf{x}_1 + \alpha_2 \mathbf{x}_2) = \mathbf{A0} = \mathbf{0}$$
$$= \alpha_1 \mathbf{A}\mathbf{x}_1 + \alpha_2 \mathbf{A}\mathbf{x}_2 \qquad (9\text{-}19)$$

But $\mathbf{A}\mathbf{x}_1 = \lambda_1 \mathbf{x}_1$ and $\mathbf{A}\mathbf{x}_2 = \lambda_2 \mathbf{x}_2$. Thus, Eq. (9-19) becomes

$$\alpha_1 \lambda_1 \mathbf{x}_1 + \alpha_2 \lambda_2 \mathbf{x}_2 = \mathbf{0} \qquad (9\text{-}20)$$

From Eq. (9-18), $\alpha_1 \mathbf{x}_1 = -\alpha_2 \mathbf{x}_2$; substituting this result into Eq. (9-20) produces

$$-\lambda_1 \alpha_2 \mathbf{x}_2 + \lambda_2 \alpha_2 \mathbf{x}_2 = \mathbf{0}$$

or

$$(\lambda_2 - \lambda_1)\alpha_2 \mathbf{x}_2 = \mathbf{0}$$

Now, $\lambda_2 - \lambda_1 \neq 0$ by hypothesis, and so it must be true that

$$\alpha_2 \mathbf{x}_2 = \mathbf{0}$$

But, since $\mathbf{x}_2$ is a nonzero vector (all eigenvectors are assumed nonzero), we must conclude that $\alpha_2 = 0$. Furthermore, if $\alpha_2 = 0$, then Eq. (9-18) becomes

$$\alpha_1 \mathbf{x}_1 = \mathbf{0}$$

and by identical reasoning we conclude that $\alpha_1 = 0$.

Hence, the only values of α_1, α_2 satisfying Eq. (9-18) are $\alpha_1 = \alpha_2 = 0$. Therefore, by definition, $\mathbf{x}_1$ and $\mathbf{x}_2$ are linearly independent. QED

Theorem 9-4 is easily generalized to:

Theorem 9-5 Eigenvectors corresponding to distinct eigenvalues of a matrix are linearly independent.

Theorem 9-6 If a matrix $\mathbf{A}$ has n linearly independent eigenvectors $\mathbf{x}_1, \mathbf{x}_2, \ldots, \mathbf{x}_n$, corresponding to eigenvalues $\lambda_1, \lambda_2, \ldots, \lambda_n$, respectively, and if $\mathbf{P} = [\mathbf{x}_1 \quad \mathbf{x}_2 \quad \cdots \quad \mathbf{x}_n]$, then $\mathbf{P}$ is nonsingular and $\mathbf{P}^{-1}\mathbf{A}\mathbf{P} = \mathbf{D}$, where $\mathbf{D}$ is a diagonal matrix whose diagonal elements are $\lambda_1, \lambda_2, \ldots, \lambda_n$.

PROOF $\mathbf{P}$ is an $n \times n$ matrix whose columns are linearly independent. Therefore, $\mathbf{P}$ is nonsingular. Let us calculate the following two products:

1.
$$\begin{aligned}\mathbf{AP} &= \mathbf{A}[\mathbf{x}_1 \quad \mathbf{x}_2 \quad \cdots \quad \mathbf{x}_n] \\ &= [\mathbf{A}\mathbf{x}_1 \quad \mathbf{A}\mathbf{x}_2 \quad \cdots \quad \mathbf{A}\mathbf{x}_n] \\ &= [\lambda_1\mathbf{x}_1 \quad \lambda_2\mathbf{x}_2 \quad \cdots \quad \lambda_n\mathbf{x}_n] \end{aligned} \tag{9-21}$$

2.
$$\begin{aligned}\mathbf{PD} &= [\mathbf{x}_1 \quad \mathbf{x}_2 \quad \cdots \quad \mathbf{x}_n]\begin{bmatrix}\lambda_1 & 0 & \cdots & 0 \\ 0 & \lambda_2 & \cdots & 0 \\ \vdots & \vdots & \ddots & \vdots \\ 0 & 0 & \cdots & \lambda_n\end{bmatrix} \\ &= [\lambda_1\mathbf{x}_1 \quad \lambda_2\mathbf{x}_2 \quad \cdots \quad \lambda_n\mathbf{x}_n]\end{aligned} \tag{9-22}$$

Upon comparing (9-21) and (9-22) we see immediately that

$$\mathbf{AP} = \mathbf{PD} \tag{9-23}$$

Premultiplying Eq. (9-23) by $\mathbf{P}^{-1}$ yields the desired result:

$$\mathbf{P}^{-1}\mathbf{AP} = \mathbf{P}^{-1}\mathbf{PD} = \mathbf{D} \qquad \text{QED}$$

EXAMPLE 9-3 Consider the matrix of Example 9-2:

$$\mathbf{A} = \begin{bmatrix} 3 & 4 \\ 4 & -3 \end{bmatrix}$$

The two eigenvalues of **A** are $\lambda_1 = 5$, $\lambda_2 = -5$, with corresponding eigenvectors, for example, of

$$\mathbf{x}_1 = \begin{bmatrix} 2 \\ 1 \end{bmatrix} \qquad \mathbf{x}_2 = \begin{bmatrix} -1 \\ 2 \end{bmatrix}$$

Thus, let

$$\mathbf{P} = \begin{bmatrix} 2 & -1 \\ 1 & 2 \end{bmatrix}$$

$$\text{Then, } \mathbf{P}^{-1} = \begin{bmatrix} \frac{2}{5} & \frac{1}{5} \\ -\frac{1}{5} & \frac{2}{5} \end{bmatrix}$$

$$\begin{aligned} \mathbf{P}^{-1}\mathbf{AP} &= \begin{bmatrix} \frac{2}{5} & \frac{1}{5} \\ -\frac{1}{5} & \frac{2}{5} \end{bmatrix} \begin{bmatrix} 3 & 4 \\ 4 & -3 \end{bmatrix} \begin{bmatrix} 2 & -1 \\ 1 & 2 \end{bmatrix} \\ &= \begin{bmatrix} \frac{2}{5} & \frac{1}{5} \\ -\frac{1}{5} & \frac{2}{5} \end{bmatrix} \begin{bmatrix} 10 & 5 \\ 5 & -10 \end{bmatrix} \\ &= \begin{bmatrix} 5 & 0 \\ 0 & -5 \end{bmatrix} = \begin{bmatrix} \lambda_1 & 0 \\ 0 & \lambda_2 \end{bmatrix} \end{aligned}$$

////

9-2 UNITARY AND ORTHOGONAL MATRICES

In this section we shall introduce several concepts which shall prove to be of extreme importance in the study of eigenvalue problems.

First, let us recall from Chap. 2 the definition of the inner product of two vectors:

Given two n-component vectors $\mathbf{x} = (x_1 \ \ x_2 \ \ \cdots \ \ x_n)^T$, $\mathbf{y} = (y_1 \ \ y_2 \ \ \cdots \ \ y_n)^T$, the inner product of $\mathbf{x}$ and $\mathbf{y}$ is defined to be

$$\mathbf{x}^*\mathbf{y} = \sum_{j=1}^{n} \bar{x}_j y_j = \bar{x}_1 y_1 + \bar{x}_2 y_2 + \cdots + \bar{x}_n y_n \qquad (9\text{-}24)$$

that is, the inner product of $\mathbf{x}$ and $\mathbf{y}$ is simply the matrix product of the $1 \times n$ row vector $\mathbf{x}^*$ with the $n \times 1$ column vector $\mathbf{y}$, resulting in a scalar.

Two vectors $\mathbf{x}$, $\mathbf{y}$ are said to be *orthogonal* if their inner product is zero:

$$\mathbf{x}^*\mathbf{y} = 0$$

(Note: If $\mathbf{x}$, $\mathbf{y}$ are three-component vectors with real components, then $\mathbf{x}^*\mathbf{y} = 0$ means that $\mathbf{x}$ and $\mathbf{y}$ are perpendicular.)

A vector $\mathbf{x}$ is said to be a *unit vector* if the inner product of $\mathbf{x}$ with itself is unity:

$$\mathbf{x}^*\mathbf{x} = \sum_{j=1}^{n} |x_j|^2 = 1$$

Any nonzero vector $\mathbf{y}$ is equal to a scalar times a unit vector: Let

$$\mathbf{y}^*\mathbf{y} = c^2 > 0$$

(Recall from Chap. 2 that for any nonzero vector $\mathbf{x}$, $\mathbf{x}^*\mathbf{x}$ is always a positive real number.) Then, if $\mathbf{x} = (1/c)\mathbf{y}$,

$$\begin{aligned} \mathbf{x}^*\mathbf{x} &= \left(\frac{1}{c}\mathbf{y}\right)^* \left(\frac{1}{c}\mathbf{y}\right) \\ &= \frac{1}{c^2}(\mathbf{y}^*\mathbf{y}) \\ &= \frac{c^2}{c^2} \\ &= 1 \end{aligned}$$

Thus, $\mathbf{x}$ is a unit vector. This process of multiplying a vector $\mathbf{y}$ by a scalar to produce a unit vector $\mathbf{x}$ is called *normalizing* the vector $\mathbf{y}$. The vector $\mathbf{x}$ is said to be *normalized.*

Observe that if any eigenvector is normalized, the resulting unit vector is also an eigenvector, since if $\mathbf{Ax} = \lambda\mathbf{x}$, for some $\mathbf{x}$, λ, it is also true that $\mathbf{Ay} = \lambda\mathbf{y}$, for $\mathbf{y} = \alpha\mathbf{x}$ (α being any scalar).

A matrix $\mathbf{U} = [\mathbf{u}_1 \quad \mathbf{u}_2 \quad \cdots \quad \mathbf{u}_n]$ whose columns are unit vectors that are mutually orthogonal is called a *unitary matrix;* by mutually orthogonal columns we mean

$$\mathbf{u}_i^*\mathbf{u}_j = 0 \qquad \text{if } i \neq j \qquad \begin{aligned} i &= 1, 2, \ldots, n \\ j &= 1, 2, \ldots, n \end{aligned} \qquad (9\text{-}25)$$

that is, each pair of columns is orthogonal.

Moreover, since each column is a unit vector, we also have

$$\mathbf{u}_i^*\mathbf{u}_i = 1 \qquad i = 1, 2, \ldots, n \tag{9-26}$$

Let us now derive some properties of unitary matrices:

Theorem 9-7 Let $\mathbf{U} = [\mathbf{u}_1 \quad \mathbf{u}_2 \quad \cdots \quad \mathbf{u}_n]$ be a unitary matrix. Then:

(*a*) $\mathbf{U}^*\mathbf{U} = \mathbf{U}\mathbf{U}^* = \mathbf{I}_n$

(*b*) $\mathbf{U}$ is nonsingular

(*c*) $\mathbf{U}^{-1} = \mathbf{U}^*$

PROOF Let us multiply $\mathbf{U}^*$ and $\mathbf{U}$ in partitioned form:

$$\begin{aligned}\mathbf{U}^*\mathbf{U} &= [\mathbf{u}_1 \quad \mathbf{u}_2 \quad \cdots \quad \mathbf{u}_n]^*[\mathbf{u}_1 \quad \mathbf{u}_2 \quad \cdots \quad \mathbf{u}_n] \\ &= \begin{bmatrix}\mathbf{u}_1^* \\ \mathbf{u}_2^* \\ \vdots \\ \mathbf{u}_n^*\end{bmatrix}[\mathbf{u}_1 \quad \mathbf{u}_2 \quad \cdots \quad \mathbf{u}_n] \\ &= \begin{bmatrix}\mathbf{u}_1^*\mathbf{u}_1 & \mathbf{u}_1^*\mathbf{u}_2 & \cdots & \mathbf{u}_1^*\mathbf{u}_n \\ \mathbf{u}_2^*\mathbf{u}_1 & \mathbf{u}_2^*\mathbf{u}_2 & \cdots & \mathbf{u}_2^*\mathbf{u}_n \\ \vdots & \vdots & & \vdots \\ \mathbf{u}_n^*\mathbf{u}_1 & \mathbf{u}_n^*\mathbf{u}_2 & \cdots & \mathbf{u}_n^*\mathbf{u}_n\end{bmatrix}\end{aligned} \tag{9-27}$$

That is, the (i, j) element of $\mathbf{U}^*\mathbf{U}$ is $\mathbf{u}_i^*\mathbf{u}_j$. But from Eqs. (9-25) and (9-26) it is immediately clear that the right-hand side of Eq. (9-27) is merely an nth-order identity matrix $\mathbf{I}_n$.

Thus,

$$\mathbf{U}^*\mathbf{U} = \mathbf{I}_n \tag{9-28}$$

Now, if we calculate the determinant of both sides of Eq. (9-28), we find that

$$\det(\mathbf{U}^*\mathbf{U}) = \det(\mathbf{I}_n) = 1$$

$$\det(\mathbf{U}^*)\det(\mathbf{U}) = 1 \tag{9-29}$$

Thus, neither det $(\mathbf{U})$ nor det $(\mathbf{U}^*)$ can be zero. Hence $\mathbf{U}$ is nonsingular and $\mathbf{U}^{-1}$ exists. Postmultiplying both sides of Eq. (9-28) by $\mathbf{U}^{-1}$ yields

$$\mathbf{U}^* = \mathbf{U}^{-1} \tag{9-30}$$

and premultiplying both sides of Eq. (9-30) by $\mathbf{U}$ yields

$$\mathbf{U}\mathbf{U}^* = \mathbf{U}\mathbf{U}^{-1} = \mathbf{I}_n \qquad \text{QED}$$

If all the elements of a unitary matrix are real, the matrix is also called an *orthogonal matrix.* If **P** is an orthogonal matrix, then

$$\mathbf{P}^T\mathbf{P} = \mathbf{I}_n = \mathbf{P}\mathbf{P}^T$$

$$\mathbf{P}^{-1} = \mathbf{P}^T$$

We conclude this section by establishing one last result.

Theorem 9-8 If $\mathbf{x}_1, \mathbf{x}_2, \ldots, \mathbf{x}_p$ is a set of mutually orthogonal nonzero vectors, then $\mathbf{x}_1, \mathbf{x}_2, \ldots, \mathbf{x}_p$ are also linearly independent.

PROOF Let

$$\alpha_1\mathbf{x}_1 + \alpha_2\mathbf{x}_2 + \cdots + \alpha_p\mathbf{x}_p = \mathbf{0}$$

We wish to show that $\alpha_1 = \alpha_2 = \cdots = \alpha_p = 0$. Suppose we select any one of the p vectors $\mathbf{x}_1, \mathbf{x}_2, \ldots, \mathbf{x}_p$, say $\mathbf{x}_k$, and calculate

$$\begin{aligned}\mathbf{x}_k^*(&\alpha_1\mathbf{x}_1 + \alpha_2\mathbf{x}_2 + \cdots + \alpha_p\mathbf{x}_p)\\ &= \alpha_1\mathbf{x}_k^*\mathbf{x}_1 + \alpha_2\mathbf{x}_k^*\mathbf{x}_2 + \cdots + \alpha_{k-1}\mathbf{x}_k^*\mathbf{x}_{k-1} + \alpha_k\mathbf{x}_k^*\mathbf{x}_k\\ &\quad + \alpha_{k+1}\mathbf{x}_k^*\mathbf{x}_{k+1} + \cdots + \alpha_p\mathbf{x}_k^*\mathbf{x}_p = 0\end{aligned} \tag{9-31}$$

Now, since vectors $\mathbf{x}_1, \mathbf{x}_2, \ldots, \mathbf{x}_p$ are mutually orthogonal, each term in the above sum is zero, except the term $\alpha_k\mathbf{x}_k^*\mathbf{x}_k$. Thus, Eq. (9-31) reduces to

$$\alpha_k\mathbf{x}_k^*\mathbf{x}_k = 0$$

But, since $\mathbf{x}_k^*\mathbf{x}_k \neq 0$, it must be true that $\alpha_k = 0$. If we repeat the above procedure for each of the vectors $\mathbf{x}_k$, $k = 1, 2, \ldots, p$, we must conclude that $\alpha_k = 0$, $k = 1, 2, \ldots, p$. QED

EXAMPLE 9-4 Consider the matrix **P** of Example 9-2:

$$\mathbf{P} = \begin{bmatrix} 2 & -1 \\ 1 & 2 \end{bmatrix} \qquad \mathbf{x}_1 = \begin{bmatrix} 2 \\ 1 \end{bmatrix} \qquad \mathbf{x}_2 = \begin{bmatrix} -1 \\ 2 \end{bmatrix} \qquad \mathbf{P} = [\mathbf{x}_1 \quad \mathbf{x}_2]$$

The columns of **P** are orthogonal:

$$\mathbf{x}_1^*\mathbf{x}_2 = [2 \quad 1]\begin{bmatrix} -1 \\ 2 \end{bmatrix} = (2)(-1) + (1)(2) = 0$$

However, $\mathbf{x}_1$ and $\mathbf{x}_2$ are not unit vectors. Let us normalize these vectors.

$$\mathbf{x}_1^*\mathbf{x}_1 = [2 \quad 1]\begin{bmatrix} 2 \\ 1 \end{bmatrix} = 5$$

$$\mathbf{x}_2^*\mathbf{x}_2 = [-1 \quad 2]\begin{bmatrix} -1 \\ 2 \end{bmatrix} = 5$$

Thus, the vectors $\mathbf{y}_1 = \left(\frac{1}{\sqrt{5}}\right)\begin{bmatrix}2\\1\end{bmatrix}$ $\mathbf{y}_2 = \left(\frac{1}{\sqrt{5}}\right)\begin{bmatrix}-1\\2\end{bmatrix}$

are the required normalized vectors. The matrix $\mathbf{Q} = [\mathbf{y}_1 \;\; \mathbf{y}_2]$ is therefore an orthogonal matrix. The reader should verify that $\mathbf{Q}^T\mathbf{Q} = \mathbf{I}$, $\mathbf{Q}^T = \mathbf{Q}^{-1}$. ////

9-3 HERMITIAN AND REAL SYMMETRIC MATRICES

We shall now investigate the properties of the eigenvalues and eigenvectors of a very special class of matrices, namely, Hermitian matrices. Such matrices occur frequently in practical situations.

Let us review briefly some properties of complex numbers which we will make use of in the following discussion.

Let $\alpha = a + bi$ be any complex number, where a, b are real numbers and $i = \sqrt{-1}$. Then, the *complex conjugate* of α, denoted by $\bar{\alpha}$, is

$$\bar{\alpha} = a - bi$$

Thus, α is a real number if and only if $b = 0$, and therefore $\alpha = \bar{\alpha}$ if and only if α is a real number. Also, α is purely imaginary $(a = 0)$ if and only if $\bar{\alpha} = -\alpha$.

Now let us consider the eigenvalues of a Hermitian matrix. Let λ_j be any eigenvalue of the Hermitian matrix $\mathbf{A}$, and let $\mathbf{x}_j$ be a corresponding unit eigenvector. Thus,

$$\mathbf{A}\mathbf{x}_j = \lambda_j \mathbf{x}_j \qquad (9\text{-}32)$$

If we premultiply both sides of Eq. (9-32) by the row vector $\mathbf{x}_j^*$, we obtain

$$\mathbf{x}_j^*\mathbf{A}\mathbf{x}_j = \mathbf{x}_j^*\lambda_j\mathbf{x}_j = \lambda_j\mathbf{x}_j^*\mathbf{x}_j = \lambda_j \qquad (9\text{-}33)$$

(since $\mathbf{x}_j$ is a unit vector, $\mathbf{x}_j^*\mathbf{x}_j = 1$).

Now, suppose we first take the transpose conjugate of both sides of Eq. (9-32) and then postmultiply both sides of the resulting equation by the eigenvector $\mathbf{x}_j$:

$$(\mathbf{A}\mathbf{x}_j)^* = (\lambda_j\mathbf{x}_j)^*$$

$$\mathbf{x}_j^*\mathbf{A}^* = \mathbf{x}_j^*\bar{\lambda}_j = \bar{\lambda}_j\mathbf{x}_j^*$$

$$\mathbf{x}_j^*\mathbf{A}^*\mathbf{x}_j = \bar{\lambda}_j\mathbf{x}_j^*\mathbf{x}_j = \bar{\lambda}_j \qquad (9\text{-}34)$$

But, since $\mathbf{A}$ is Hermitian, $\mathbf{A} = \mathbf{A}^*$, and Eq. (9-34) becomes

$$\mathbf{x}_j^*\mathbf{A}\mathbf{x}_j = \bar{\lambda}_j \qquad (9\text{-}35)$$

We now note that the left-hand sides of Eqs. (9-33) and (9-35) are identical, and therefore their respective right-hand sides must be equal:

$$\lambda_j = \bar{\lambda}_j$$

From our earlier discussion, $\lambda_j = \bar{\lambda}_j$ only if λ_j is a real number. Furthermore, the above argument must hold for *all* eigenvalues of **A** (since λ_j was assumed to be any such eigenvalue). We have just proved:

Theorem 9-9 The eigenvalues of a Hermitian matrix are all real numbers.

A Hermitian matrix **A** has only real eigenvalues even if the elements of **A** are complex numbers!

We shall now state and prove another interesting and very important result, concerning the eigenvectors of a Hermitian matrix:

Theorem 9-10 If λ_j and λ_k are distinct eigenvalues ($\lambda_j \neq \lambda_k$) of a Hermitian matrix **A**, with corresponding eigenvectors $\mathbf{x}_j$, $\mathbf{x}_k$, respectively, then $\mathbf{x}_j$ and $\mathbf{x}_k$ are orthogonal.

PROOF We must show that $\mathbf{x}_j^*\mathbf{x}_k = 0$. First, from our hypothesis, we have that

$$\mathbf{A}\mathbf{x}_j = \lambda_j \mathbf{x}_j \qquad (9\text{-}36)$$

$$\mathbf{A}\mathbf{x}_k = \lambda_k \mathbf{x}_k \qquad (9\text{-}37)$$

and of course that $\mathbf{A} = \mathbf{A}^*$. Suppose we premultiply Eq. (9-36) by $\mathbf{x}_k^*$ and Eq. (9-37) by $\mathbf{x}_j^*$. We then obtain:

$$\mathbf{x}_k^*\mathbf{A}\mathbf{x}_j = \lambda_j \mathbf{x}_k^*\mathbf{x}_j \qquad (9\text{-}38)$$

$$\mathbf{x}_j^*\mathbf{A}\mathbf{x}_k = \lambda_k \mathbf{x}_j^*\mathbf{x}_k \qquad (9\text{-}39)$$

If we now calculate the transpose conjugate of both sides of Eq. (9-38), the result will be

$$(\mathbf{x}_k^*\mathbf{A}\mathbf{x}_j)^* = (\lambda_j \mathbf{x}_k^*\mathbf{x}_j)^*$$

$$[\mathbf{x}_k^*(\mathbf{A}\mathbf{x}_j)]^* = \bar{\lambda}_j(\mathbf{x}_j^*\mathbf{x}_k)$$

$$(\mathbf{A}\mathbf{x}_j)^*(\mathbf{x}_k^*)^* = \bar{\lambda}_j(\mathbf{x}_j^*\mathbf{x}_k)$$

$$\mathbf{x}_j^*\mathbf{A}^*\mathbf{x}_k = \bar{\lambda}_j(\mathbf{x}_j^*\mathbf{x}_k) \qquad (9\text{-}40)$$

However, we know from Theorem 9-9 that $\bar{\lambda}_j = \lambda_j$ and by hypothesis, $\mathbf{A} = \mathbf{A}^*$. Substituting these relations into Eq. (9-40) yields

$$\mathbf{x}_j^*\mathbf{A}\mathbf{x}_k = \lambda_j \mathbf{x}_j^*\mathbf{x}_k \qquad (9\text{-}41)$$

Now, we observe that the left-hand sides of Eqs. (9-39) and (9-41) are identical, and therefore

$$\lambda_j \mathbf{x}_j^* \mathbf{x}_k = \lambda_k \mathbf{x}_j^* \mathbf{x}_k$$

or

$$\lambda_j \mathbf{x}_j^* \mathbf{x}_k - \lambda_k \mathbf{x}_j^* \mathbf{x}_k = 0$$

$$(\lambda_j - \lambda_k)(\mathbf{x}_j^* \mathbf{x}_k) = 0 \qquad (9\text{-}42)$$

Since $\lambda_j \neq \lambda_k$ by assumption, Eq. (9-42) implies that $\mathbf{x}_j^* \mathbf{x}_k = 0$.
QED

We can now combine Theorems 9-6, 9-7, 9-8, and 9-10 to arrive at:

Theorem 9-11 If a Hermitian matrix $\mathbf{A}$ has n distinct eigenvalues, $\lambda_1, \lambda_2, \ldots, \lambda_n$, with corresponding unit eigenvectors $\mathbf{x}_1, \mathbf{x}_2, \ldots, \mathbf{x}_n$, and if $\mathbf{U} = [\mathbf{x}_1 \quad \mathbf{x}_2 \quad \cdots \quad \mathbf{x}_n]$, then $\mathbf{U}^*\mathbf{A}\mathbf{U} = \mathbf{D}$, where $\mathbf{D}$ is a diagonal matrix whose diagonal elements are $\lambda_1, \lambda_2, \ldots, \lambda_n$.

PROOF By Theorem 9-10, the eigenvectors $\mathbf{x}_1, \mathbf{x}_2, \ldots, \mathbf{x}_n$ are mutually orthogonal, and therefore by Theorem 9-8 $\mathbf{x}_1, \mathbf{x}_2, \ldots, \mathbf{x}_n$ are linearly independent. Hence, the matrix $\mathbf{U}$ is nonsingular. By Theorem 9-6, we have $\mathbf{U}^{-1}\mathbf{A}\mathbf{U} = \mathbf{D}$; however, $\mathbf{U}$ is a unitary matrix, and by Theorem 9-7, $\mathbf{U}^{-1} = \mathbf{U}^*$. Thus, $\mathbf{U}^*\mathbf{A}\mathbf{U} = \mathbf{D}$. QED

Corollary 9-12 For any matrix $\mathbf{A}$ for which there exists a nonsingular matrix $\mathbf{P}$ such that $\mathbf{P}^{-1}\mathbf{A}\mathbf{P} = \mathbf{D}$: $\mathbf{A}^k = \mathbf{P}\mathbf{D}^k\mathbf{P}^{-1}$, for all positive integers k, and $\mathbf{D}^k$ is a diagonal matrix with diagonal elements $\lambda_1{}^k, \lambda_2{}^k, \ldots, \lambda_n{}^k$.

The proof is left as an exercise for the reader.

Since a real symmetric matrix is merely a special case of a Hermitian matrix, we can also state:

Corollary 9-13 Let $\mathbf{A}$ be a real symmetric matrix of order n. Then:

1 The eigenvalues of $\mathbf{A}$ are all real numbers.

2 Eigenvectors of $\mathbf{A}$ corresponding to distinct eigenvalues are orthogonal.

3 If $\mathbf{A}$ has n distinct eigenvalues, then there exists an orthogonal matrix $\mathbf{P}$ such that $\mathbf{P}^T\mathbf{A}\mathbf{P} = \mathbf{D}$, where $\mathbf{D}$ is a diagonal matrix with diagonal elements $\lambda_1, \lambda_2, \ldots, \lambda_n$ (the eigenvalues of $\mathbf{A}$); moreover, $\mathbf{P} = [\mathbf{x}_1 \quad \mathbf{x}_2 \quad \cdots \quad \mathbf{x}_n]$ where $\mathbf{x}_1, \mathbf{x}_2, \ldots, \mathbf{x}_n$ are unit eigenvectors corresponding to $\lambda_1, \lambda_2, \ldots, \lambda_n$, respectively.

EXAMPLE 9-5 Let $\mathbf{A} = \begin{bmatrix} 5 & 0 & 1 \\ 0 & -1 & 0 \\ 1 & 0 & 5 \end{bmatrix}$ **A** is real and symmetric

The eigenvalues of **A** are the roots of

$$\begin{vmatrix} 5-\lambda & 0 & 1 \\ 0 & -1-\lambda & 0 \\ 1 & 0 & 5-\lambda \end{vmatrix} = 0$$

or

$$(5-\lambda)(-1-\lambda)(5-\lambda) + 1(1+\lambda) = 0$$

$$(\lambda+1)[-\lambda^2 + 10\lambda - 25 + 1] = 0$$

$$(\lambda+1)(-\lambda^2 + 10\lambda - 24) = 0$$

$$(\lambda+1)(\lambda-4)(\lambda-6) = 0$$

Hence, $\lambda_1 = -1$, $\lambda_2 = 4$, $\lambda_3 = 6$.

An eigenvector corresponding to $\lambda_1 = -1$ is found from

$$(\mathbf{A} - \lambda_1\mathbf{I})\mathbf{x} = \mathbf{0}$$

$$\begin{bmatrix} 6 & 0 & 1 \\ 0 & 0 & 0 \\ 1 & 0 & 6 \end{bmatrix}\begin{bmatrix} \alpha_1 \\ \alpha_2 \\ \alpha_3 \end{bmatrix} = \begin{bmatrix} 0 \\ 0 \\ 0 \end{bmatrix}$$

$$6\alpha_1 + \alpha_3 = 0$$

$$\alpha_1 + 6\alpha_3 = 0$$

$\alpha_1 = \alpha_3 = 0$, α_2 arbitrary. Thus, let us choose $\mathbf{x}_1 = \begin{bmatrix} 0 \\ 1 \\ 0 \end{bmatrix}$ as the unit eigenvector.

Similarly, we determine eigenvectors for λ_2 and λ_3:

$$(\mathbf{A} - \lambda_2\mathbf{I})\mathbf{x} = \mathbf{0}$$

$$\begin{bmatrix} 1 & 0 & 1 \\ 0 & -5 & 0 \\ 1 & 0 & 1 \end{bmatrix}\begin{bmatrix} \alpha_1 \\ \alpha_2 \\ \alpha_3 \end{bmatrix} = \begin{bmatrix} 0 \\ 0 \\ 0 \end{bmatrix}$$

$$\alpha_1 + \alpha_3 = 0$$

$$-5\alpha_2 = 0$$

$$\alpha_1 + \alpha_3 = 0$$

$\alpha_1 = -\alpha_3$, $\alpha_2 = 0$. If we let $\alpha_3 = 1$, we obtain the eigenvector $\mathbf{y} = \begin{bmatrix} -1 \\ 0 \\ 1 \end{bmatrix}$;

normalizing $\mathbf{y}$ yields $\mathbf{x}_2 = \begin{bmatrix} -1/\sqrt{2} \\ 0 \\ 1/\sqrt{2} \end{bmatrix}$ a unit eigenvector corresponding to $\lambda_2 = 4$.

$$(\mathbf{A} - \lambda_3 \mathbf{I})\mathbf{x} = \mathbf{0}$$

$$\begin{bmatrix} -1 & 0 & 1 \\ 0 & -7 & 0 \\ 1 & 0 & -1 \end{bmatrix} \begin{bmatrix} \alpha_1 \\ \alpha_2 \\ \alpha_3 \end{bmatrix} = \begin{bmatrix} 0 \\ 0 \\ 0 \end{bmatrix}$$

$$-\alpha_1 + \alpha_3 = 0$$

$$-7\alpha_2 = 0$$

$$-\alpha_1 - \alpha_3 = 0$$

$\alpha_1 = \alpha_3$, $\alpha_2 = 0$. Letting $\alpha_3 = 1$ yields the eigenvector $\mathbf{z} = \begin{bmatrix} 1 \\ 0 \\ 1 \end{bmatrix}$

Upon normalizing $\mathbf{z}$, we obtain $\mathbf{x}_3 = \begin{bmatrix} 1/\sqrt{2} \\ 0 \\ 1/\sqrt{2} \end{bmatrix}$

a unit vector corresponding to $\lambda_3 = 6$. Now, if we let $\mathbf{P} = [\mathbf{x}_1 \quad \mathbf{x}_2 \quad \mathbf{x}_3]$, we easily verify that $\mathbf{P}$ is orthogonal, by calculating $\mathbf{P}^T\mathbf{P} = \mathbf{I}_3$. Moreover,

$$\mathbf{P}^T\mathbf{A}\mathbf{P} = \begin{bmatrix} 0 & 1 & 0 \\ -1/\sqrt{2} & 0 & 1/\sqrt{2} \\ 1/\sqrt{2} & 0 & 1/\sqrt{2} \end{bmatrix} \begin{bmatrix} 5 & 0 & 1 \\ 0 & -1 & 0 \\ 1 & 0 & 5 \end{bmatrix} \begin{bmatrix} 0 & -1/\sqrt{2} & 1/\sqrt{2} \\ 1 & 0 & 0 \\ 0 & 1/\sqrt{2} & 1/\sqrt{2} \end{bmatrix}$$

$$= \begin{bmatrix} 0 & -1 & 0 \\ -4/\sqrt{2} & 0 & 4/\sqrt{2} \\ 6/\sqrt{2} & 0 & 6/\sqrt{2} \end{bmatrix} \begin{bmatrix} 0 & -1/\sqrt{2} & 1/\sqrt{2} \\ 1 & 0 & 0 \\ 0 & 1/\sqrt{2} & 1/\sqrt{2} \end{bmatrix}$$

$$= \begin{bmatrix} -1 & 0 & 0 \\ 0 & 4 & 0 \\ 0 & 0 & 6 \end{bmatrix} = \begin{bmatrix} \lambda_1 & 0 & 0 \\ 0 & \lambda_2 & 0 \\ 0 & 0 & \lambda_3 \end{bmatrix}$$

////

We now turn to the question of determining the dimensions d_j of the null spaces $N(\mathbf{A} - \lambda_j \mathbf{I})$ of Hermitian matrices $\mathbf{A}$ for each distinct λ_j. In Sec. 9-2 we learned that $1 \le d_j \le m_j$, where m_j is the algebraic multiplicity of λ_j. We shall now establish that if $\mathbf{A}$ is Hermitian, $d_j = m_j$ for each distinct eigenvalue λ_j, $j = 1, 2, \ldots, p$. In order to simplify the notation somewhat, we shall restrict our attention to the special case of a real symmetric matrix, and prove†

Theorem 9-14 If $\mathbf{A}$ is a real symmetric matrix of order n, with eigenvalues $\lambda_1, \lambda_2, \ldots, \lambda_n$ (not necessarily distinct), then there exists an orthogonal matrix $\mathbf{U}$ such that $\mathbf{U}^T\mathbf{A}\mathbf{U} = \mathbf{D}$, where $\mathbf{D}$ is a diagonal matrix with diagonal elements $\lambda_1, \lambda_2, \ldots, \lambda_n$.

PROOF First, note that if $\mathbf{A}$ is a real symmetric matrix, then for any real matrix $\mathbf{B}$, the matrix $\mathbf{B}^T\mathbf{A}\mathbf{B}$ is also a real symmetric matrix:

$$(\mathbf{B}^T\mathbf{A}\mathbf{B})^T = \mathbf{B}^T\mathbf{A}^T(\mathbf{B}^T)^T = \mathbf{B}^T\mathbf{A}\mathbf{B}$$

We shall make use of this fact in the ensuing proof.

Now, we shall proceed by induction on n. Let us consider the case $n = 2$. Then, we may represent any 2×2 real symmetric matrix $\mathbf{A}$ by

$$\mathbf{A} = \begin{bmatrix} a & b \\ b & c \end{bmatrix}$$

Moreover, we know that $\mathbf{A}$ has at least one eigenvalue with a corresponding unit eigenvector; denote these by λ_1 and $\mathbf{x}_1 = (\alpha_1 \ \alpha_2)^T$ respectively.

Now, define a vector $\mathbf{x}_2 = (\beta_1 \quad \beta_2)^T$ and let

$$\mathbf{U} = [\mathbf{x}_1 \quad \mathbf{x}_2] = \begin{bmatrix} \alpha_1 & \beta_1 \\ \alpha_2 & \beta_2 \end{bmatrix}$$

Thus, the first column of $\mathbf{U}$ is the eigenvector $\mathbf{x}_1$. If $\mathbf{U}$ is to be orthogonal, then $\mathbf{x}_1$ and $\mathbf{x}_2$ must be unit vectors and $\mathbf{x}_1$, $\mathbf{x}_2$ must be orthogonal:

$$\mathbf{x}_1{}^T\mathbf{x}_1 = \alpha_1{}^2 + \alpha_2{}^2 = 1 \qquad (9\text{-}43)$$

$$\mathbf{x}_2{}^T\mathbf{x}_2 = \beta_1{}^2 + \beta_2{}^2 = 1 \qquad (9\text{-}44)$$

$$\mathbf{x}_2{}^T\mathbf{x}_1 = \alpha_1\beta_1 + \alpha_2\beta_2 = 0 \qquad (9\text{-}45)$$

† The proof of Theorem 9-14 is extremely tedious. However, it is not difficult to follow, and the author felt it would be instructive to include this proof in detail.

Let us calculate $\mathbf{U}^T\mathbf{A}\mathbf{U}$:

$$\mathbf{U}^T\mathbf{A}\mathbf{U} = \begin{bmatrix} \mathbf{x}_1{}^T \\ \mathbf{x}_2{}^T \end{bmatrix} \mathbf{A}[\mathbf{x}_1 \quad \mathbf{x}_2]$$

$$= \begin{bmatrix} \mathbf{x}_1{}^T \\ \mathbf{x}_2{}^T \end{bmatrix} [\mathbf{A}\mathbf{x}_1 \quad \mathbf{A}\mathbf{x}_2]$$

$$= \begin{bmatrix} \mathbf{x}_1{}^T \\ \mathbf{x}_2{}^T \end{bmatrix} [\lambda_1\mathbf{x}_1 \quad \mathbf{A}\mathbf{x}_2]$$

$$= \begin{bmatrix} \lambda_1\mathbf{x}_1{}^T\mathbf{x}_1 & \mathbf{x}_1{}^T\mathbf{A}\mathbf{x}_2 \\ \lambda_1\mathbf{x}_2{}^T\mathbf{x}_1 & \mathbf{x}_2{}^T\mathbf{A}\mathbf{x}_2 \end{bmatrix} \qquad (9\text{-}46)$$

But, the two elements in column 1 of the right-hand side above are equal to λ_1 and 0, respectively, by Eqs. (9-43) and (9-45). Hence, Eq. (9-46) becomes:

$$\mathbf{U}^T\mathbf{A}\mathbf{U} = \begin{bmatrix} \lambda_1 & \mathbf{x}_1{}^T\mathbf{A}\mathbf{x}_2 \\ 0 & \mathbf{x}_2{}^T\mathbf{A}\mathbf{x}_2 \end{bmatrix} \qquad (9\text{-}47)$$

Moreover, since by our comment at the beginning of this proof $\mathbf{U}^T\mathbf{A}\mathbf{U}$ is a real symmetric matrix, it must also be true that $\mathbf{x}_1{}^T\mathbf{A}\mathbf{x}_2 = 0$. Thus,

$$\mathbf{U}^T\mathbf{A}\mathbf{U} = \begin{bmatrix} \lambda_1 & 0 \\ 0 & \mathbf{x}_2{}^T\mathbf{A}\mathbf{x}_2 \end{bmatrix} \qquad (9\text{-}48)$$

We have thus far shown that $\mathbf{U}^T\mathbf{A}\mathbf{U}$ is a diagonal matrix whose first diagonal element is λ_1; we must now show that $\mathbf{x}_2{}^T\mathbf{A}\mathbf{x}_2 = \lambda_2$. From Eq. (9-45) we have that $\alpha_1\beta_1 + \alpha_2\beta_2 = 0$. Suppose we choose $\beta_1 = -\alpha_2$, $\beta_2 = \alpha_1$. Then, Eq. (9-45) is clearly satisfied; that is, $\mathbf{x}_1 = (\alpha_1 \quad \alpha_2)^T$ and $\mathbf{x}_2 = (-\alpha_2 \quad \alpha_1)^T$ are orthogonal. Moreover, since α_1, α_2 must satisfy Eq. (9-43), it is clear that Eq. (9-44) is also satisfied by the above choice for β_1, β_2.

Next, we observe that

$$\mathbf{A}\mathbf{x}_2 = \begin{bmatrix} a & b \\ b & c \end{bmatrix} \begin{bmatrix} \beta_1 \\ \beta_2 \end{bmatrix}$$

$$= \begin{bmatrix} a & b \\ b & c \end{bmatrix} \begin{bmatrix} -\alpha_2 \\ \alpha_1 \end{bmatrix}$$

$$= \begin{bmatrix} -a\alpha_2 + b\alpha_1 \\ -b\alpha_2 + c\alpha_1 \end{bmatrix} \qquad (9\text{-}49)$$

However, from $\mathbf{Ax}_1 = \lambda_1 \mathbf{x}_1$, we obtain

$$\begin{bmatrix} a & b \\ b & c \end{bmatrix} \begin{bmatrix} \alpha_1 \\ \alpha_2 \end{bmatrix} = \lambda_1 \begin{bmatrix} \alpha_1 \\ \alpha_2 \end{bmatrix}$$

$$a\alpha_1 + b\alpha_2 = \lambda_1\alpha_1 \qquad b\alpha_1 + c\alpha_2 = \lambda_1\alpha_2$$

$$b\alpha_2 = \lambda_1\alpha_1 - a\alpha_1 \qquad b\alpha_1 = \lambda_1\alpha_2 - c\alpha_2 \tag{9-50}$$

Upon substituting Eqs. (9-50) into Eq. (9-49) we obtain

$$\begin{aligned} \mathbf{Ax}_2 &= \begin{bmatrix} -a\alpha_2 + \lambda_1\alpha_2 - c\alpha_2 \\ -\lambda_1\alpha_1 + a\alpha_1 + c\alpha_1 \end{bmatrix} \\ &= (a + c - \lambda_1)\begin{bmatrix} -\alpha_2 \\ \alpha_1 \end{bmatrix} \\ &= (a + c - \lambda_1)\mathbf{x}_2 \end{aligned} \tag{9-51}$$

Thus, Eq. (9-51) states that $\mathbf{x}_2$ is an eigenvector of $\mathbf{A}$ corresponding to the eigenvalue $(a + c - \lambda_1)$. Since the 2×2 matrix $\mathbf{A}$ has only two eigenvalues, it must be true that

$$\lambda_2 = a + c - \lambda_1$$

Premultiplying $\mathbf{Ax}_2 = \lambda_2 \mathbf{x}_2$ by $\mathbf{x}_2{}^T$ yields the desired result,

$$\mathbf{x}_2{}^T\mathbf{Ax}_2 = \lambda_2$$

Inserting this result into Eq. (9-48) yields

$$\mathbf{U}^T\mathbf{AU} = \begin{bmatrix} \lambda_1 & 0 \\ 0 & \lambda_2 \end{bmatrix}$$

Note that it is possible that $\lambda_1 = \lambda_2$.

Now, we must show that if the theorem is true for a matrix of order $(n - 1)$, then it is also true for a matrix of order n. The induction hypothesis is:

Assume that for every real symmetric matrix $\mathbf{S}$ of order $(n - 1)$ there exists an orthogonal matrix $\mathbf{V}$ of order $(n - 1)$ such that

$$\mathbf{V}^T\mathbf{SV} = \hat{\mathbf{D}}$$

where, as usual, $\hat{\mathbf{D}}$ is a diagonal matrix with the $(n - 1)$ eigenvalues of $\mathbf{S}$ along its diagonal.

We proceed along the same lines as in the proof of the $n = 2$ case. Let λ_1 be any eigenvalue of $\mathbf{A}$ with corresponding unit eigenvector $\mathbf{x}_1$.

Then let $\mathbf{W}$ be an orthogonal matrix whose first column is $\mathbf{x}_1$:

$$\mathbf{W} = [\mathbf{x}_1 \quad \mathbf{x}_2 \quad \cdots \quad \mathbf{x}_n]$$

Now, we calculate $\mathbf{W}^T\mathbf{A}\mathbf{W}$:

$$\mathbf{W}^T\mathbf{A}\mathbf{W} = \begin{bmatrix} \mathbf{x}_1{}^T \\ \mathbf{x}_2{}^T \\ \vdots \\ \mathbf{x}_n{}^T \end{bmatrix} \mathbf{A}[\mathbf{x}_1 \quad \mathbf{x}_2 \quad \cdots \quad \mathbf{x}_n]$$

$$= \begin{bmatrix} \mathbf{x}_1{}^T \\ \mathbf{x}_2{}^T \\ \vdots \\ \mathbf{x}_n{}^T \end{bmatrix} [\mathbf{A}\mathbf{x}_1 \quad \mathbf{A}\mathbf{x}_2 \quad \cdots \quad \mathbf{A}\mathbf{x}_n]$$

$$= \begin{bmatrix} \mathbf{x}_1{}^T \\ \mathbf{x}_2{}^T \\ \vdots \\ \mathbf{x}_n{}^T \end{bmatrix} [\lambda_1\mathbf{x}_1 \quad \mathbf{A}\mathbf{x}_2 \quad \cdots \quad \mathbf{A}\mathbf{x}_n]$$

$$= \begin{bmatrix} \lambda_1\mathbf{x}_1{}^T\mathbf{x}_1 & \mathbf{x}_1{}^T\mathbf{A}\mathbf{x}_2 & \mathbf{x}_1{}^T\mathbf{A}\mathbf{x}_n \\ \lambda_1\mathbf{x}_2{}^T\mathbf{x}_1 & \mathbf{x}_2{}^T\mathbf{A}\mathbf{x}_2 & \mathbf{x}_2{}^T\mathbf{A}\mathbf{x}_n \\ \vdots & \vdots & \vdots \\ \lambda_1\mathbf{x}_n{}^T\mathbf{x}_1 & \mathbf{x}_n{}^T\mathbf{A}\mathbf{x}_2 & \mathbf{x}_n{}^T\mathbf{A}\mathbf{x}_n \end{bmatrix} \qquad (9\text{-}52)$$

If we denote $\mathbf{W}^T\mathbf{A}\mathbf{W} = \mathbf{F} = [f_{ij}]$, then by Eq. (9-52),

$$f_{11} = \lambda_1\mathbf{x}_1{}^T\mathbf{x}_1 = \lambda_1 \qquad \text{since } \mathbf{x}_1 \text{ is a unit vector}$$

$$\left.\begin{aligned} f_{21} &= \lambda_1\mathbf{x}_2{}^T\mathbf{x}_1 = 0 \\ f_{31} &= \lambda_1\mathbf{x}_3{}^T\mathbf{x}_1 = 0 \\ &\cdots\cdots\cdots\cdots \\ f_{n1} &= \lambda_1\mathbf{x}_n{}^T\mathbf{x}_1 = 0 \end{aligned}\right\} \qquad \begin{array}{l}\text{since the columns of} \\ \text{the orthogonal matrix} \\ \mathbf{W} \text{ are mutually orthogonal}\end{array} \qquad (9\text{-}53)$$

Moreover, by the observation made at the very beginning of this proof, $\mathbf{F} = \mathbf{W}^T\mathbf{A}\mathbf{W}$ is a real symmetric matrix. Therefore, by Eq. (9-53),

$$f_{12} = f_{13} = \cdots f_{1n} = 0$$

Thus, we may write Eq. (9-52) in partitioned form as follows:

$$\mathbf{W}^T\mathbf{A}\mathbf{W} = \left[\begin{array}{c|c} \lambda_1 & \mathbf{0} \\ \hline \mathbf{0} & \mathbf{S} \end{array}\right] \qquad (9\text{-}54)$$

where $\mathbf{S}$ is a real symmetric matrix of order $(n - 1)$.

Moreover, by our induction hypothesis, there exists an orthogonal matrix $\mathbf{V}$, of order $(n-1)$, such that $\mathbf{V}^T\mathbf{S}\mathbf{V} = \hat{\mathbf{D}}$.

Consider the nth-order matrix

$$\mathbf{P} = \left[\begin{array}{c:c} 1 & \mathbf{0} \\ \hdashline \mathbf{0} & \mathbf{V} \end{array}\right] \tag{9-55}$$

Observe that, by Eq. (9-54)

$$\begin{aligned} \mathbf{P}^T(\mathbf{W}^T\mathbf{A}\mathbf{W})\mathbf{P} &= \begin{bmatrix} 1 & \mathbf{0} \\ \mathbf{0} & \mathbf{V}^T \end{bmatrix} \begin{bmatrix} \lambda_1 & \mathbf{0} \\ \mathbf{0} & \mathbf{S} \end{bmatrix} \begin{bmatrix} 1 & \mathbf{0} \\ \mathbf{0} & \mathbf{V} \end{bmatrix} \\ &= \begin{bmatrix} \lambda_1 & \mathbf{0} \\ \mathbf{0} & \mathbf{V}^T\mathbf{S} \end{bmatrix} \begin{bmatrix} 1 & \mathbf{0} \\ \mathbf{0} & \mathbf{V} \end{bmatrix} \\ &= \begin{bmatrix} \lambda_1 & \mathbf{0} \\ \mathbf{0} & \mathbf{V}^T\mathbf{S}\mathbf{V} \end{bmatrix} \\ &= \begin{bmatrix} \lambda_1 & \mathbf{0} \\ \mathbf{0} & \hat{\mathbf{D}} \end{bmatrix} \end{aligned} \tag{9-56}$$

But the left-hand side of Eq. (9-56) may be expressed as

$$\mathbf{P}^T(\mathbf{W}^T\mathbf{A}\mathbf{W})\mathbf{P} = (\mathbf{W}\mathbf{P})^T\mathbf{A}(\mathbf{W}\mathbf{P}) \tag{9-57}$$

Letting $\mathbf{U} = \mathbf{W}\mathbf{P}$, we see that Eqs. (9-56) and (9-57) combine to produce

$$\mathbf{U}^T\mathbf{A}\mathbf{U} = \begin{bmatrix} \lambda_1 & \mathbf{0} \\ \mathbf{0} & \hat{\mathbf{D}} \end{bmatrix} \tag{9-58}$$

The right-hand side of Eq. (9-58) is clearly a diagonal matrix. But two questions still remain:

1 Is $\mathbf{U}$ orthogonal?

2 Are the diagonal elements of $\begin{bmatrix} \lambda_1 & \mathbf{0} \\ \mathbf{0} & \hat{\mathbf{D}} \end{bmatrix}$ the eigenvalues of $\mathbf{A}$?

These questions are easily resolved:

1 Recall that, $\mathbf{U} = \mathbf{W}\mathbf{P}$ and that $\mathbf{W}$ is orthogonal. We now observe that $\mathbf{P}$ is also orthogonal, by calculating $\mathbf{P}^T\mathbf{P}$:

$$\begin{aligned} \mathbf{P}^T\mathbf{P} &= \begin{bmatrix} 1 & \mathbf{0} \\ \mathbf{0} & \mathbf{V}^T \end{bmatrix} \begin{bmatrix} 1 & \mathbf{0} \\ \mathbf{0} & \mathbf{V} \end{bmatrix} \\ &= \begin{bmatrix} 1 & \mathbf{0} \\ \mathbf{0} & \mathbf{I}_{n-1} \end{bmatrix} \qquad \text{since } \mathbf{V} \text{ is orthogonal} \end{aligned}$$

Now, we calculate

$$\begin{aligned}\mathbf{U}^T\mathbf{U} &= (\mathbf{WP})^T\mathbf{WP} \\ &= \mathbf{P}^T\mathbf{W}^T\mathbf{WP} \\ &= \mathbf{P}^T\mathbf{IP} \\ &= \mathbf{P}^T\mathbf{P} \\ &= \mathbf{I}_n\end{aligned}$$

Thus, **U** is indeed orthogonal.

2 To show that the diagonal elements of $\begin{bmatrix}\lambda_1 & \mathbf{0} \\ \mathbf{0} & \hat{\mathbf{D}}\end{bmatrix}$ are in fact the eigenvalues of **A**, we need only to examine Eq. (9-58) carefully. Since **U** is orthogonal, $\mathbf{U}^T = \mathbf{U}^{-1}$ and we may write the left-hand side of Eq. (9-58) as

$$\mathbf{U}^{-1}\mathbf{AU} = \begin{bmatrix}\lambda_1 & \mathbf{0} \\ \mathbf{0} & \hat{\mathbf{D}}\end{bmatrix} \tag{9-59}$$

Now, Theorem 9-2 tells us that the matrices **A** and $\mathbf{U}^{-1}\mathbf{AU}$ have the same characteristic polynomial (and hence the same eigenvalues). But from the right-hand side of Eq. (9-59) we clearly see that the eigenvalues of $\mathbf{U}^{-1}\mathbf{AU}$ (and hence of **A**) are merely the diagonal elements of $\begin{bmatrix}\lambda_1 & \mathbf{0} \\ \mathbf{0} & \hat{\mathbf{D}}\end{bmatrix}$ QED

By a completely analogous procedure (substituting the transpose conjugate for the transpose and unitary matrices for orthogonal matrices where appropriate) one can prove:

Theorem 9-15 If **A** is a Hermitian matrix of order n, with eigenvalues $\lambda_1, \lambda_2, \ldots, \lambda_n$ (not necessarily distinct), then there exists a unitary matrix **U**, such that $\mathbf{U}^*\mathbf{AU} = \mathbf{D}$, where **D** is a diagonal matrix with diagonal elements $\lambda_1, \lambda_2, \ldots, \lambda_n$.

Now, let us recall from Chap. 6 that if a matrix **A** is pre- and/or postmultiplied by a nonsingular matrix, the resulting matrix and **A** have the same rank. In particular, the matrices $\mathbf{U}^T\mathbf{AU} = \mathbf{D}$, and **A** of Eq. (9-58) have the same rank. Now, suppose we investigate the null spaces of $(\mathbf{A} - \lambda_j\mathbf{I})$, for each distinct λ_j. In Chap. 7 we learned that the dimension of the null space of any matrix **B** of order n is equal to $n - r$, where r is the rank of **B**.

But it is obvious from the right-hand side of Eq. (9-58) that the rank of $(\mathbf{U}^T\mathbf{AU} - \lambda_j\mathbf{I})$ is equal to $(n - m_j)$, where m_j is the algebraic multiplicity of λ_j [since the matrix $(\mathbf{U}^T\mathbf{AU} - \lambda_j\mathbf{I})$ will contain exactly m_j rows of zeros and $(n - m_j)$ rows with exactly one nonzero diagonal element].

Thus, we have reached the conclusion that the dimension of $N(\mathbf{A} - \lambda_j\mathbf{I})$ is equal to m_j for any real symmetric matrix $\mathbf{A}$. More generally, we have:

Theorem 9-16 If $\mathbf{A}$ is a Hermitian matrix with distinct eigenvalues $\lambda_1, \lambda_2, \ldots, \lambda_p$, then the dimension of $N(\mathbf{A} - \lambda_j\mathbf{I})$ is equal to the algebraic multiplicity m_j of λ_j, for $j = 1, 2, \ldots, p$.

Theorem 9-16 now leads directly to:

Theorem 9-17 If $\mathbf{A}$ is a Hermitian matrix of order n, then $\mathbf{A}$ has a set of n linearly independent, mutually orthogonal unit eigenvectors.

PROOF Let $\lambda_1, \lambda_2, \ldots, \lambda_p$ be the distinct eigenvalues of $\mathbf{A}$.

For each λ_j, the dimension of $N(\mathbf{A} - \lambda_j\mathbf{I}) = m_j$, and for each of these null spaces we can find a set of m_j mutually orthogonal unit eigenvectors. But since eigenvectors corresponding to distinct eigenvalues are also orthogonal (Theorem 9-10), the entire set of $m_1 + m_2 + \cdots + m_p$ eigenvectors must be mutually orthogonal (and thus linearly independent by Theorem 9-8). But, of course, it is clear that $m_1 + m_2 + \cdots + m_p = n$. QED

Finally, if we combine Theorems 9-15 and 9-17, we will have proved:

Theorem 9-18 If $\mathbf{A}$ is a Hermitian matrix of order n, and if $\mathbf{x}_1, \mathbf{x}_2, \ldots, \mathbf{x}_n$ is a mutually orthogonal set of unit eigenvectors, corresponding to eigenvalues $\lambda_1, \lambda_2, \ldots, \lambda_n$, respectively, and if $\mathbf{U} = [\mathbf{x}_1 \quad \mathbf{x}_2 \quad \cdots \quad \mathbf{x}_n]$, then $\mathbf{U}^*\mathbf{AU} = \mathbf{D}$, where $\mathbf{D}$ is a diagonal matrix whose diagonal elements are $\lambda_1, \lambda_2, \ldots, \lambda_n$, respectively.

9-4 THE JORDAN CANONICAL FORM

In the previous section we observed that any Hermitian matrix can be diagonalized by a unitary matrix; that is, if $\mathbf{A}$ is a Hermitian matrix, then there exists a unitary matrix $\mathbf{U}$ such that $\mathbf{U}^*\mathbf{AU} = \mathbf{D}$. Moreover, since $\mathbf{U}^* = \mathbf{U}^{-1}$, we can also write $\mathbf{U}^{-1}\mathbf{AU} = \mathbf{D}$. Theorem 9-6 states that any matrix **A** (of order n) which has n linearly independent eigenvectors can be diagonalized by a nonsingular matrix **P**; that is, $\mathbf{P}^{-1}\mathbf{A}\mathbf{P} = \mathbf{D}$.

If **A** is *any* matrix and there exists a nonsingular matrix **P** such that $\mathbf{P}^{-1}\mathbf{A}\mathbf{P} = \mathbf{D}$ (where as usual, **D** is a diagonal matrix with the n eigenvalues of **A** along its diagonal), then **A** is said to be *diagonalizable.*

Unfortunately, not every matrix is diagonalizable. To see this, consider the matrix

$$\mathbf{A} = \begin{bmatrix} 2 & 1 \\ 0 & 2 \end{bmatrix} \tag{9-60}$$

The eigenvalues of **A** are clearly $\lambda_1 = \lambda_2 = 2$. Thus, if **A** is diagonalizable, then there exists a nonsingular matrix **P** such that

$$\mathbf{P}^{-1}\mathbf{A}\mathbf{P} = \begin{bmatrix} 2 & 0 \\ 0 & 2 \end{bmatrix} \tag{9-61}$$

If we let $\mathbf{P} = \begin{bmatrix} a & b \\ c & d \end{bmatrix}$

then

$$\mathbf{P}^{-1} = \alpha \begin{bmatrix} d & -b \\ -c & a \end{bmatrix}$$

where

$$\alpha = \frac{1}{ad - bc} = \frac{1}{\det(\mathbf{P})}$$

Then,

$$\begin{aligned}
\mathbf{P}^{-1}\mathbf{A}\mathbf{P} &= \alpha \begin{bmatrix} d & -b \\ -c & a \end{bmatrix} \begin{bmatrix} 2 & 1 \\ 0 & 2 \end{bmatrix} \begin{bmatrix} a & b \\ c & d \end{bmatrix} \\
&= \alpha \begin{bmatrix} 2d & d - 2b \\ -2c & -c + 2a \end{bmatrix} \begin{bmatrix} a & b \\ c & d \end{bmatrix} \\
&= \alpha \begin{bmatrix} 2ad + dc - 2bc & 2bd + d^2 - 2bd \\ -2ac - c^2 + 2ac & -2bc - dc + 2ad \end{bmatrix} \\
&= \alpha \begin{bmatrix} dc + 2/\alpha & d^2 \\ -c^2 & -dc + 2/\alpha \end{bmatrix} \\
&= \begin{bmatrix} \alpha dc + 2 & \alpha d^2 \\ -\alpha c^2 & -\alpha dc + 2 \end{bmatrix}
\end{aligned} \tag{9-62}$$

Equating the right-hand sides of Eqs. (9-61) and (9-62) and comparing the corresponding elements yields

$$\begin{aligned} \alpha dc + 2 &= 2 \\ \alpha d^2 &= 0 \\ -\alpha c^2 &= 0 \\ -\alpha dc + 2 &= 2 \end{aligned} \tag{9-63}$$

But the second and third equations of (9-63) imply that either $\alpha = 0$ or $c = d = 0$. But either of these results means that det $(\mathbf{P}) = 0$. This is a contradiction, since $\mathbf{P}$ was assumed to be nonsingular. Thus, there is no nonsingular matrix $\mathbf{P}$ satisfying Eq. (9-61), and therefore $\mathbf{A}$ is not diagonalizable.

If two matrices $\mathbf{A}$ and $\mathbf{B}$ are related by

$$\mathbf{P}^{-1}\mathbf{A}\mathbf{P} = \mathbf{B}$$

where $\mathbf{P}$ is any nonsingular matrix (and $\mathbf{B}$ is not necessarily diagonal), then $\mathbf{A}$ and $\mathbf{B}$ are said to be *similar* matrices. Thus, if $\mathbf{A}$ is diagonalizable, then $\mathbf{A}$ is similar to a diagonal matrix.

We shall now present without proof the very important result that every matrix is similar to an "almost" diagonal matrix with zeros everywhere except along its diagonal and along its "superdiagonal"†:

Theorem 9-19 (The Jordan canonical form) Let $\mathbf{A}$ be a matrix of order n, and let $\lambda_1, \lambda_2, \ldots, \lambda_p$ denote the distinct eigenvalues of $\mathbf{A}$, with respective algebraic multiplicities $m_1, m_2, \ldots, m_p$ and geometric multiplicities $d_1, d_2, \ldots, d_p$. Then there exists a nonsingular matrix $\mathbf{P}$ such that

$$\mathbf{P}^{-1}\mathbf{A}\mathbf{P} = \begin{bmatrix} \mathbf{J}_1 & \mathbf{O} & \cdots & \mathbf{O} \\ \mathbf{O} & \mathbf{J}_2 & \cdots & \mathbf{O} \\ \vdots & \vdots & \ddots & \vdots \\ \mathbf{O} & \mathbf{O} & \cdots & \mathbf{J}_p \end{bmatrix} \tag{9-64}$$

where each $\mathbf{J}_k$ is a square matrix of order m_k, $k = 1, 2, \ldots, p$. Moreover, each $\mathbf{J}_k$ has the form

$$\mathbf{J}_k = \begin{bmatrix} \lambda_k & \delta_1 & 0 & 0 & \cdots & 0 \\ 0 & \lambda_k & \delta_2 & 0 & \cdots & 0 \\ 0 & 0 & \lambda_k & \delta_3 & \cdots & 0 \\ 0 & 0 & 0 & \lambda_k & \cdots & 0 \\ \vdots & \vdots & \vdots & \vdots & \ddots & \vdots \\ 0 & 0 & 0 & 0 & \cdots & \delta_{m_k - 1} \\ 0 & 0 & 0 & 0 & \cdots & \lambda_k \end{bmatrix}$$

† The *superdiagonal* of a square matrix $\mathbf{A} = [a_{ij}]$ consists of the elements $a_{i,i+1}$, $i = 1, 2, \ldots, n-1$; that is, the elements directly above the diagonal elements.

and $(d_k - 1)$ of the δ_i, $i = 1, 2, \ldots, m_k - 1$, are equal to zero, and the number of δ_i's equaling 1 is $(m_k - d_k)$.

The right-hand side of Eq. (9-64) is called the *Jordan canonical form* of **A**. There are several alternate forms which are sometimes used, the difference being in the ordering of the δ_i's. However, for the purposes of this text, the existence of the form in (9-64) is sufficient. An excellent detailed discussion and derivation of the Jordan canonical form can be found in Bronson [1]. However, the reader should keep in mind that a matrix **A** has more than one Jordan canonical form.

The primary application of Theorem 9-19 occurs in the proof of convergence of various iterative algorithms, such as the Jacobi and Gauss-Seidel methods discussed in Chap. 4. In such algorithms, convergence is dependent on the iteration matrix **Q** having the property that

$$\lim_{k \to \infty} \mathbf{Q}^k = \mathbf{O}$$

Theorem 9-19 leads directly to:

Theorem 9-20 If **Q** is any nth-order matrix with distinct eigenvalues $\lambda_1, \lambda_2, \ldots, \lambda_p$, the $\lim_{k \to \infty} \mathbf{Q}^k = \mathbf{O}$ if and only if $\max_{j=1, \ldots, p} |\lambda_j| < 1$.

PROOF If we denote a Jordan canonical form of **Q** by **J**, then Theorem 9-19 states that there exists a nonsingular matrix **P** such that

$$\mathbf{P}^{-1}\mathbf{Q}\mathbf{P} = \mathbf{J}$$

Moreover,

$$\mathbf{Q} = \mathbf{P}\mathbf{J}\mathbf{P}^{-1}$$

and

$$\mathbf{Q}^k = \mathbf{P}\mathbf{J}^k\mathbf{P}^{-1} \tag{9-65}$$

But $\mathbf{J}^k$ is easily shown to be an upper triangular matrix whose diagonal elements are $\lambda_1{}^k, \lambda_2{}^k, \ldots, \lambda_p{}^k$, and whose off-diagonal elements are terms involving the λ_j's raised to different powers less than k. The reader should verify this statement. Thus

$$\lim_{k \to \infty} \mathbf{J}^k = \mathbf{O}$$

if and only if all the λ_j's are strictly less than 1 in absolute value; i.e., if and only if

$$\max_{j=1, 2, \ldots, p} |\lambda_j| < 1 \tag{9-66}$$

From (9-65) we see then that

$$\lim_{k\to\infty} \mathbf{Q}^k = \mathbf{P}[\lim_{k\to\infty} \mathbf{J}^k]\mathbf{P}^{-1}$$
$$= \mathbf{POP}^{-1} = \mathbf{O}$$

if and only if (9-66) holds. QED

EXAMPLE 9-6 A Jordan canonical form of

$$\mathbf{A} = \begin{bmatrix} 4 & 0 & 1 & 0 \\ 2 & 2 & 3 & 0 \\ -1 & 0 & 2 & 0 \\ 4 & 0 & 1 & 2 \end{bmatrix}$$

is

$$\mathbf{J} = \left[\begin{array}{cc:cc} 3 & 1 & 0 & 0 \\ 0 & 3 & 0 & 0 \\ \hdashline 0 & 0 & 2 & 0 \\ 0 & 0 & 0 & 2 \end{array}\right]$$

and

$$\mathbf{P} = \begin{bmatrix} 1 & 1 & 0 & 0 \\ 3 & -1 & 1 & 0 \\ 0 & -1 & 0 & 0 \\ 1 & 3 & 0 & 1 \end{bmatrix}$$

The reader should verify that $\mathbf{P}^{-1}\mathbf{AP} = \mathbf{J}$. ////

REFERENCES

1. BRONSON, RICHARD: "Matrix Methods," Academic, New York, 1969.
2. HOHN, FRANZ E.: "Elementary Matrix Algebra," 2d ed., Macmillan, New York, 1964.
3. LANCASTER, PETER: "Theory of Matrices," Academic, New York, 1969.

EXERCISES

1 For each of the following matrices, find the eigenvalues and a basis for the null spaces $(\mathbf{A} - \lambda_j\mathbf{I})$ for each distinct eigenvalue:

(*a*) $\mathbf{A} = \begin{bmatrix} 3 & 0 & 0 \\ 0 & 3 & 4 \\ 0 & 4 & 3 \end{bmatrix}$

(*b*) $\mathbf{A} = \begin{bmatrix} -3 & 0 & 4 \\ 6 & 2 & 5 \\ 4 & 0 & 3 \end{bmatrix}$

(c)
$$\mathbf{A} = \begin{bmatrix} 3 & 2 & 4 \\ 0 & 5 & 0 \\ 4 & 6 & -3 \end{bmatrix}$$

(d)
$$\mathbf{A} = \begin{bmatrix} 3 & 1 & 0 \\ 0 & 3 & 1 \\ 0 & 0 & 3 \end{bmatrix}$$

(e)
$$\mathbf{A} = \begin{bmatrix} 3 & 1 & 1 \\ 0 & 3 & 1 \\ 0 & 0 & 3 \end{bmatrix}$$

(f)
$$\mathbf{A} = \begin{bmatrix} -10 & \sqrt{2} & \sqrt{6} \\ \sqrt{2} & -11 & \sqrt{3} \\ \sqrt{6} & \sqrt{3} & -9 \end{bmatrix}$$

(g)
$$\mathbf{A} = \begin{bmatrix} 3 & 0 & -1 & 0 \\ 0 & 3 & 0 & -1 \\ -1 & 0 & 3 & 0 \\ 0 & -1 & 0 & 3 \end{bmatrix}$$

(h)
$$\mathbf{A} = \begin{bmatrix} 5 & 1 & 1 & 1 \\ 1 & 5 & 1 & 1 \\ 1 & 1 & 5 & 1 \\ 1 & 1 & 1 & 5 \end{bmatrix}$$

2 Prove that $\lambda = 0$ is an eigenvalue of a matrix $\mathbf{A}$ if and only if $\mathbf{A}$ is singular.

3 Prove that if λ_j is an eigenvalue of a nonsingular matrix $\mathbf{A}$ with corresponding eigenvector $\mathbf{x}_j$, then $1/\lambda_j$ is an eigenvalue of $\mathbf{A}^{-1}$ with corresponding eigenvector $\mathbf{x}_j$, $j = 1, 2, \ldots, n$.

4 Prove that the characteristic polynomials of $\mathbf{A}$ and $\mathbf{A}^T$ are identical, and hence $\mathbf{A}$ and $\mathbf{A}^T$ have the same eigenvalues. What, if anything, can be said about the eigenvectors of $\mathbf{A}$ and $\mathbf{A}^T$?

5 Prove by induction that if λ is an eigenvalue of a matrix $\mathbf{A}$ with corresponding eigenvector $\mathbf{x}$, then for all positive integers k, λ^k is an eigenvalue of $\mathbf{A}^k$ with corresponding eigenvector $\mathbf{x}$.

6 Prove by induction: The product of k unitary matrices is a unitary matrix.

7 Suppose a matrix $\mathbf{A}$ has the property that $\mathbf{A}^5 = \mathbf{A}$. What can be said about the eigenvalues of $\mathbf{A}$?

8 Use Eq. (9-6) to show that, for any square matrix $\mathbf{A}$, $\det(\mathbf{A}) = \lambda_1\lambda_2 \cdots \lambda_n$.

9 Expand Eq. (9-4) and compare it with Eq. (9-5) to show that $\alpha_1 = \lambda_1 + \lambda_2 + \cdots + \lambda_n = a_{11} + a_{22} + \cdots + a_{nn} \equiv \text{trace}(\mathbf{A})$.

10 Prove that the eigenvalues of a skew-symmetric matrix ($\mathbf{A}^T = -\mathbf{A}$) are all pure imaginary.

11 Prove that if λ_j is an eigenvalue of a matrix $\mathbf{A}$ with corresponding eigenvector $\mathbf{x}_j$, then $(\lambda_j + \mu)$ is an eigenvalue of the matrix $(\mathbf{A} + \mu\mathbf{I})$, with corresponding eigenvector $\mathbf{x}_j$.

12 If $f(\beta) = \sum_{k=0}^{p} \alpha_k \beta^k$ is any pth-degree polynomial in the scalar β, then we define the *matrix polynomial* $f(\mathbf{A})$ obtained by replacing the scalar β by the square matrix $\mathbf{A}$: $f(\mathbf{A}) = \sum_{k=0}^{p} \alpha_k \mathbf{A}^k$ (where $\mathbf{A}^0 \equiv \mathbf{I}_n$). Note that $f(\mathbf{A})$ is also a square matrix. Prove that if λ_j is an eigenvalue of $\mathbf{A}$ with corresponding eigenvector $\mathbf{x}_j$, then $f(\lambda_j)$ is an eigenvalue of $f(\mathbf{A})$ with corresponding eigenvector $\mathbf{x}_j$.

13 Prove that the determinant of an orthogonal matrix is either $+1$ or -1.

14 A square matrix $\mathbf{N}$ is called *nilpotent* if there exists a positive integer k such that $\mathbf{N}^k = \mathbf{O}$. Prove that, if $\mathbf{N}$ is nilpotent:

(a) $\det(\mathbf{N}) = 0$

(b) All the eigenvalues of $\mathbf{N}$ are zero.

(c) The trace of $\mathbf{N}$ is zero.

15 Construct an example to show that, given two square matrices **A** and **B** of the same order, if λ is an eigenvalue of **A** and μ is an eigenvalue of **B**, then in general $(\lambda + \mu)$ is *not* an eigenvalue of $(\mathbf{A} + \mathbf{B})$.

16 Construct an example to show that, given two square matrices **A** and **B** of the same order, if λ is an eigenvalue of **A** and μ is an eigenvalue of **B**, then in general $\lambda\mu$ is *not* an eigenvalue of **AB**.

17 Prove that if **A** and **B** commute, and if λ is an eigenvalue of **A** with corresponding eigenvector **x**, then **Bx** is also an eigenvector of **A** corresponding to λ, provided $\mathbf{Bx} \neq \mathbf{0}$.

18 Find a matrix **P** which diagonalizes

$$\mathbf{A} = \begin{bmatrix} 3 & 0 & 0 \\ 0 & 3 & 4 \\ 0 & 4 & 3 \end{bmatrix}$$

19 Find an orthogonal matrix **P** which diagonalizes

$$\mathbf{A} = \begin{bmatrix} 3 & 0 & -1 & 0 \\ 0 & 3 & 0 & -1 \\ -1 & 0 & 3 & 0 \\ 0 & -1 & 0 & 3 \end{bmatrix}$$

20 Find an orthogonal matrix **P** which diagonalizes

$$\mathbf{A} = \begin{bmatrix} 7 & 1 & 1 \\ 1 & 7 & 1 \\ 1 & 1 & 7 \end{bmatrix}$$

given that the eigenvalues of **A** are $\lambda_1 = \lambda_2 = 6$, $\lambda_3 = 9$.

21 Prove Corollary 9-12.

22 Use Theorem 9-2 to prove that given any nonsingular matrix **P**, the matrix $\mathbf{P}^{-1}\mathbf{AP}$ has eigenvalues λ_j with corresponding eigenvectors $\mathbf{P}^{-1}\mathbf{x}_j$, where λ_j and $\mathbf{x}_j$ are eigenvalues and corresponding eigenvectors of **A**.

23 Find an orthogonal matrix **P** such that $\mathbf{P}^T\mathbf{AP}$ is a diagonal matrix, where

$$\mathbf{A} = \begin{bmatrix} 5 & 1 & 1 & 1 \\ 1 & 5 & 1 & 1 \\ 1 & 1 & 5 & 1 \\ 1 & 1 & 1 & 5 \end{bmatrix}$$

if the eigenvalues of **A** are $\lambda_1 = \lambda_2 = \lambda_3 = 4$, $\lambda_4 = 8$.

24 Find a matrix **P** which diagonalizes

$$\mathbf{A} = \begin{bmatrix} 3 & 0 & 1 & 2 \\ 0 & 2 & 0 & 1 \\ 0 & 0 & 1 & 4 \\ 0 & 0 & 0 & -1 \end{bmatrix}$$

25 Prove that all the eigenvalues of **A*A** are nonnegative real numbers for any $m \times n$ matrix **A**, m not necessarily equal to n. [*Hint*: First note that **A*A** is Hermitian.]

26 To find a unitary matrix **U** which diagonalizes a given Hermitian matrix **A**, we need to find a basis consisting of n linearly independent, mutually orthogonal unit vectors. This involves finding such a basis for the null spaces $(\mathbf{A} - \lambda_j \mathbf{I})$ for each distinct λ_j. If a particular λ_j has dimension p, then we can easily find a basis for the corresponding space, by Theorem 7-10.

Suppose the vectors $\mathbf{x}_1, \mathbf{x}_2, \ldots, \mathbf{x}_p$ constitute a basis for a null space. Show that the vectors $\mathbf{u}_1, \mathbf{u}_2, \ldots, \mathbf{u}_p$ are mutually orthogonal unit vectors and hence a desired basis for this null space, where:

$$\alpha_1 = (\mathbf{x}_1^* \mathbf{x}_1)^{1/2} \qquad \mathbf{u}_1 = \frac{1}{\alpha_1} \mathbf{x}_1$$

$$\mathbf{v}_2 = \mathbf{x}_2 - (\mathbf{u}_1^* \mathbf{x}_2)\mathbf{u}_1 \qquad \alpha_2 = (\mathbf{v}_2^* \mathbf{v}_2)^{1/2} \qquad \mathbf{u}_2 = \frac{1}{\alpha_3} \mathbf{v}_3$$

$$\mathbf{v}_3 = \mathbf{x}_3 - \{(\mathbf{u}_1^* \mathbf{x}_3)\mathbf{u}_1 + (\mathbf{u}_2^* \mathbf{x}_3)\mathbf{u}_2\} \qquad \alpha_3 = (\mathbf{v}_3^* \mathbf{v}_3)^{1/2} \qquad \mathbf{u}_3 = \frac{1}{\alpha_3} \mathbf{v}_3$$

$$\mathbf{v}_k = \mathbf{x}_k - \sum_{i=1}^{k-1} (\mathbf{u}_i^* \mathbf{x}_k)\mathbf{u}_i \qquad \alpha_k = (\mathbf{v}_k^* \mathbf{v}_k)^{1/2} \qquad \mathbf{u}_k = \frac{1}{\alpha_k} \mathbf{v}_k$$

$$k = 2, 3, \ldots, p$$

The process of computing a mutually orthogonal set of unit vectors from a given set of linearly independent vectors, by means of the above equations, is called the *Gram-Schmidt process.*

27 Use the Gram-Schmidt process (see Exercise 26) to find a mutually orthogonal set of unit vectors which forms a basis for the vector space spanned by the following set of linearly independent vectors:

$$\mathbf{x}_1 = \begin{bmatrix} 1 \\ 0 \\ -1 \\ 2 \\ 0 \end{bmatrix} \qquad \mathbf{x}_2 = \begin{bmatrix} 1 \\ 1 \\ 0 \\ -3 \\ 2 \end{bmatrix} \qquad \mathbf{x}_3 = \begin{bmatrix} 1 \\ 0 \\ 0 \\ 0 \\ 0 \end{bmatrix}$$

28 Prove Theorem 9-5: eigenvectors corresponding to distinct eigenvalues of a matrix are linearly independent.

10

NUMERICAL CALCULATION OF EIGENVALUES

10-1 INTRODUCTION

In our discussion in Chap. 9 on the theory of eigenvalue problems, we learned among other things, that the eigenvalues of a square matrix **A** are the roots of an nth-degree polynomial. However, for large matrices, it is highly impractical to try to calculate this polynomial, let alone to find its roots. Moreover, even if we knew the characteristic polynomial of a matrix, the determination of its roots is also a difficult problem, for large matrices. In this chapter we shall consider some computational aspects of finding eigenvalues. In some cases we shall be concerned only with finding–or estimating–the largest eigenvalue, as in the test for convergence of the Jacobi iteration method for solving simultaneous equations; in other instances we shall desire all the eigenvalues.

10-2 COMPUTING BOUNDS FOR EIGENVALUES

In some applications it is sufficient to determine only a rough idea of the values of the eigenvalues of a matrix **A**. For example, if we have the characteristic polynomial of **A** and wish to use a root-finding numerical method,† we usually need an initial approximation for each root (eigenvalue).

Consider then the following: If λ is an eigenvalue of $\mathbf{A} = [a_{ij}]_{n\times n}$ with corresponding eigenvector $\mathbf{x} = (x_1\ x_2 \cdots x_n)^T$, then we can write

$$\sum_{j=1}^{n} a_{ij}x_j = \lambda x_i \qquad i = 1, 2, \ldots, n \tag{10-1}$$

or, equivalently,

$$\sum_{\substack{j=1\\ j\neq i}}^{n} a_{ij}x_j = (\lambda - a_{ii})x_i \qquad i = 1, 2, \ldots, n \tag{10-2}$$

Now, since **x** is an eigenvector, it is nonzero. Suppose its kth component is the largest in absolute value:

$$|x_k| = \max_{j=1,\,2,\,\ldots\,n} \{|x_j|\} \tag{10-3}$$

Let us divide both sides of the $i = k$ equation of (10-2) by x_k (which is obviously nonzero):

$$\sum_{\substack{j=1\\ j\neq k}}^{n} a_{kj}\frac{x_j}{x_k} = \lambda - a_{kk} \tag{10-4}$$

If we now take the absolute value of both sides of Eq. (10-4), we obtain

$$|\lambda - a_{kk}| = \left|\sum_{\substack{j=1\\ j\neq k}}^{n} a_{kj}\frac{x_j}{x_k}\right| \tag{10-5}$$

and by the triangle inequality (see Chap. 1) the right-hand side of Eq. (10-5) satisfies the inequality

$$\left|\sum_{\substack{j=1\\ j\neq k}}^{n} a_{kj}\frac{x_j}{x_k}\right| \leq \sum_{\substack{j=1\\ j\neq k}}^{n} |a_{kj}|\frac{|x_j|}{|x_k|} \tag{10-6}$$

Moreover, by Eq. (10-3), it is apparent that

$$\frac{|x_j|}{|x_k|} \leq 1 \qquad j = 1, 2, \ldots, n$$

† Discussion of such methods is beyond the scope of this text, but may be found in most books on numerical analysis, such as reference [1].

and therefore

$$\left| \sum_{\substack{j=1 \\ j \neq k}}^{n} a_{kj} \frac{x_j}{x_k} \right| \leq \sum_{\substack{j=1 \\ j \neq k}}^{n} |a_{kj}| \frac{|x_j|}{|x_k|}$$

$$\leq \sum_{\substack{j=1 \\ j \neq k}}^{n} |a_{kj}| \cdot 1 \qquad (10\text{-}7)$$

Upon combining (10-5) and (10-7) we obtain

$$|\lambda - a_{kk}| \leq \sum_{\substack{j=1 \\ j \neq k}}^{n} |a_{kj}| \qquad (10\text{-}8)$$

Let us examine (10-8) more closely, to see just exactly what information it provides. First of all, we note that for *some* k, $k = 1, 2, \ldots, n$, the inequality (10-8) must hold for *each* eigenvalue of $\mathbf{A}$. However, we do not know which k to choose for each eigenvalue, since the eigenvector $\mathbf{x}$ is unknown.

Secondly, for each k, the set of λ which satisfies (10-8) is a disk† with center at a_{kk} and radius r_k, where

$$r_k = \sum_{\substack{j=1 \\ j \neq k}}^{n} |a_{kj}| \qquad (10\text{-}9)$$

Thus, if we denote each of these n disks by D_k,

$$D_k = \{\lambda \mid |\lambda - a_{kk}| \leq r_k\} \qquad k = 1, 2, \ldots, n \qquad (10\text{-}10)$$

then each (and therefore every) eigenvalue of $\mathbf{A}$ must lie in the union S of these disks:

$$S = \bigcup_{k=1}^{n} D_k \qquad (10\text{-}11)$$

The above discussion may be summarized by:

Theorem 10-1 Gerschgorin's theorem If $\mathbf{A} = [a_{ij}]_{n \times n}$ is any square matrix, then its eigenvalues all lie in the set S, defined by Eq. (10-11), along with Eqs. (10-9) and (10-10).

This theorem was first proved by the mathematician S. Gerschgorin in 1931. Each of the disks D_k of Eq. (10-10) is known as a *Gerschgorin disk.*

† The term *disk* means a circle and its interior. Here, the disk is in the complex plane (set of all complex numbers.)

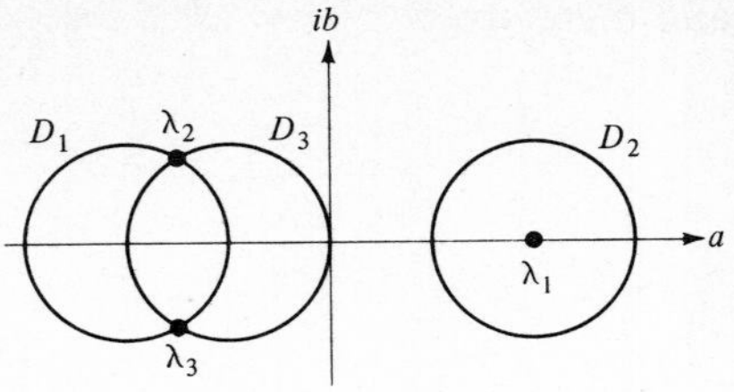

FIGURE 10-1

EXAMPLE 10-1 Consider the matrix

$$A = \begin{bmatrix} -2 & 0 & -1 \\ 0 & 2 & 1 \\ 1 & 0 & -1 \end{bmatrix}$$

The eigenvalues of **A** are

$$\lambda_1 = 2 \qquad \lambda_2 = \tfrac{1}{2}(-3 + \sqrt{3}\, i) \qquad \lambda_3 = \tfrac{1}{2}(-3 - \sqrt{3}\, i)$$

The Gerschgorin disks are

$$D_1 = \{\lambda \mid |\lambda + 2| \le 1\}$$
$$D_2 = \{\lambda \mid |\lambda - 2| \le 1\}$$
$$D_3 = \{\lambda \mid |\lambda + 1| \le 1\}$$

These disks and the eigenvalues λ_1, λ_2, λ_3 are shown in Fig. 10-1. ////

Observe that if we apply Theorem 10-1 to a real symmetric matrix, then we obtain even better estimates of the eigenvalues, since we know from Chap. 9 that the eigenvalues must be real numbers. Thus, instead of a disk in the complex plane, the set D_k of Eq. (10-10) is merely an interval on the real line:

$$D_k = \{\lambda \mid (a_{kk} - r_k) \le \lambda \le (a_{kk} + r_k)\}$$

Note also that, since the eigenvalues of the transpose of a matrix **A** are the same as those of **A**, we can apply Theorem 10-1 to $\mathbf{A}^T$, yielding a different set, say $\hat{S}$, of Gerschgorin disks. We then know that the eigenvalues of **A** must lie in the intersection of S and $\hat{S}$.

We turn now to a method for estimating the largest (in magnitude) eigenvalue of a matrix **A**. From Eq. (10-1), we have that

$$\sum_{j=1}^{n} a_{ij} x_j = \lambda x_i \qquad i = 1, 2, \ldots, n$$

where as before $\mathbf{x} = (x_1\ x_2 \cdots x_n)^T$ is an eigenvector corresponding to the eigenvalue λ. Choosing x_k by Eq. (10-3), we can divide both sides of Eq. (10-2), for $i = k$, by x_k and obtain

$$\sum_{j=1}^{n} a_{kj} \frac{x_j}{x_k} = \lambda \qquad (10\text{-}12)$$

Taking the absolute value of both sides of (10-12) yields

$$\left| \sum_{j=1}^{n} a_{kj} \frac{x_j}{x_k} \right| = |\lambda|$$

and, since $|x_j/x_k| \leq 1$, we have that

$$|\lambda| \leq \sum_{j=1}^{n} |a_{kj}| \qquad (10\text{-}13)$$

Equation (10-13) implies only that for a specific eigenvalue λ, $|\lambda|$ is less than or equal to the sum of the absolute values of the elements in *some* row (i.e., row k) of $\mathbf{A}$. Unfortunately, we don't know *which* row, since we don't know $\mathbf{x}$. However, if we calculate the sum of the absolute values of the elements for each row, and take the largest of these, then we certainly can state that

$$|\lambda| \leq \max_{k=1,2,\ldots,n} \left\{ \sum_{j=1}^{n} |a_{kj}| \right\} \qquad (10\text{-}14)$$

In general, (10-14) doesn't provide as good an estimate of each λ as the Gerschgorin disks do. However, if we are interested only in getting an upper bound for the largest (in magnitude) eigenvalue, then (10-14) is quite useful. We summarize in:

Theorem 10-2 If $\mathbf{A} = [a_{ij}]_{n \times n}$, and if λ_1 is the eigenvalue of $\mathbf{A}$ largest in magnitude, then

$$|\lambda_1| \leq \max_{k=1,2,\ldots,n} \left\{ \sum_{j=1}^{n} |a_{kj}| \right\}$$

As in the case of the Gerschgorin disks, we can state a similar result for the columns of $\mathbf{A}$:

Corollary 10-3 If $\mathbf{A} = [a_{ij}]_{n \times n}$ and if λ_1 is the eigenvalue of $\mathbf{A}$ largest in magnitude, then

$$|\lambda_1| \leq \max_{j=1,2,\ldots,n} \left\{ \sum_{k=1}^{n} |a_{kj}| \right\} \qquad (10\text{-}15)$$

Moreover, we can combine these last two results in:

Corollary 10-4 If $\mathbf{A} = [a_{ij}]_{n \times n}$ and if λ_1 is the eigenvalue of $\mathbf{A}$ largest in magnitude, then $|\lambda_1|$ is less than or equal to the minimum of the right-hand sides of (10-14) and (10-15).

10-3 THE POWER METHOD

In this section we shall discuss a method for calculating the eigenvalue largest in magnitude of a given matrix. This method, known as the *power method*, is quite simple in concept and is very easy to apply. It is probably the most widely used method of finding the largest (in magnitude) eigenvalue of a matrix $\mathbf{A}$. After our discussion of the power method, we shall show how it can also be applied to find the other eigenvalues of $\mathbf{A}$.

We begin by considering the special case where $\mathbf{A}$ is a real symmetric matrix, of order n. Let us also assume for the moment that, if λ_1 is the eigenvalue of $\mathbf{A}$ largest in magnitude, then

$$|\lambda_1| > |\lambda_j| \qquad j = 2, 3, \ldots, n \tag{10-16}$$

Equation (10-16) implies that λ_1 is an eigenvalue of multiplicity one, and also that $\lambda = -\lambda_1$ is not an eigenvalue. We shall see later that these restrictions can be removed.

Since $\mathbf{A}$ is a real symmetric matrix, we know that all the eigenvalues of $\mathbf{A}$ are real and that $\mathbf{A}$ has n linearly independent eigenvectors. Let us denote these latter by $\mathbf{x}_1, \mathbf{x}_2, \ldots, \mathbf{x}_n$, where as usual, $\mathbf{x}_j$ is an eigenvector corresponding to λ_j, $j = 1, 2, \ldots, n$: $\mathbf{A}\mathbf{x}_j = \lambda\mathbf{x}_j$, $j = 1, 2, \ldots, n$.

Thus, these n eigenvectors form a basis for the set of all n-component vectors. This means that, given any vector $\mathbf{y}_0$, there exists a unique set of scalars $\alpha_1, \alpha_2, \ldots, \alpha_n$ such that

$$\mathbf{y}_0 = \alpha_1\mathbf{x}_1 + \alpha_2\mathbf{x}_2 + \cdots + \alpha_n\mathbf{x}_n = \sum_{j=1}^{n} \alpha_j\mathbf{x}_j \tag{10-17}$$

Suppose we arbitrarily choose a nonzero vector of $\mathbf{y}_0$, and calculate successively vectors $\mathbf{y}_1, \mathbf{y}_2, \ldots, \mathbf{y}_k, \ldots$ according to the formulas

$$\begin{aligned} \mathbf{y}_1 &= \mathbf{A}\mathbf{y}_0 \\ \mathbf{y}_2 &= \mathbf{A}\mathbf{y}_1 = \mathbf{A}^2\mathbf{y}_0 \\ &\vdots \\ \mathbf{y}_k &= \mathbf{A}\mathbf{y}_{k-1} = \mathbf{A}^k\mathbf{y}_0 \end{aligned} \tag{10-18}$$

Substitution of Eq. (10-17) into (10-18) yields

$$\mathbf{y}_k = \mathbf{A}^k\left\{\sum_{j=1}^{n} \alpha_j \mathbf{x}_j\right\}$$

$$= \sum_{j=1}^{n} \alpha_j \mathbf{A}^k \mathbf{x}_j \tag{10-19}$$

Moreover, since $\mathbf{A}^k\mathbf{x}_j = \lambda_j^k\mathbf{x}_j$, $j = 1, 2, \ldots, n$, Eq. (10-19) becomes

$$\mathbf{y}_k = \sum_{j=1}^{n} \alpha_j \lambda_j^k \mathbf{x}_j \tag{10-20}$$

If we divide both sides of Eq. (10-20) by the scalar $(1/\lambda_1^k)$, we obtain

$$\left(\frac{1}{\lambda_1^k}\right)\mathbf{y}_k = \sum_{j=1}^{n} \alpha_j \left(\frac{\lambda_j}{\lambda_1}\right)^k \mathbf{x}_j$$

$$= \alpha_1 \mathbf{x}_1 + \sum_{j=2}^{n} \alpha_j \left(\frac{\lambda_j}{\lambda_1}\right)^k \mathbf{x}_j \tag{10-21}$$

However, since $(\lambda_j/\lambda_1) < 1, j = 2, 3, \ldots, n$, by assumption, the terms in the summation in the right-hand side of Eq. (10-21) are becoming smaller and smaller as k increases. In particular, in the limit, as $k \to \infty$, we have

$$\lim_{k\to\infty}\left\{\left(\frac{1}{\lambda_1^k}\right)\mathbf{y}_k\right\} = \lim_{k\to\infty}\left\{\alpha_1\mathbf{x}_1 + \sum_{j=2}^{n}\alpha_j\left(\frac{\lambda_j}{\lambda_1}\right)^k\right\}\mathbf{x}_j$$

$$= \alpha_1 \mathbf{x}_1 \tag{10-22}$$

Thus, the limit of the right-hand side of Eq. (10-21) is seen to be $\alpha_1\mathbf{x}_1$, which is an eigenvector of $\mathbf{A}$ corresponding to λ_1. However, in order to calculate this limit, since the right-hand side of Eq. (10-21) is unknown, we would need to know λ_1 [i.e., the left-hand side of Eq. (10-21)]. But, of course, this is what we are trying to find in the first place!

Let us see how we can make use of the above result to calculate λ_1.

If we let $\mathbf{x}_j = (x_{j1}, x_{j2}, \ldots, x_{jn})^T$ and $\mathbf{y}_k = (y_{k1} \quad y_{k2} \quad \cdots \quad y_{kn})^T$, then by Eq. (10-20)

$$y_{kp} = \sum_{j=1}^{n} \alpha_j \lambda_j^k x_{jp} \qquad p = 1, 2, \ldots, n$$

Moreover, the components of $\mathbf{y}_{k+1}$, again by Eq. (10-20), are seen to be

$$y_{k+1,\,p} = \sum_{j=1}^{n} \alpha_j \lambda_j^{k+1} x_{jp}$$

Now, for each p for which $y_{kp} \neq 0$, suppose we compute

$$\begin{aligned}\frac{y_{k+1,p}}{y_{kp}} &= \frac{\sum_{j=1}^{n} \alpha_j \lambda_j^{k+1} x_{jp}}{\sum_{j=1}^{n} \alpha_j \lambda_j^{k} x_{jp}} \\ &= \frac{\alpha_1 \lambda_1^{k+1} x_{1p} + \sum_{j=2}^{n} \alpha_j \lambda_j^{k+1} x_{jp}}{\alpha_1 \lambda_1^{k} x_{1p} + \sum_{j=2}^{n} \alpha_j \lambda_j^{k} x_{jp}} \\ &= \frac{\lambda_1^{k+1}[\alpha_1 x_{1p} + \sum_{j=2}^{n} \alpha_j (\lambda_j/\lambda_1)^{k+1} x_{jp}]}{\lambda_1^{k}[\alpha_1 x_{1p} + \sum_{j=2}^{n} \alpha_j (\lambda_j/\lambda_1)^{k} x_{jp}]} \\ &= \lambda_1 \frac{[\alpha_1 x_{1p} + \sum_{j=2}^{n} \alpha_j (\lambda_j/\lambda_1)^{k+1} x_{jp}]}{[\alpha_1 x_{1p} + \sum_{j=2}^{n} \alpha_j (\lambda_j/\lambda_1)^{k} x_{jp}]} \end{aligned} \qquad (10\text{-}23)$$

If we now take the limit, as $k \to \infty$, of each of the ratios in (10-23), we obtain†

$$\lim_{k \to \infty} \left\{ \frac{y_{k+1,p}}{y_{kp}} \right\} = \lim_{k \to \infty} \left\{ \frac{\lambda_1[\alpha_1 x_{1p} + \sum_{j=2}^{n} \alpha_j (\lambda_j/\lambda_1)^{k+1} x_{jp}]}{[\alpha_1 x_{1p} + \sum_{j=2}^{n} \alpha_j (\lambda_j/\lambda_1)^{k} x_{jp}]} \right\} = \lambda_1 \qquad (10\text{-}24)$$

since each term in the summations again goes to zero as $k \to \infty$.

Now, we do have a relation which is computable, namely, the ratios of the corresponding components of successive $\mathbf{y}_k$'s. Equation (10-24) implies that for large enough k these ratios will approach λ_1.

In order to determine what value of k is "large enough," we observe that for each nonzero y_{kp}, the ratio $y_{k+1,p}/y_{kp}$ must equal λ_1. Thus, each time a new vector $\mathbf{y}_{k+1}$ is computed, we calculate these ratios for each nonzero y_{kp}. Whenever all these ratios are approximately equal to within an acceptable tolerance level (e.g., to four decimal places), then this common value is λ_1.

Moreover, the corresponding $\mathbf{y}_{k+1}$ (the most recently calculated) is in fact an eigenvector corresponding to λ_1, since by Eq. (10-24)

$$\mathbf{y}_{k+1} \approx \lambda_1 \mathbf{y}_k$$

and by definition, $\mathbf{y}_{k+1} = \mathbf{A}\mathbf{y}_k$.

There is one additional computational detail which should be mentioned: In the calculation of the successive vectors $\mathbf{y}_k$, we note from Eq. (10-20) that if λ_1 is larger than 1 in magnitude, the right-hand side of Eq. (10-20)—and thus the components of $\mathbf{y}_k$—will tend to grow rather large as k increases. On the other hand, if $|\lambda_1| < 1$, then this expression will tend toward 0 as k increases. To avoid the resulting computational difficulties, the following variation is often

† We assume for the moment that $\alpha_1 \neq 0$ and $x_{1p} \neq 0$. We shall discuss the $\alpha_1 = 0$ case later. Note also that since $\mathbf{y}_k$ is tending toward $\mathbf{x}_1$ in the limit, for k large enough it will be true that $x_{1p} \neq 0$ if $y_{kp} \neq 0$, and the limit in Eq. (10-24) is computed only for $y_{kp} \neq 0$.

employed: Instead of computing just the sequence $\mathbf{y}_0, \mathbf{y}_1, \ldots, \mathbf{y}_k$, we compute in addition the sequence $\hat{\mathbf{y}}_0, \hat{\mathbf{y}}_1, \ldots, \hat{\mathbf{y}}_k$, where

$$\hat{\mathbf{y}}_k = \beta_k \mathbf{y}_k \tag{10-25}$$

$$\frac{1}{\beta_k} = \max_{p=1,2,\ldots,n} \{|y_{kp}|\} \tag{10-26}$$

and

$$\mathbf{y}_{k+1} = \mathbf{A}\hat{\mathbf{y}}_k \tag{10-27}$$

Thus, the components of the vectors $\hat{\mathbf{y}}_{k+1}$, $\hat{\mathbf{y}}_k$ will always be less than or equal to 1 in magnitude, and at least one component will be equal to 1. Therefore, the difficulty discussed in the preceding paragraph is avoided.

Observe now that the ratios of corresponding components of $\mathbf{y}_{k+1}$ (not $\hat{\mathbf{y}}_{k+1}$) to $\hat{\mathbf{y}}_k$ yield λ_1 in the limit.

We now summarize the power method for calculating λ_1:

1 Choose arbitrarily $\mathbf{y}_0$; set $k = 1$.
2 Compute $\hat{\mathbf{y}}_k = \beta_k \mathbf{y}_k$ by Eq. (10-26) and compute $\mathbf{y}_{k+1} = \mathbf{A}\hat{\mathbf{y}}_k$.
3 For each nonzero $\hat{y}_{kp}$, $p = 1, 2, \ldots, n$, compute the ratios:

$$\frac{y_{k+1,p}}{\hat{y}_{kp}} = r_p \tag{10-28}$$

4 Compare the ratios r_p computed in step 3. If all are approximately equal (to within the desired accuracy), then

$$\lambda_1 = r_p$$

$$\mathbf{x}_1 = \mathbf{y}_{k+1}$$

and the method terminates. If these ratios are not approximately equal, increase k by 1 and return to step 2.

Let us consider now some of the restrictive assumptions employed in the above derivation of the power method. First of all, we have assumed that $\alpha_1 \neq 0$. It is easily seen by Eq. (10-23) that if $\alpha_1 = 0$, the ratios $y_{k+1,p}/y_{kp}$ will tend to λ_2 (the second largest eigenvalue in magnitude) in the limit. Obviously, whether or not α_1 is zero depends upon the choice of $\mathbf{y}_0$. However, in practice the possibility of $\alpha_1 = 0$ causes no difficulties, since the roundoff errors introduced in the successive calculation of each $\mathbf{y}_k$ and $\hat{\mathbf{y}}_k$ usually perturb the problem enough so that even if α_1 is theoretically zero, it will actually be a small but nonzero number (at least after several iterations).

We have also assumed [Eq. (10-16)] that λ_1 is larger in magnitude than the other eigenvalues. We now eliminate this restriction. *If λ_1 is a multiple root, then the power method will still converge to λ_1, but it will yield only* *one eigenvector corresponding to* λ_1. To see this, we need only to modify Eq. (10-23) appropriately. The details are left as an exercise for the reader.

If $\lambda = -\lambda_1$ is an eigenvalue, then the ratios computed at *every other* step will approach the constant $\lambda_1{}^2$, and two successive $\mathbf{y}_k$'s approach the corresponding two eigenvectors $\mathbf{x}_1$, $\mathbf{x}_2$.

All of the above discussion still depends on the assumption that **A** is a real symmetric matrix. If this is not the case, two new difficulties arise: (1) The eigenvalues of **A** may not be real, and (2) **A** may not possess n linearly independent eigenvectors.

The latter difficulty means that our initial choice of $\mathbf{y}_0$ is no longer completely arbitrary. For some choices of $\mathbf{y}_0$, the power method might fail (if $\mathbf{y}_0$ is not in the span of the eigenvectors of **A**). However, a trial-and-error approach may be employed, using different choices for $\mathbf{y}_0$.

If λ_1 is not real, then $\bar{\lambda}_1$, its complex conjugate, is also an eigenvalue provided **A** is real. If we let $\lambda_1 = r + qi$, then λ_1 and $\lambda_2 = \bar{\lambda}_1$ are roots of some quadratic equation:

$$(\lambda - \lambda_1)(\lambda - \lambda_2) = \lambda^2 - (\lambda_1 + \lambda_2)\lambda + \lambda_1\lambda_2$$

If we denote this quadratic by $\lambda^2 + a\lambda + b$, then

$$\begin{aligned} a &= \lambda_1 + \lambda_2 = \lambda_1 + \bar{\lambda}_1 \\ b &= \lambda_1\lambda_2 = \lambda_1\bar{\lambda}_1 \end{aligned} \tag{10-29}$$

Thus, if we can somehow use the power method to find a, b, then we can immediately calculate

$$\lambda_1 = -\tfrac{1}{2}[a - \sqrt{4b^2 - a^2}\, i]$$
$$\lambda_2 = -\tfrac{1}{2}[a + \sqrt{4b^2 - a^2}\, i]$$

To modify the power method appropriately, we observe that if $\mathbf{x}_1$ is an eigenvector corresponding to λ_1, then $\mathbf{x}_2 = \bar{\mathbf{x}}_1$ is an eigenvector corresponding to $\lambda_2 = \bar{\lambda}_1$, so long as **A** is a real matrix:

$$\mathbf{A}\mathbf{x}_1 = \lambda_1\mathbf{x}_1$$
$$\bar{\mathbf{A}}\bar{\mathbf{x}}_1 = \bar{\lambda}_1\bar{\mathbf{x}}_1$$

Since **A** is real, $\mathbf{A} = \bar{\mathbf{A}}$, and

$$\mathbf{A}\bar{\mathbf{x}}_1 = \bar{\lambda}_1\bar{\mathbf{x}}_1$$

or

$$\mathbf{A}\mathbf{x}_2 = \lambda_2\mathbf{x}_2$$

Moreover, note that if we choose $\mathbf{y}_0$ as a real vector, then $\alpha_2 = \bar{\alpha}_1$, where as before

$$\mathbf{y}_0 = \sum_{j=1}^{n} \alpha_j \mathbf{x}_j$$

since otherwise $\mathbf{y}_0$ will have some complex components. Thus, we can write

$$\mathbf{y}_0 = (\alpha_1 \mathbf{x}_1 + \bar{\alpha}_1 \bar{\mathbf{x}}_1) + \sum_{j=3}^{n} \alpha_j \mathbf{x}_j \tag{10-30}$$

Again, we obtain the sequence $\mathbf{y}_k = \mathbf{A}^k \mathbf{y}_0$:

$$\begin{aligned} \mathbf{y}_k &= (\alpha_1 \mathbf{A}^k \mathbf{x}_1 + \bar{\alpha}_1 \mathbf{A}^k \bar{\mathbf{x}}_1) + \sum_{j=3}^{n} \alpha_j \mathbf{A}^k \mathbf{x}_j \\ &= (\alpha_1 \lambda_1{}^k \mathbf{x}_1 + \bar{\alpha}_1 \bar{\lambda}_1{}^k \bar{\mathbf{x}}_1) + \sum_{j=3}^{n} \alpha_j \lambda_j{}^k \mathbf{x}_j \end{aligned} \tag{10-31}$$

Now, if $|\lambda_1| > |\lambda_j|$, $j = 3, 4, \ldots, n$, then, as before, the terms in the summation will become successively smaller compared to $(\alpha_1 \lambda_1{}^k \mathbf{x}_1 + \bar{\alpha}_1 \bar{\lambda}_1{}^k \bar{\mathbf{x}}_1)$, for increasing k. For very large k, then, let us write

$$C_k \simeq \alpha_1 \lambda_1{}^k g + \bar{\alpha}_1 \bar{\lambda}_1{}^k \bar{g} \tag{10-32}$$

where g is the largest component of $\mathbf{x}_1$ and C_k is the corresponding component of $\mathbf{y}_k$.

Next, we calculate the quantities

$$\begin{aligned} \mu_k &\equiv C_{k+1} C_{k-1} - C_k{}^2 \\ \eta_k &\equiv C_{k+2} C_{k-1} - C_{k+1} C_k \end{aligned} \tag{10-33}$$

Using Eq. (10-32), it can be shown† that

$$\begin{aligned} \mu_k &= |\alpha_1 g|^2 (\lambda_1 - \bar{\lambda}_1)^2 (\lambda_1 \bar{\lambda}_1)^{k-1} \\ \eta_k &= |\alpha_1 g|^2 (\lambda_1 - \bar{\lambda}_1)^2 (\lambda_1 \bar{\lambda}_1)^{k-1} (\lambda_1 + \bar{\lambda}_1) \end{aligned} \tag{10-34}$$

Thus, the two ratios

$$R_k \equiv \frac{\mu_{k+1}}{\mu_k} = \lambda_1 \bar{\lambda}_1 \tag{10-35}$$

$$P_k \equiv \frac{\eta_k}{\mu_k} = \lambda_1 + \bar{\lambda}_1 \tag{10-36}$$

tend in the limit (since we have been working with large k) to $R_k = b$, $P_k = a$, where a, b are defined by Eq. (10-29).

† The reader should verify the equations (10-34).

Hence, the power method can indeed be modified to find complex eigenvalues. At each iteration, we must use four successive vectors $\mathbf{y}_{k-1}$, $\mathbf{y}_k$, $\mathbf{y}_{k+1}$, $\mathbf{y}_{k+2}$ [see Eqs. (10-33)]. This is not unreasonable, however, since we will obtain two eigenvalues simultaneously. (For λ_1 real, we obtained only one eigenvalue and used two successive $\mathbf{y}_k$'s.) Moreover, at each succeeding iteration after the first we are calculating only one additional $\mathbf{y}_k$ vector.

EXAMPLE 10-2 Consider the matrix of Example 9-5:

$$\mathbf{A} = \begin{bmatrix} 5 & 0 & 1 \\ 0 & -1 & 0 \\ 1 & 0 & 5 \end{bmatrix}$$

The eigenvalues of $\mathbf{A}$ were found to be $\lambda_1 = 6$, $\lambda_2 = 4$, $\lambda_3 = -1$, with corresponding eigenvectors

$$\mathbf{x}_1 = \begin{bmatrix} 1 \\ 0 \\ 1 \end{bmatrix} \qquad \mathbf{x}_2 = \begin{bmatrix} -1 \\ 0 \\ 1 \end{bmatrix} \qquad \mathbf{x}_3 = \begin{bmatrix} 0 \\ 1 \\ 0 \end{bmatrix}$$

We shall apply the power method with two different initial vectors $\mathbf{y}_0$, for purposes of illustration.

First, suppose we choose

$$\mathbf{y}_0 = \begin{bmatrix} 1 \\ 1 \\ 1 \end{bmatrix}$$

and round our calculations to three decimal places. Then

$$\mathbf{y}_1 = \mathbf{A}\mathbf{y}_0 = \begin{bmatrix} 6 \\ -1 \\ 6 \end{bmatrix} \qquad \hat{\mathbf{y}}_1 = \begin{bmatrix} 1 \\ -0.167 \\ 1 \end{bmatrix} \qquad \mathbf{y}_2 = \mathbf{A}\hat{\mathbf{y}}_1 = \begin{bmatrix} 6 \\ 0.167 \\ 6 \end{bmatrix}$$

$$\hat{\mathbf{y}}_2 = \begin{bmatrix} 1 \\ 0.028 \\ 1 \end{bmatrix} \qquad \mathbf{y}_3 = \mathbf{A}\hat{\mathbf{y}}_2 = \begin{bmatrix} 6 \\ -0.028 \\ 6 \end{bmatrix} \qquad \hat{\mathbf{y}}_3 = \begin{bmatrix} 1 \\ +0.005 \\ 1 \end{bmatrix}$$

$$\mathbf{y}_4 = \mathbf{A}\hat{\mathbf{y}}_3 = \begin{bmatrix} 6 \\ -0.005 \\ 6 \end{bmatrix} \qquad \hat{\mathbf{y}}_4 = \begin{bmatrix} 1 \\ -0.000 \\ 1 \end{bmatrix} \qquad \mathbf{y}_5 = \mathbf{A}\hat{\mathbf{y}}_4 = \begin{bmatrix} 6 \\ 0 \\ 6 \end{bmatrix} \qquad \hat{\mathbf{y}}_5 = \begin{bmatrix} 1 \\ 0 \\ 1 \end{bmatrix}$$

Since $\hat{\mathbf{y}}_4 = \hat{\mathbf{y}}_5$, we have converged to the eigenvector of λ_1, and it is clear that $\lambda_1 = 6$. Thus, we required only five iterations to obtain the exact result.

However, life is not always so wonderful. Suppose we had initially chosen $\mathbf{y}_0 = \begin{bmatrix} 1 \\ 0 \\ 0 \end{bmatrix}$

Table 10-1 indicates the sequence of $\hat{\mathbf{y}}_k$'s; again, calculations are rounded to three decimal places.

Thus, $\hat{\mathbf{y}}_{19} = \hat{\mathbf{y}}_{18}$ and the method has converged, in 19 iterations. The resulting estimate of the eigenvector of λ_1 is thus $\begin{bmatrix} 1.000 \\ 0 \\ 0.999 \end{bmatrix}$

Moreover, since

$$\mathbf{y}_{19} = \begin{bmatrix} 5.999 \\ 0 \\ 5.995 \end{bmatrix}$$

our estimate of λ_1 is $\lambda_1 = 5.999$ (i.e., $\mathbf{A}\hat{\mathbf{y}}_{19} = 5.999\hat{\mathbf{y}}_{19}$).

We have seen above that the rate of convergence of the power method is dependent upon our initial choice for $\mathbf{y}_0$. However, there is usually no a priori method for choosing a "good" $\mathbf{y}_0$. In the above two choices for $\mathbf{y}_0$, both seem equally reasonable, particularly in light of the additional information that the true eigenvector is $\begin{bmatrix} 1 \\ 0 \\ 1 \end{bmatrix}$. ////

10-4 DEFLATION AND THE POWER METHOD

In this section we shall proceed to show how the power method can be extended to obtain the other eigenvalues and eigenvectors of a real matrix **A**.

Suppose we have applied the power method to a matrix **A** and have obtained the largest eigenvalue λ_1 and corresponding eigenvector $\mathbf{x}_1$. For convenience, we assume that the first component of $\mathbf{x}_1$ is equal to 1.†

Table 10-1 ITERATIONS FOR EXAMPLE 10-2

	Components of $\hat{\mathbf{y}}_k$				Components of $\hat{\mathbf{y}}_k$		
k	1	2	3	k	1	2	3
0	1.000	0	0	10	1.000	0	0.966
1	1.000	0	0.200	11	1.000	0	0.977
2	1.000	0	0.385	12	1.000	0	0.985
3	1.000	0	0.543	13	1.000	0	0.990
4	1.000	0	0.670	14	1.000	0	0.993
5	1.000	0	0.767	15	1.000	0	0.995
6	1.000	0	0.838	16	1.000	0	0.997
7	1.000	0	0.889	17	1.000	0	0.998
8	1.000	0	0.925	18	1.000	0	0.999
9	1.000	0	0.949	19	1.000	0	0.999

† The eigenvector $\mathbf{x}_1$ can always be scaled so that its first component is 1, unless this component is 0. In this case, the discussion above must be modified. The reader is asked to provide the details of this modification in the exercises.

We begin by partitioning **A** as follows:

$$\mathbf{A} = \begin{bmatrix} \mathbf{a}_1{}^T \\ \mathbf{a}_2{}^T \\ \vdots \\ \mathbf{a}_n{}^T \end{bmatrix}$$

where $\mathbf{a}_i{}^T$ denotes the ith row of **A**. Thus, since $\mathbf{A}\mathbf{x}_1 = \lambda_1 \mathbf{x}_1$ by assumption, we have

$$\mathbf{a}_1{}^T \mathbf{x}_1 = \lambda_1 \tag{10-37}$$

Moreover, if λ_2 is any other eigenvalue of **A** with corresponding eigenvector $\mathbf{x}_2$, then if the first component of $\mathbf{x}_2$ is also 1,

$$\mathbf{a}_1{}^T \mathbf{x}_2 = \lambda_2 \tag{10-38}$$

Now, let us define the matrix $\hat{\mathbf{A}}$:

$$\hat{\mathbf{A}} = \mathbf{A} - \mathbf{x}_1 \mathbf{a}_1{}^T \tag{10-39}$$

We observe that λ_2 *is an eigenvalue of* $\hat{\mathbf{A}}$ *with corresponding eigenvector* $(\mathbf{x}_1 - \mathbf{x}_2)$:

$$\begin{aligned} \hat{\mathbf{A}}(\mathbf{x}_1 - \mathbf{x}_2) &= \hat{\mathbf{A}}\mathbf{x}_1 - \hat{\mathbf{A}}\mathbf{x}_2 \\ &= (\mathbf{A} - \mathbf{x}_1\mathbf{a}_1{}^T)\mathbf{x}_1 - (\mathbf{A} - \mathbf{x}_1\mathbf{a}_1{}^T)\mathbf{x}_2 \\ &= \mathbf{A}\mathbf{x}_1 - \mathbf{x}_1(\mathbf{a}_1{}^T\mathbf{x}_1) - \mathbf{A}\mathbf{x}_2 + \mathbf{x}_1(\mathbf{a}_1{}^T\mathbf{x}_2) \\ &= \lambda_1\mathbf{x}_1 - \mathbf{x}_1\lambda_1 - \lambda_2\mathbf{x}_2 + \mathbf{x}_1\lambda_2 \\ &= \lambda_2(\mathbf{x}_1 - \mathbf{x}_2) \end{aligned}$$

In going from the third line to the fourth line in the above sequence, we have made use of Eqs. (10-37) and (10-38), as well as the relations $\mathbf{A}\mathbf{x}_1 = \lambda_1\mathbf{x}_1$, $\mathbf{A}\mathbf{x}_2 = \lambda_2\mathbf{x}_2$.

The above discussion indicates that $\hat{\mathbf{A}}$ has eigenvalues $\lambda_2, \lambda_3, \ldots, \lambda_n$ (i.e., the eigenvalues of **A** excluding λ_1). However, since $\hat{\mathbf{A}}$ is of order n, we must determine its nth eigenvalue. But this last eigenvalue is clearly zero, since the first row of $\hat{\mathbf{A}}$ consists entirely of zeros: From Eq. (10-39),

$$\hat{\mathbf{a}}_1{}^T = \mathbf{a}_1{}^T - \alpha_1\mathbf{a}_1{}^T \tag{10-40}$$

where† $\alpha_1 = 1$; Eq. (10-40) yields $\hat{\mathbf{a}}_1{}^T = \mathbf{0}$.

† Here, α_1 is the first component of $\mathbf{x}_1$, assumed to be 1.

Thus, we may partition $\hat{\mathbf{A}}$ as follows:

$$\hat{\mathbf{A}} = \left[\begin{array}{c:c} 0_{1 \times 1} & \mathbf{0}_{1 \times (n-1)} \\ \hdashline \mathbf{b}_{(n-1) \times 1} & \mathbf{B}_{(n-1) \times (n-1)} \end{array}\right]$$

The eigenvalues of the $(n-1) \times (n-1)$ matrix $\mathbf{B}$ are $\lambda_2, \lambda_3, \ldots, \lambda_n$. If we apply the power method to $\mathbf{B}$, then, we will obtain λ_2 and $\mathbf{y}_2$ (λ_2 being the eigenvalue of $\mathbf{A}$ second largest in magnitude), where $\mathbf{y}_2$ is an eigenvector of $\mathbf{B}$ corresponding to λ_2.

One question remains, however: How do we find $\mathbf{x}_2$, the eigenvector of $\mathbf{A}$ corresponding to λ_2? Recalling that the first components of $\mathbf{x}_1$ and $\mathbf{x}_2$ were assumed to be 1, let us write

$$\mathbf{x}_1 = \begin{bmatrix} 1 \\ \mathbf{z}_1 \end{bmatrix} \qquad \mathbf{x}_2 = \begin{bmatrix} 1 \\ \mathbf{z}_2 \end{bmatrix} \qquad \mathbf{w} = \begin{bmatrix} 0 \\ \mathbf{y}_2 \end{bmatrix} \tag{10-41}$$

where $\mathbf{z}_1$, $\mathbf{z}_2$ are $(n-1)$-component vectors (as is $\mathbf{y}_2$).

Consider the equation

$$\mathbf{x}_2 = \mathbf{x}_1 + \beta \mathbf{w} \tag{10-42}$$

$$\begin{bmatrix} 1 \\ \mathbf{z}_2 \end{bmatrix} = \begin{bmatrix} 1 \\ \mathbf{z}_1 \end{bmatrix} + \beta \begin{bmatrix} 0 \\ \mathbf{y}_2 \end{bmatrix}$$

Now, is there a value for the scalar β which satisfies Eq. (10-42)?

If we premultiply both sides of Eq. (10-42) by $\mathbf{a}_1^T$, the first row of $\mathbf{A}$, we obtain

$$\mathbf{a}_1^T \mathbf{x}_2 = \mathbf{a}_1^T \mathbf{x}_1 + \beta \mathbf{a}_1^T \mathbf{w} \tag{10-43}$$

and, by Eqs. (10-37) and (10-38), Eq. (10-43) becomes

$$\lambda_2 = \lambda_1 + \beta \mathbf{a}_1^T \mathbf{w}$$

Thus, if $\mathbf{a}_1^T \mathbf{w}$ (which is a scalar) is nonzero,† we have

$$\beta = \frac{\lambda_2 - \lambda_1}{\mathbf{a}_1^T \mathbf{w}} \tag{10-44}$$

The process of finding a matrix $\mathbf{B}$ of order $(n-1)$ whose eigenvalues are identical with those of $\mathbf{A}$, except λ_1, is called *deflation*.

This process can then be repeated, until all the eigenvalues of $\mathbf{A}$ have been found.

† $\mathbf{a}_1^T \mathbf{w} = 0$ if $\lambda_2 = 0$ (in which case all the remaining eigenvalues are also zero) or if the first component of $\mathbf{x}_2$ is zero.

EXAMPLE 10-3 In Example 10-2, we found that the largest eigenvalue of

$$\mathbf{A} = \begin{bmatrix} 5 & 0 & 1 \\ 0 & -1 & 0 \\ 1 & 0 & 5 \end{bmatrix}$$

is $\lambda_1 = 6$, with corresponding eigenvector

$$\mathbf{x}_1 = \begin{bmatrix} 1 \\ 0 \\ 1 \end{bmatrix}$$

Thus,

$$\hat{\mathbf{A}} = \mathbf{A} - \mathbf{x}_1 \mathbf{a}_1{}^T$$

$$= \begin{bmatrix} 5 & 0 & 1 \\ 0 & -1 & 0 \\ 1 & 0 & 5 \end{bmatrix} - \begin{bmatrix} 1 \\ 0 \\ 1 \end{bmatrix} [5 \quad 0 \quad 1]$$

$$= \begin{bmatrix} 5 & 0 & 1 \\ 0 & -1 & 0 \\ 1 & 0 & 5 \end{bmatrix} - \begin{bmatrix} 5 & 0 & 1 \\ 0 & 0 & 0 \\ 5 & 0 & 1 \end{bmatrix}$$

$$= \left[\begin{array}{c:cc} 0 & 0 & 0 \\ \hdashline 0 & -1 & 0 \\ -4 & 0 & 4 \end{array}\right]$$

Thus,

$$\mathbf{B} = \begin{bmatrix} -1 & 0 \\ 0 & 4 \end{bmatrix}$$

Since $\mathbf{B}$ is diagonal, we see immediately that $\lambda_2 = 4$, with corresponding eigenvector $\mathbf{y}_2 = \begin{bmatrix} 0 \\ 1 \end{bmatrix}$.

To find the corresponding eigenvector $\mathbf{x}_2$ of $\mathbf{A}$, we employ Eqs. (10-41), (10-42), and (10-44):

Letting

$$\mathbf{w} = \begin{bmatrix} 0 \\ \mathbf{y}_2 \end{bmatrix} = \begin{bmatrix} 0 \\ 0 \\ 1 \end{bmatrix}$$

we have

$$\mathbf{x}_2 = \mathbf{x}_1 + \beta \mathbf{w}$$

$$\beta = \frac{\lambda_2 - \lambda_1}{\mathbf{a}_1{}^T \mathbf{w}}$$

$$= (4 - 6) \Bigg/ \left[(5 \quad 0 \quad 1) \begin{pmatrix} 0 \\ 0 \\ 1 \end{pmatrix} \right]$$

$$= -2$$

$$\mathbf{x}_2 = \begin{bmatrix} 1 \\ 0 \\ 1 \end{bmatrix} + (-2) \begin{bmatrix} 0 \\ 0 \\ 1 \end{bmatrix} = \begin{bmatrix} 1 \\ 0 \\ -1 \end{bmatrix}$$ ////

10-5 JACOBI'S METHOD

In this section we shall consider a method for finding simultaneously all the eigenvalues and eigenvectors of a real symmetric matrix. This method has proved to be computationally quite successful.

Recall from Chap. 9 that if $\mathbf{A}$ is a real symmetric matrix, then there exists an orthogonal matrix $\mathbf{P}$ ($\mathbf{P}^T\mathbf{P} = \mathbf{I}$) such that $\mathbf{P}^T\mathbf{A}\mathbf{P} = \mathbf{D}$, where $\mathbf{D}$ is a diagonal matrix whose diagonal elements are the eigenvalues of $\mathbf{A}$. Moreover, the columns of $\mathbf{P}$ are the corresponding eigenvectors of $\mathbf{A}$. We shall now derive an iterative scheme for finding $\mathbf{P}$.

Let us first consider the case $n = 2$:

Theorem 10-5 Let

$$\mathbf{A} = \begin{bmatrix} a_{11} & a_{12} \\ a_{12} & a_{22} \end{bmatrix} \qquad \mathbf{P} = \begin{bmatrix} p_{11} & p_{12} \\ p_{21} & p_{22} \end{bmatrix}$$

where $p_{11} = \cos\theta = p_{22}$ and $p_{12} = \sin\theta = -p_{21}$. Then, there exists a value of θ such that:

(*a*) $\mathbf{P}^T\mathbf{P} = \mathbf{I}_2$

(*b*) $\mathbf{P}^T\mathbf{A}\mathbf{P} = \begin{bmatrix} \lambda_1 & 0 \\ 0 & \lambda_2 \end{bmatrix}$

where λ_1 and λ_2 are the eigenvalues of $\mathbf{A}$.

PROOF

(*a*)

$$\mathbf{P} = \begin{bmatrix} \cos\theta & \sin\theta \\ -\sin\theta & \cos\theta \end{bmatrix}$$

$$\mathbf{P}^T\mathbf{P} = \begin{bmatrix} \cos\theta & -\sin\theta \\ \sin\theta & \cos\theta \end{bmatrix} \begin{bmatrix} \cos\theta & \sin\theta \\ -\sin\theta & \cos\theta \end{bmatrix}$$

$$= \begin{bmatrix} \cos^2\theta + \sin^2\theta & \cos\theta\sin\theta - \cos\theta\sin\theta \\ \cos\theta\sin\theta - \cos\theta\sin\theta & \sin^2\theta + \cos^2\theta \end{bmatrix}$$

$$= \begin{bmatrix} 1 & 0 \\ 0 & 1 \end{bmatrix} \qquad \text{for all } \theta$$

(*b*)

$$\mathbf{P}^T\mathbf{A}\mathbf{P} = \begin{bmatrix} \cos\theta & -\sin\theta \\ \sin\theta & \cos\theta \end{bmatrix} \begin{bmatrix} a_{11} & a_{12} \\ a_{12} & a_{22} \end{bmatrix} \begin{bmatrix} \cos\theta & \sin\theta \\ -\sin\theta & \cos\theta \end{bmatrix}$$

$$\begin{bmatrix} a_{11}\cos\theta - a_{12}\sin\theta & a_{12}\cos\theta - a_{22}\sin\theta \\ a_{11}\sin\theta + a_{12}\cos\theta & a_{12}\sin\theta + a_{22}\cos\theta \end{bmatrix} \begin{bmatrix} \cos\theta & \sin\theta \\ -\sin\theta & \cos\theta \end{bmatrix}$$

$$\begin{bmatrix} a_{11}\cos^2\theta - 2a_{12}\sin\theta\cos\theta + a_{22}\sin^2\theta & (a_{11} - a_{22})\cos\theta\sin\theta + a_{12}(\cos^2\theta - \sin^2\theta) \\ (a_{11} - a_{22})\cos\theta\sin\theta + a_{12}(\cos^2\theta - \sin^2\theta) & a_{11}\sin^2\theta + 2a_{12}\sin\theta\cos\theta + a_{22}\cos^2\theta \end{bmatrix}$$

Thus, if we let $\mathbf{D} = \mathbf{P}^T\mathbf{A}\mathbf{P} = [d_{ij}]$, then

$$d_{12} = d_{21} = (a_{11} - a_{22})\cos\theta\sin\theta + a_{12}(\cos^2\theta - \sin^2\theta) \tag{10-45}$$

$$d_{11} = a_{11}\cos^2\theta - 2a_{12}\sin\theta\cos\theta + a_{22}\sin^2\theta \tag{10-46}$$

$$d_{22} = a_{11}\sin^2\theta + 2a_{12}\sin\theta\cos\theta + a_{22}\cos^2\theta \tag{10-47}$$

If $\mathbf{D}$ is to be diagonal, then $d_{12} = d_{21} = 0$. Invoking the well-known trigonometric identities

$$\begin{aligned} \cos^2\theta - \sin^2\theta &= \cos 2\theta \\ \cos\theta\sin\theta &= \tfrac{1}{2}\sin 2\theta \end{aligned} \tag{10-48}$$

and setting $d_{12} = d_{21} = 0$ yields, from Eq. (10-45):

$$\tfrac{1}{2}(a_{11} - a_{22})\sin 2\theta + a_{12}\cos 2\theta = 0 \tag{10-49}$$

Or

$$\frac{\sin 2\theta}{\cos 2\theta} = \frac{2a_{12}}{a_{22} - a_{11}} \qquad a_{22} \neq a_{11}$$

$$\tan 2\theta = \frac{2a_{12}}{a_{22} - a_{11}} \qquad a_{22} \neq a_{11} \tag{10-50}$$

Thus, if $a_{11} \neq a_{22}$, then θ is the angle which satisfies Eq. (10-49); if $a_{11} = a_{22}$, then Eq. (10-48) becomes

$$a_{12} \cos 2\theta = 0$$

and therefore we can choose $\theta = \pi/4$. QED

For any 2×2 real symmetric matrix, we have shown how to immediately calculate $\mathbf{P}$ such that $\mathbf{P}^T\mathbf{AP} = \mathbf{D}$.

Suppose now that we form an nth-order matrix $\mathbf{P} = [p_{ij}]$, by modifying an identity matrix as follows:

Let all the elements of $\mathbf{P}$ be the corresponding elements of $\mathbf{I}_n$, except that for some row r and column q, we let

$$\begin{aligned} p_{rr} &= p_{qq} = \cos\theta \\ p_{rq} &= \sin\theta = -p_{qr} \end{aligned} \tag{10-51a}$$

Thus, for all other p_{ij},

$$p_{ij} = \begin{cases} 1 & i = j \\ 0 & i \neq j \end{cases} \tag{10-51b}$$

Then, we have:

Theorem 10-6 If $\mathbf{A}$ is a real symmetric matrix and $\mathbf{P}$ is the matrix defined by Eqs. (10-51), then:

(*a*) $\mathbf{P}^T\mathbf{P} = \mathbf{I}_n$ (i.e., $\mathbf{P}$ is orthogonal).

(*b*) There exists a θ such that the (r, q) element and the (q, r) element of $\mathbf{P}^T\mathbf{AP}$ are zero.

(*c*) A value of θ which satisfies the criterion of (*b*) is found from

$$\begin{aligned} \tan 2\theta &= \frac{2a_{rq}}{a_{qq} - a_{rr}} & a_{qq} &\neq a_{rr} \\ \theta &= \pi/4 & a_{qq} &= a_{rr} \end{aligned} \tag{10-52}$$

(*d*) If $\mathbf{B} = [b_{ij}] = \mathbf{P}^T\mathbf{AP}$, then $\mathbf{B}$ is symmetric and

(i) For $i \neq r, q$ and $j \neq r, q$, $b_{ij} = a_{ij}$ (10-53a)

(ii) For $i \neq r, q$, $b_{ir} = b_{ri} = a_{ir} \cos\theta - a_{iq} \sin\theta$ (10-53b)

$$b_{iq} = b_{qi} = a_{ir} \sin\theta + a_{iq} \cos\theta \tag{10-53c}$$

(iii) $b_{rr} = a_{rr} \cos^2\theta + a_{qq} \sin^2\theta - 2a_{rq} \sin\theta \cos\theta$ (10-53d)

$$b_{qq} = a_{rr} \sin^2\theta + a_{qq} \cos^2\theta + 2a_{rq} \sin\theta \cos\theta$$

PROOF Let $\mathbf{P} = [\mathbf{p}_1 \quad \mathbf{p}_2 \quad \cdots \quad \mathbf{p}_n]$

Then, by Eqs. (10-51):

for $j \neq r,\ q$: $\mathbf{p}_j = \mathbf{e}_j =$ vector whose jth component is 1, all other components being 0.

$$\mathbf{p}_r = \begin{bmatrix} 0 \\ 0 \\ \vdots \\ \cos\theta \\ \vdots \\ -\sin\theta \\ \vdots \\ 0 \end{bmatrix} \begin{matrix} \\ \\ \\ r\text{th component} \\ \\ q\text{th component} \\ \\ \\ \end{matrix} \qquad \mathbf{p}_q = \begin{bmatrix} 0 \\ 0 \\ \vdots \\ \sin\theta \\ \vdots \\ \cos\theta \\ \vdots \\ 0 \end{bmatrix} \begin{matrix} \\ \\ \\ r\text{th component} \\ \\ q\text{th component} \\ \\ \\ \end{matrix}$$

Thus,

$$\mathbf{P}^T\mathbf{P} = \begin{bmatrix} \mathbf{p}_2{}^T \\ \vdots \\ \mathbf{p}_2{}^T \\ \mathbf{p}_n{}^T \end{bmatrix} [\mathbf{p}_1 \mathbf{p}_2 \quad \cdots \quad \mathbf{p}_n]$$

$$= \begin{bmatrix} \mathbf{p}_1{}^T\mathbf{p}_1 & \mathbf{p}_1{}^T\mathbf{p}_2 & \cdots & \mathbf{p}_1{}^T\mathbf{p}_n \\ \mathbf{p}_2{}^T\mathbf{p}_1 & \mathbf{p}_2{}^T\mathbf{p}_2 & \cdots & \mathbf{p}_2{}^T\mathbf{p}_n \\ \vdots & \vdots & & \vdots \\ \mathbf{p}_n{}^T\mathbf{p}_1 & \mathbf{p}_n{}^T\mathbf{p}_2 & \cdots & \mathbf{p}_n{}^T\mathbf{p}_n \end{bmatrix}$$

The (i, j) element of $\mathbf{P}^T\mathbf{P}$ is $\mathbf{p}_i{}^T\mathbf{p}_j$. For i and j not equal to r or q:

$$\mathbf{p}_i{}^T\mathbf{p}_j = \mathbf{e}_i{}^T\mathbf{e}_j = \begin{cases} 1 & \text{if } i = j \\ 0 & \text{if } i \neq j \end{cases}$$

For $i = r$ and $j \neq r,\ q$:

$$\mathbf{p}_i{}^T\mathbf{p}_j = \mathbf{p}_r{}^T\mathbf{e}_j = 0$$

for $i = r$ and $j = r$:

$$\mathbf{p}_r{}^T\mathbf{p}_r = \cos^2\theta + \sin^2\theta = 1$$

for $i = q$ and $j = q$:

$$\mathbf{p}_q{}^T\mathbf{p}_q = \sin^2\theta + \cos^2\theta = 1$$

for $i = r$ and $j = q$:

$$\mathbf{p}_r{}^T\mathbf{p}_q = \cos\theta\sin\theta - \sin\theta\cos\theta = 0$$

Moreover, $\mathbf{P}^T\mathbf{P}$ is symmetric. Hence, we have included all cases and have shown that for all i and j,

$$\mathbf{p}_i{}^T\mathbf{p}_j = \begin{cases} 1 & \text{if } i = j \\ 0 & \text{if } i \neq j \end{cases}$$

Hence, $\mathbf{P}^T\mathbf{P} = \mathbf{I}_n$, and $\mathbf{P}$ is therefore an orthogonal matrix.

To prove the remainder of the theorem, let us compute $\mathbf{B} = [b_{ij}]$, letting $\mathbf{A} = [\mathbf{a}_1 \quad \mathbf{a}_2 \quad \cdots \quad \mathbf{a}_n]$:

$$\begin{aligned}
\mathbf{B} &\equiv \mathbf{P}^T\mathbf{A}\mathbf{P} \\
&= \begin{bmatrix} \mathbf{p}_1{}^T \\ \mathbf{p}_2{}^T \\ \vdots \\ \mathbf{p}_n{}^T \end{bmatrix} \mathbf{A}[\mathbf{p}_1 \quad \mathbf{p}_2 \quad \cdots \quad \mathbf{p}_n] \\
&= \begin{bmatrix} \mathbf{p}_1{}^T \\ \mathbf{p}_2{}^T \\ \vdots \\ \mathbf{p}_n{}^T \end{bmatrix} [\mathbf{A}\mathbf{p}_1 \quad \mathbf{A}\mathbf{p}_2 \quad \cdots \quad \mathbf{A}\mathbf{p}_n] \\
&= \begin{bmatrix} \mathbf{p}_1{}^T\mathbf{A}\mathbf{p}_1 & \mathbf{p}_1{}^T\mathbf{A}\mathbf{p}_2 & \cdots & \mathbf{p}_1{}^T\mathbf{A}\mathbf{p}_n \\ \mathbf{p}_2{}^T\mathbf{A}\mathbf{p}_1 & \mathbf{p}_2{}^T\mathbf{A}\mathbf{p}_2 & \cdots & \mathbf{p}_2{}^T\mathbf{A}\mathbf{p}_n \\ \vdots & \vdots & & \vdots \\ \mathbf{p}_n{}^T\mathbf{A}\mathbf{p}_1 & \mathbf{p}_n{}^T\mathbf{A}\mathbf{p}_2 & \cdots & \mathbf{p}_n{}^T\mathbf{A}\mathbf{p}_n \end{bmatrix}
\end{aligned}$$

Hence, $b_{ij} = \mathbf{p}_i{}^T\mathbf{A}\mathbf{p}_j$.

Again let us consider separately the cases:

<u>$i \neq r, q$ and $j \neq r, q$:</u>

$$b_{ij} = \mathbf{e}_i{}^T\mathbf{A}\mathbf{e}_j = \mathbf{e}_i{}^T\mathbf{a}_j = a_{ij}$$

<u>$i \neq r, q$ and $j = r$:</u>

$$\begin{aligned}
b_{ir} = \mathbf{e}_i{}^T\mathbf{A}\mathbf{p}_r = e_i{}^T &\begin{bmatrix} a_{1r}\cos\theta - a_{1q}\sin\theta \\ a_{2r}\cos\theta - a_{2q}\sin\theta \\ \vdots \\ a_{nr}\cos\theta - a_{nq}\sin\theta \end{bmatrix} \\
&= a_{ir}\cos\theta - a_{iq}\sin\theta
\end{aligned}$$

<u>$i \neq r, q$ and $j = q$:</u>

$$\begin{aligned}
b_{iq} = \mathbf{e}_i{}^T\mathbf{A}\mathbf{p}_q = e_i{}^T &\begin{bmatrix} a_{1r}\sin\theta + a_{1q}\cos\theta \\ a_{2r}\sin\theta + a_{2q}\cos\theta \\ \vdots \\ a_{nr}\sin\theta + a_{nq}\cos\theta \end{bmatrix} \\
&= a_{ir}\sin\theta + a_{iq}\cos\theta
\end{aligned}$$

$\underline{i = j = r}$:

$$b_{rr} = \mathbf{p}_r{}^T\mathbf{A}\mathbf{p}_r = [0 \cdots \overset{(r)}{\cos\theta} \cdots \overset{(q)}{-\sin\theta} \cdots 0]\begin{bmatrix} a_{1r}\cos\theta - a_{1q}\sin\theta \\ a_{2r}\cos\theta - a_{2q}\sin\theta \\ \vdots \\ a_{nr}\cos\theta - a_{nq}\sin\theta \end{bmatrix}$$

$$= a_{rr}\cos^2\theta - a_{rq}\cos\theta\sin\theta - a_{qr}\sin\theta\cos\theta + a_{qq}\sin^2\theta$$

$$= a_{rr}\cos^2\theta + a_{qq}\sin^2\theta - 2a_{rq}\sin\theta\cos\theta$$ (since **A** is symmetric, $a_{rq} = a_{qr}$)

$\underline{i = j \doteq q}$:

$$b_{qq} = \mathbf{p}_q{}^T\mathbf{A}\mathbf{p}_q = [0 \cdots \overset{(r)}{\sin\theta} \cdots \overset{(q)}{\cos\theta} \cdots 0]\begin{bmatrix} a_{1r}\sin\theta + a_{1q}\cos\theta \\ a_{2r}\sin\theta + a_{2q}\cos\theta \\ \vdots \\ a_{nr}\sin\theta + a_{nq}\cos\theta \end{bmatrix}$$

$$= a_{rr}\sin^2\theta + a_{rq}\sin\theta\cos\theta + a_{qr}\cos\theta\sin\theta + a_{qq}\cos^2\theta$$

$$= a_{rr}\sin^2\theta + a_{qq}\cos^2\theta + 2a_{rq}\sin\theta\cos\theta$$

$\underline{i = r \text{ and } j = q}$:

$$b_{rq} = \mathbf{p}_r{}^T\mathbf{A}\mathbf{p}_q = [0 \cdots \overset{(r)}{\cos\theta} \cdots \overset{(q)}{-\sin\theta} \cdots 0]\begin{bmatrix} a_{1r}\sin\theta + a_{1q}\cos\theta \\ a_{2r}\sin\theta + a_{2q}\cos\theta \\ \vdots \\ a_{nr}\sin\theta + a_{nq}\cos\theta \end{bmatrix}$$

$$= a_{rr}\cos\theta\sin\theta + a_{rq}\cos^2\theta - a_{qr}\sin^2\theta - a_{qq}\sin\theta\cos\theta$$

$$= a_{rq}(\cos^2\theta - \sin^2\theta) + (a_{rr} - a_{qq})\sin\theta\cos\theta$$

$$= a_{rq}\cos 2\theta + \tfrac{1}{2}(a_{rr} - a_{qq})\sin 2\theta \qquad \text{by Eq. (10-48)}$$

Now we observe that **B** is symmetric, since **A** is symmetric and $\mathbf{B}^T = (\mathbf{P}^T\mathbf{A}\mathbf{P})^T = \mathbf{P}^T\mathbf{A}^T\mathbf{P} = \mathbf{P}^T\mathbf{A}\mathbf{P} = \mathbf{B}$. Thus, we have covered all cases above. Moreover, we see that we can make $b_{rq} = b_{qr} = 0$ by a proper choice of θ:

$$0 = a_{rq}\cos 2\theta + \tfrac{1}{2}(a_{rr} - a_{qq})\sin 2\theta$$

If $a_{rr} - a_{qq} \neq 0$, then

$$\frac{\sin 2\theta}{\cos 2\theta} = \tan 2\theta = \frac{2a_{rq}}{a_{qq} - a_{rr}}$$

If $a_{rr} - a_{qq} = 0$, then

$$a_{rq}\cos 2\theta = 0$$

which is satisfied by $\theta = \pi/4$.

Hence we have verified all the statements of the theorem.

QED

To summarize, Theorem 10-6 tells us that, given a symmetric matrix **A**, we can find an orthogonal matrix **P** such that $\mathbf{P}^T\mathbf{AP} = \mathbf{B}$, where for some chosen r and q, $b_{rq} = b_{qr} = 0$. Moreover:

Theorem 10-7 If **P** is any orthogonal matrix, then $\mathbf{P}^T\mathbf{AP} = \mathbf{B}$ and **A** have the same eigenvalues.

PROOF The proof follows immediately from Theorems 9-2 and 9-7. The details are left as an exercise for the reader.

Thus, **B** and **A** have the same eigenvalues, and at least two off-diagonal elements of **B** are zero; namely, $b_{rq} = b_{qr} = 0$. However, **B** is not necessarily diagonal, which is our ultimate goal.

Suppose we generate a sequence of matrices $\mathbf{B}_1, \mathbf{B}_2, \ldots$ by computing a sequence of orthogonal matrices $\mathbf{P}_1, \mathbf{P}_2, \ldots$ each of which is forcing a different pair of elements to zero. Specifically, consider the following:

$$\begin{aligned}
\mathbf{B}_1 &= \mathbf{P}_1{}^T\mathbf{A}\mathbf{P}_1 \\
\mathbf{B}_2 &= \mathbf{P}_2{}^T\mathbf{B}_1\mathbf{P}_2 \\
&\vdots \\
\mathbf{B}_k &= \mathbf{P}_k{}^T\mathbf{B}_{k-1}\mathbf{P}_k
\end{aligned}$$

By induction, each $\mathbf{B}_k$ has the same eigenvalues as **A**; if the sequence $\mathbf{B}_1, \mathbf{B}_2, \ldots$ converges to a diagonal matrix, then, we will have accomplished our goal of finding all the eigenvalues of **A**.

Unfortunately, there is no known finite sequence which leads to a diagonal matrix. Moreover, although at one step of the procedure we may have forced, say $b_{rq} = b_{qr} = 0$, at a subsequent step these same two elements may become nonzero, as the following example will demonstrate:

EXAMPLE 10-4

$$\text{Let } \mathbf{A} = \begin{bmatrix} 4 & 6 & 9 & 7 \\ 6 & 5 & 6 & 8 \\ 9 & 6 & 9 & 7 \\ 7 & 8 & 7 & 9 \end{bmatrix}$$

Suppose we choose $p = 3$, $q = 4$. Then, since $a_{33} = a_{44}$, we find that $\theta = \pi/4$ and $\cos\theta = \sin\theta = \sqrt{2}/2$. Thus,

$$\mathbf{P}_1 = \begin{bmatrix} 1 & 0 & 0 & 0 \\ 0 & 1 & 0 & 0 \\ 0 & 0 & \sqrt{2}/2 & -\sqrt{2}/2 \\ 0 & 0 & \sqrt{2}/2 & \sqrt{2}/2 \end{bmatrix}$$

and

$$\mathbf{B}_1 = \mathbf{P}_1{}^T\mathbf{A}\mathbf{P}_1 = \begin{bmatrix} 4 & 6 & 8\sqrt{2} & -\sqrt{2} \\ 6 & 5 & 7\sqrt{2} & \sqrt{2} \\ 8\sqrt{2} & 7\sqrt{2} & 16 & 0 \\ -\sqrt{2} & \sqrt{2} & 0 & 2 \end{bmatrix}$$

Next, suppose we choose $r = 1$, $q = 3$. Then we find that $\tan 2\theta = 2(8\sqrt{2})/(16 - 4) \cong 1.8856$, and $\theta \cong -31.0308$; $\sin\theta = 0.515499$, $\cos\theta = 0.85689$.
Hence,

$$\mathbf{P}_2 = \begin{bmatrix} 0.85689 & 0 & 0.515499 & 0 \\ 0 & 1 & 0 & 0 \\ -0.515499 & 0 & 0.85689 & 0 \\ 0 & 0 & 0 & 1 \end{bmatrix}$$

and

$$\mathbf{B}_2 = \mathbf{P}_2{}^T\mathbf{B}_1\mathbf{P}_2 = \begin{bmatrix} -2.806 & 0.3816 & 0 & -1.2118 \\ 0.3816 & 5 & 11.576 & 1.4142 \\ 0 & 11.576 & 22.806 & -0.7290 \\ -1.2118 & 1.4142 & -0.7290 & 2 \end{bmatrix}$$

Observe that, although the (1, 3) and (3, 1) elements of $\mathbf{B}_2$ are now zero, the formerly zero (in $\mathbf{B}_1$) elements (3, 4) and (4, 3) are no longer zero. ////

The above example indicates that the sequence $\mathbf{B}_1, \mathbf{B}_2, \ldots$ generated may not be finite, even though at each step we force two off-diagonal elements to zero and there are of course only $(n^2 - n)$ such elements. However, it can be shown (see reference [5]) that if we choose the pair (r, q) properly at each step, then the method will in fact converge, producing the desired diagonal matrix. We shall indicate the proof of the above statement in the Exercises. There are several different methods for selecting (r, q), one of which is to choose (r, q) such that

$$|b_{rq}| = \max_{i \neq j} \{|b_{ij}|\}$$

that is, by forcing the off-diagonal element largest in magnitude to zero.

The method we have just described is called *Jacobi's method* for determining the eigenvalues and eigenvectors of a real symmetric matrix. Some variations of Jacobi's method may be found in references [1] and [5].

10-6 OTHER METHODS

In this section we shall present a discussion of several other useful methods for computing eigenvalues. We shall not develop these methods in detail, but rather shall provide a general description of them. Our purpose here is merely to familiarize the reader with the existence of the techniques and to provide the necessary references so that he can obtain detailed information should he desire.

The methods we shall describe below are generally known as Givens' method, Householder's method, the LR algorithm, and the QR algorithm. These methods are described more fully in references [1], [4], [5], and [6].

Givens' Method

As we noted in the discussion of Jacobi's method in Sec. 10-5, there is, in general, no finite sequence of transformations of the type $\mathbf{B}_k = \mathbf{P}_k^T \mathbf{B}_{k-1} \mathbf{P}_k$ which leads to a diagonal matrix. However, there is such a finite sequence which leads to a *tridiagonal matrix;* that is, a matrix of the form

$$\hat{\mathbf{A}} = \begin{bmatrix} \alpha_1 & \beta_1 & 0 & 0 & \cdots & 0 & 0 \\ \beta_1 & \alpha_2 & \beta_2 & 0 & \cdots & 0 & 0 \\ 0 & \beta_2 & \alpha_3 & \beta_3 & \cdots & 0 & 0 \\ 0 & 0 & \beta_3 & \alpha_4 & \cdots & 0 & 0 \\ \vdots & \vdots & \vdots & \vdots & & \vdots & \vdots \\ 0 & 0 & 0 & 0 & \cdots & \alpha_{n-1} & \beta_{n-1} \\ 0 & 0 & 0 & 0 & \cdots & \beta_{n-1} & \alpha_n \end{bmatrix} \qquad (10\text{-}54)$$

In Givens' method, we begin with a real symmetric matrix $\mathbf{A}$ and, using orthogonal matrices $\mathbf{P}_1, \mathbf{P}_2, \ldots,$ of the type defined by Eqs. (10-51), we successively drive to zero the elements: $a_{13}, a_{14}, \ldots, a_{1n}; a_{24}, a_{25}, \ldots, a_{2n}; \ldots; a_{n-3,n-1}, a_{n-3,n}; a_{n-2,n}$. Since $\mathbf{A}$ is symmetric, we will be simultaneously driving two elements to zero at each step, resulting in the symmetric tridiagonal matrix $\hat{\mathbf{A}}$ of Eq. (10-54). Thus, there is a total of $\sum_{k=1}^{n-2} = \frac{1}{2}(n-2)(n-1)$ such matrices $\mathbf{P}_k$ and $\mathbf{B}_k = \mathbf{P}_k^T \mathbf{B}_{k-1} \mathbf{P}_k$ to compute (where $\mathbf{B}_0 = \mathbf{A}$). There is one minor difference between the matrices $\mathbf{P}_k$ of Jacobi's method and those of Givens' method, and this difference occurs in the computation of the angle θ: In Givens' method, if $i = r$, $j = q$ in Eqs. (10-51), then θ is chosen so that the $(r-1, q)$ and $(q, r-1)$ elements are forced to zero [not the (r, q), (q, r) elements, as in Jacobi's method].

After the tridiagonal matrix $\hat{\mathbf{A}}$ is obtained, Givens' method then employs a recursive technique which estimates the intervals in which the eigenvalues of $\hat{\mathbf{A}}$ lie. (Again, recall that $\hat{\mathbf{A}}$ and $\mathbf{A}$ have the same eigenvalues.)

Once these intervals are known, any standard root-finding numerical method may be used to determine the actual values $\lambda_1, \lambda_2, \ldots, \lambda_n$. The reader will find several such methods described in the references cited previously in this section.

Finally, to obtain the eigenvectors of **A**, we must solve the systems of homogeneous equations $(\mathbf{A} - \lambda_j \mathbf{I}_n)\mathbf{x} = \mathbf{0}$, $j = 1, 2, \ldots, n$. For a good discussion of the computational difficulties involved in determining the eigenvectors see Wilkinson [6].

We note also that Givens' method can be generalized to the case of finding the eigenvalues and eigenvectors of a general $n \times n$ matrix.

Householder's Method

As we indicated in the discussion of Givens' method, it is possible to reduce the real symmetric matrix **A** to the tridiagonal form $\hat{\mathbf{A}}$ of Eq. (10-54) by $\frac{1}{2}(n-1)(n-2)$ transformations of the form $\mathbf{B}_k = \mathbf{P}_k{}^T\mathbf{B}_{k-1}\mathbf{P}_k$, $k = 0, 1, 2, \ldots,$ where $\mathbf{B}_0 = \mathbf{A}$.

Householder's method employs a sequence of only $(n-2)$ transformations of the same form to obtain $\hat{\mathbf{A}}$; however, the orthogonal matrices $\mathbf{P}_k$ are constructed quite differently from those used in Givens' method or Jacobi's method.

Briefly, each orthogonal matrix **P** used in Householder's method is of the form

$$\mathbf{P} = \mathbf{I}_n - \mathbf{w}\mathbf{w}^T$$

where **w** is an n-component vector of unit length:

$$\mathbf{w}^T\mathbf{w} = 1$$

It can be shown that if the components of **w** are properly chosen, then only $(n-2)$ transformations of the above type are necessary to obtain $\hat{\mathbf{A}}$. From this point on, Givens' method and Householder's method are identical. Householder's method also can be generalized to the case of a general square matrix **A**.

Wilkinson [6] has shown that Givens' method and Householder's method are essentially equivalent theoretically, even though they appear to generate quite a different sequence of transformations. However, because each does involve a different detailed set of calculations, it is possible that one will achieve more accurate results for a given matrix than the other. For this reason, one should be familiar with both methods.

The LR Algorithm

The LR algorithm, which is due to H. Rutishauser, utilizes the following facts:

1 The eigenvalues of an upper triangular matrix are equal to its diagonal elements.

2 Let **B** be a square nonsingular matrix and let **R** be the upper triangular matrix obtained by performing the forward pass of Gaussian elimination on **B**. Furthermore, let **L** be the lower triangular matrix whose diagonal elements are equal to 1 and whose elements below the diagonal are the multipliers employed in the above Gaussian elimination; that is, for $i > j$, the (i, j) element of **L** is m_{ij}, where m_{ij} is the multiple of row i subtracted from row j (in the ith step of the forward pass). Then,†

$$\mathbf{B} = \mathbf{LR} \tag{10-55}$$

3 If **L** and **R** are as defined above, then **L** is nonsingular. Moreover, by Theorem 9-2, **B** and $\mathbf{L}^{-1}\mathbf{BL}$ have the same eigenvalues. Note that

$$\begin{aligned}\mathbf{L}^{-1}\mathbf{BL} &= \mathbf{L}^{-1}(\mathbf{LR})\mathbf{L} \\ &= \mathbf{RL}\end{aligned} \tag{10-56}$$

Hence, $\mathbf{B} = \mathbf{LR}$ and **RL** have the same eigenvalues.

4 Suppose we define the sequence of matrices $\mathbf{B}_k$, $\mathbf{L}_k$, $\mathbf{R}_k$ by:

$$\left.\begin{aligned}\mathbf{B}_{k-1} &= \mathbf{L}_{k-1}\mathbf{R}_{k-1} \\ \mathbf{R}_{k-1}\mathbf{L}_{k-1} &= \mathbf{B}_k\end{aligned}\right\} \qquad k = 1, 2, \ldots \tag{10-57}$$

where $\mathbf{B}_0 = \mathbf{A}$. Then Rutishauser showed that under certain conditions,

$$\lim_{k\to\infty} (\mathbf{L}_k) = \mathbf{I}$$

$$\lim_{k\to\infty} (\mathbf{R}_k) = \lim_{k\to\infty} (\mathbf{A}_k) = \mathbf{U}$$

where **U** is an upper triangular matrix. By comments 1 and 3 above, the diagonal elements of **U** are the eigenvalues of **A**.

Thus, the LR algorithm is defined by comment 4 above; that is, the sequence of matrices $\mathbf{B}_0 = \mathbf{A}, \mathbf{B}_1, \mathbf{B}_2, \ldots$ is generated until a step p is reached at which the elements below the diagonal of $\mathbf{B}_p$ are sufficiently close to zero so that $\mathbf{B}_p$ is essentially upper triangular. Note that the calculation of each $\mathbf{B}_k$ requires the performance of a complete forward pass of Gaussian elimination.

† Equation (10-55) holds only if no row interchanges were necessary in performing the forward pass; if row interchanges were performed, the procedure described here must be modified.

The QR Algorithm

The QR algorithm (due to J. G. F. Francis) is conceptually similar to the LR algorithm discussed above. The key difference is that instead of "factoring" **B** into the product of the lower triangular matrix **L** and the upper triangular matrix **R**, we factor **B** into the product of a unitary matrix **Q** and an upper triangular matrix **R**:

$$\mathbf{B} = \mathbf{QR}$$

Again, we generate the sequence

$$\begin{aligned} \mathbf{B}_{k-1} &= \mathbf{Q}_{k-1}\mathbf{R}_{k-1} \\ \mathbf{B}_k &= \mathbf{R}_{k-1}\mathbf{Q}_{k-1} \end{aligned} \qquad k = 1, 2, \ldots$$

where $\mathbf{B}_0 = \mathbf{A}$.

Then, as in the LR algorithm, the sequence $\mathbf{B}_0$, $\mathbf{B}_1$, ... approaches an upper triangular matrix whose diagonal elements are the eigenvalues of **A**.

We have given just the barest of outlines of the LR and QR algorithms, and a few additional comments are in order. First, both of these algorithms require, in general, a great deal of computational effort. However, such effort may be justified, depending on how badly one needs to know the eigenvalues of a given matrix! Moreover, the computational effort can be reduced considerably in certain special cases. Second, the QR algorithm seems to display better "stability" properties then the LR algorithm in most cases. By this we mean that numerical roundoff difficulties tend to be far less severe with the QR algorithm. An excellent and very comprehensive analysis of these algorithms, including the above points, can be found in Wilkinson [6]. It is strongly recommended that the reader consult this source should he need to use one of these algorithms.

REFERENCES

1. CARNAHAN, BRICE, H. A. LUTHER, and JAMES O. WILKES: "Applied Numerical Methods," Wiley, New York, 1969.
2. FORSYTHE, GEORGE, and CLEVE B. MOLER: "Computer Solution of Linear Algebraic Systems," Prentice-Hall, Englewood Cliffs, N.J., 1967.
3. FRANKLIN, J. N.: "Matrix Theory," Prentice-Hall, Englewood Cliffs, N.J., 1968.
4. FRÖBERG, CARL-ERIK: "Introduction to Numerical Analysis," 2d ed., Addison-Wesley, Reading, Mass., 1969.
5. ISAACSON, EUGENE, and HERBERT B. KELLER: "Analysis of Numerical Methods, Wiley, New York, 1966.
6. WILKINSON, J. H.: "The Algebraic Eigenvalue Problem," Oxford University Press, London, 1965.

EXERCISES

1 Use Gerschgorin's theorem to obtain bounds on the magnitude of the eigenvalues for each of the following matrices, and graph the Gerschgorin disks.

(*a*) $\begin{bmatrix} -1 & 1 & 0 \\ 1 & 5 & -1 \\ 0 & -1 & 2 \end{bmatrix}$ (*b*) $\begin{bmatrix} 0 & -1 & -1 \\ 1 & 1 & 1 \\ 1 & 0 & -3 \end{bmatrix}$

(*c*) $\begin{bmatrix} 2+i & 1-i & 0 \\ 3 & 2-3i & 0 \\ 0 & 1 & 1 \end{bmatrix}$

2 For each of the matrices of Exercise 1, sketch the Gerschgorin disks of the transpose matrix to obtain bounds on the eigenvalues of the latter.

3 Use Corollary 10-4 to obtain an upper bound on the magnitude of the largest eigenvalue of each of the matrices in Exercise 1.

4 Prepare a detailed flow chart for the power method of Sec. 10-3.

5 Find the largest eigenvalue of

$$\mathbf{A} = \begin{bmatrix} -1 & 1 & 0 \\ 1 & 5 & -1 \\ 0 & -1 & 2 \end{bmatrix}$$

by the power method.

6 Use Eqs. (10-32) and (10-33) to verify Eqs. (10-34).

7 Use the method of deflation to find the remaining eigenvalues of the matrix of Exercise 5.

8 Prepare a flow chart describing Jacobi's method.

9 Note that at each step of Jacobi's method, we do not actually require the value of θ, but merely the values of $\sin\theta$ and $\cos\theta$ [see Eqs. (10-53)]. If $a_{pp} = a_{qq}$, then $\theta = \pi/4$ and $\sin\theta = \cos\theta = \sqrt{2}/2$. If $a_{pp} \neq a_{qq}$, then $\tan 2\theta = 2a_{rq}/(a_{qq} - a_{rr})$. Let

$$\alpha = \pm 2|a_{rq}| \qquad \beta = |a_{qq} - a_{rr}| \tag{10-58}$$

where the sign of α is chosen so that $\tan 2\theta = \alpha/\beta$.

(*a*) Show that

$$\cos^2 2\theta = \beta^2/(\alpha^2 + \beta^2)$$
$$\cos 2\theta = \beta/\sqrt{\alpha^2 + \beta^2}$$

(*b*) Use the trigonometric identity $\cos 2\theta = 2\cos^2\theta - 1$ to show that

$$\cos\theta = \{1/2[1 + \beta/(\alpha^2 + \beta^2)]\}^{1/2} \tag{10-59}$$

Thus, since $\sin\theta = \sqrt{1 - \cos^2\theta}$, we can compute the desired quantities successively from Eqs. (10-58) and (10-59).

10 Prove Theorem 10-7.

11 To prove that Jacobi's method does indeed converge a diagonal matrix, we must show that the sum of squares of the off-diagonal elements of each matrix $\mathbf{B}_k$

decreases at each step. Equivalently, we can show that if $\mathbf{B} = \mathbf{P}^T\mathbf{A}\mathbf{P}$ (where $\mathbf{P}$ is chosen according to Theorem 10-6), then

$$\sum_{i=1}^{n}\sum_{j=1}^{n} b_{ij}^2 - \sum_{i=1}^{n} b_{ii}^2 < \sum_{i=1}^{n}\sum_{j=1}^{n} a_{ij}^2 - \sum_{i=1}^{n} a_{ii}^2$$

Note that

$$\text{trace}\,(\mathbf{B}^T\mathbf{B}) = \sum_{i=1}^{n}\sum_{j=1}^{n} b_{ij}^2$$

$$\text{trace}\,(\mathbf{A}^T\mathbf{A}) = \sum_{i=1}^{n}\sum_{j=1}^{n} a_{ij}^2$$

and that

$$\begin{aligned}\mathbf{B}^T\mathbf{B} &= (\mathbf{P}^T\mathbf{A}\mathbf{P})^T(\mathbf{P}^T\mathbf{A}\mathbf{P})\\ &= \mathbf{P}^T\mathbf{A}^T\mathbf{P}\mathbf{P}^T\mathbf{A}\mathbf{P}\\ &= \mathbf{P}^T(\mathbf{A}^T\mathbf{A})\mathbf{P}\end{aligned}$$

Thus, by Theorem 10-7, $\mathbf{B}^T\mathbf{B}$ and $\mathbf{A}^T\mathbf{A}$ have the same eigenvalues. According to Exercise 9, Chap. 9, then

$$\text{trace}\,(\mathbf{B}^T\mathbf{B}) = \text{trace}\,(\mathbf{A}^T\mathbf{A})$$

Hence, to complete the desired proof, we need only to show that

$$\sum_{i=1}^{n} b_{ii}^2 > \sum_{i=1}^{n} a_{ii}^2 \qquad (10\text{-}60)$$

Use Eqs. (10-53) to show that

$$\sum_{i=1}^{n} b_{ii}^2 = \sum_{i=1}^{n} a_{ii}^2 + 2a_{rq}^2$$

and hence that (10-60) is satisfied.

12 Write a computer program for the power method, and test it by finding the largest eigenvalue for the matrices of Exercises 1*a, f, g, h*, of Chapter 9.

13 Write a computer program for Jacobi's method, making use of the results of Exercise 9. Test your program by finding the eigenvalues and eigenvectors for the matrices of Exercises 1*a, f, g, h*, of Chapter 9.

14 In the discussion of the method of deflation in Sec. 10-4, it is assumed that the first components of the eigenvectors $\mathbf{x}_1$, $\mathbf{x}_2$ are nonzero.

(*a*) Discuss how to modify the method of deflation if the first component of $\mathbf{x}_1$ is zero.

(*b*) Assume that the first component of $\mathbf{x}_1$ is nonzero, but that the first component of $\mathbf{x}_2$ is zero. Note that, before the method of deflation (and the power method) are applied to find λ_2, the eigenvector $\mathbf{x}_2$ is unknown. Discuss how the method should be modified to handle this case.

15 Suppose the power method is applied to a real symmetric matrix $\mathbf{A}$, whose largest eigenvalue λ_1 is a root of multiplicity m. Show that the power method will still converge to λ_1, but will yield only one eigenvector corresponding to λ_1.

INDEX